Introduction to Avionics

Microwave Technology Series

The *Microwave Technology Series* publishes authoritative works for professional engineers, researchers and advanced students across the entire range of microwave devices, sub-systems, systems and applications. The series aims to meet the reader's needs for relevant information useful in practical applications. Engineers involved in microwave devices and circuits, antennas, broadcasting communications, radar, infra-red and avionics will find the series an invaluable source of design and reference information.

Series editors:
Michel-Henri Carpentier
Professor in 'Grandes Écoles', France
Fellow of the IEEE, and President of the French SEE

Bradford L. Smith
International Patents Consultant and Engineer
with the Alcatel group in Paris, France,
and a Senior Member of the IEEE and French SEE

Titles available

Introduction to Avionics

R.P.G. Collinson BScEng., CEng., FIEE., FRAeS

Formerly of GEC Marconi-Avionics Ltd, Rochester, Kent, UK.

CHAPMAN & HALL

London · Weinheim · New York · Tokyo · Melbourne · Madras

Published by Chapman & Hall, 2–6 Boundary Row, London SE1 8HN, UK

Chapman & Hall, 2–6 Boundary Row, London SE1 8HN, UK

Chapman & Hall GmbH, Pappelallee 3, 69469 Weinheim, Germany

Chapman & Hall USA, 115 Fifth Avenue, New York, NY 10003, USA

Chapman & Hall Japan, ITP-Japan, Kyowa Building, 3F, 2-2-1 Hirakawacho, Chiyoda-ku, Tokyo 102, Japan

Chapman & Hall Australia, 102 Dodds Street, South Melbourne, Victoria 3205, Australia

Chapman & Hall India, R. Seshadri, 32 Second Main Road, CIT East, Madras 600 035, India

First edition 1996
Reprinted 1997

© 1996 R.P.G. Collinson

Printed in Great Britain by St. Edmundsbury Press, Suffolk

ISBN 0 412 48250 9

A catalogue record for this book is available from the British Library

♾ Printed on permanent acid-free text paper, manufactured in accordance with ANSI/NISO Z39.48-1992 (Permanence of Paper).

Contents

Foreword

Avionics covers the diverse topics of computing, electronics, control and communications. The dramatic advances in these disciplines in recent years have been successfully applied to aircraft systems including flight control and guidance systems, aircraft navigation, displays, fault tolerance, sensors and monitoring. In the systems engineering approach to design, the system designer needs a sound grasp of the principles of operation of avionics systems, an understanding of signals and error sources and knowledge of aircraft systems to be able to design, develop and integrate aircraft systems.

Dick Collinson's book brings these threads together. It is a difficult task to cover navigation systems, displays, flight control and the related topics which underpin the operation of modern civil and military aircraft. The author brings knowledge of avionics systems design based on his close involvement with many of the major international aircraft projects of recent years. His rigorous treatment of topics which are difficult to extract from the literature and his insights into the systems design approach make this book a fresh and useful addition for engineers in industry and for university courses covering aeronautical systems engineering.

This book is appropriate for the aeronautical engineer wishing to understand aircraft systems, for the electronics engineer needing to grasp the fundamentals of flight dynamics and navigation and for the control engineer who lacks knowledge of aircraft operation and aircraft systems. Although the book is ambitious in attempting to cover a very wide range of topics in modern avionics, Dick Collinson has managed to bring a practical perspective to many otherwise dry and theoretical topics. His approach has a sufficient blend of rigor and practical discussion to serve as an essential reference for designers, manufacturers and operators of modern avionic equipment.

Professor D.J.Allerton
Head of Department of Avionics,
College of Aeronautics,
Cranfield University.
United Kingdom.

Preface

My reason for writing this book is to meet the need for a graduate level text book which explains the basic principles underlying the key avionic systems in a modern civil or military aircraft from a systems engineering standpoint, and their implementation using modern technology.

A particular aim, based on my experience over many years, is to meet the needs of graduates (or equivalent) entering the avionics industry who have been educated in a wide variety of disciplines, for example, electronic engineering, computer science, mathematics, physics, mechanical and aeronautical engineering. The book also aims to meet the needs of engineers at all levels working in particular areas of avionics who require an appreciation and understanding of other areas and disciplines and the latest technology implementations.

A further objective is to show the very wide range of disciplines which are involved in avionic systems, as this makes the subject an interesting and challenging field. Apart from the interest inherent in aircraft, the range of disciplines and technologies which are exploited covers satellite navigation, optical gyroscopes, man–machine interaction, speech recognition, advanced display systems, holographic optics, intelligent knowledge based systems, closed loop control systems, high integrity failure survival systems, high integrity software, VLSI design and data bus systems.

The technology used to implement avionic systems is subject to a process of continual change as improved performance and lower cost is sought. New technologies and systems which offer significant improvements are introduced over relatively short time scales which, in fact, are tending to get shorter particularly in respect of electronic implementation and integrated circuit technology.

The basic philosophy adopted in this book has, therefore, been to concentrate on the underlying principles and basic theory involved in the avionic systems as this is unchanging. The latest means, however, for implementing the avionic systems have been explained. Future technologies and systems which are still in the advanced development stages and not yet in service have been reviewed. My recent background and responsibility for advanced avionic systems research has also enabled me to comment from the basis of first hand experience.

Personally, I have found avionics a fascinating subject throughout my career and I hope the reader will be able to share this interest.

Dick Collinson
July 1995

Acknowledgements

I would like to thank the management of GEC-Marconi Avionics Ltd., Rochester for their assistance in writing this book and their permission to use a number of the illustrations showing GEC-Marconi Avionics systems and concepts. In particularly, I would like to express my appreciation to Brian Tucker, Director and General Manager, and Robin Sleight, Director of Marketing and Technology of the Displays and New Ventures Division, for their encouragement and assistance. I would also like to thank my former boss, Ron Howard, formerly Managing Director and later Chairman of GEC Avionics (now retired) for his help and encouragement. My thanks and appreciation to my former colleagues in the Company, Chris Bartlett, Gordon Belcher, John Corney, Andrew Gibson, Derek Hamlin and Dave Jibb for their help in obtaining information, checking the draft chapters and providing helpful and constructive comments.

I would also like to thank Professor David Allerton, of Cranfield University for his help and support.

My thanks and appreciation also to Peter Potocki, Department Manager/Cockpit Avionics, and Gerard Peiller, Senior Engineer/Automatic Flight Systems of Airbus Industrie for their advice and assistance on flight management systems.

The excellent work carried out by Bob and Jean Ellwood and Andy Poad in producing this book is gratefully acknowledged. Bob and Jean for producing the typed and formatted text including all the equations to the 'camera ready copy' stage and Andy for producing most of the computer-generated illustrations from hand drawn diagrams.

Grateful acknowledgement is made to the following companies for permission to use their illustrations.

Airbus Industrie,
British Aerospace Defence Ltd,
GEC-Marconi Avionics Ltd,
GEC-Marconi Defence Systems Ltd,
Litton Industries, Canada,
Lockheed Advanced Development Company,
Rockwell International, Collins Avionics Division,
Schlumberger Industries,
Smiths Industries.

Finally, I would like to thank my wife and family for their whole-hearted support and encouragement in writing this book.

1

INTRODUCTION

1.1 IMPORTANCE AND ROLE OF AVIONICS

'Avionics' is a word derived from the combination of aviation and electronics. It was first used in the USA in the early 1950s and has since gained wide scale usage and acceptance although it must be said that it may still be necessary to explain what it means to the lay person on occasions.

The term 'avionic system' or 'avionic sub-system' is used in this book to mean any system in the aircraft which is dependent on electronics for its operation, although the system may contain electro-mechanical elements. For example, a Fly-by-wire (FBW) flight control system depends on electronic digital computers for its effective operation, but there are also other equally essential elements in the system. These are basically electro-mechanical devices although they may have associated electronics: for instance, the pilot's control stick sensor assembly, rate gyros and accelerometers to measure the aircraft's motion, air data sensors to measure the aircraft height, airspeed and airstream incidence, and electro-hydraulic servo actuators to control the angular positions of the control surfaces.

The avionics industry is now a major multi-billion dollar industry world-wide and the avionics equipment on a modern military or civil aircraft can account for around 30% of the total cost of the aircraft. This figure for the avionics content is more like 40% in the case of a maritime patrol/anti-submarine aircraft (or helicopter) and can be over 75% of the total cost in the case of an airborne early warning aircraft such as an AWACS.

Even on modern general aviation category aircraft the avionic equipment is now relatively significant (eg colour head down displays) and can be in the region of 10% of the total cost of the aircraft.

The avionic systems are essential to enable the flight crew to carry out the aircraft mission safely and efficiently. Whether the mission is the carrying of the passengers to their destination in the case of a civil airliner, or, in the case of a military aircraft, intercepting a hostile aircraft, attacking a ground target, reconnaissance or maritime patrol.

A major driver in the development and introduction of avionic systems has been the need to meet the mission requirements with the minimum flight crew. In the case of a modern civil airliner, this means a crew of two only, namely the First Pilot (or Captain) and the Second Pilot. This is only made possible by reducing the crew workload by automating the tasks which used to be carried out by the Navigator and

Flight Engineer. The achievement of safe two crew operation has very considerable economic benefits for the airline in a highly competitive market with the consequent saving of crew salaries, expenses and training costs. The reduction in weight is also significant and can be translated into more passengers or longer range on less fuel, as unnecessary weight is geared up ten to one, as will be explained later. In the military case, a single seat fighter or strike (attack) aircraft is lighter and costs less than an equivalent two seat version. The elimination of the second crew member (Navigator/Observer/Radar Operator) has also significant economic benefits in terms of reduction in training costs. (The cost of training and selection of aircrew for fast jet operation is very high.)

Other very important drivers for avionic systems are increased safety, air traffic control requirements, all weather operation, reduction in fuel consumption, improved aircraft performance and control and handling and reduction in maintenance costs.

In the military case, the avionic systems are also being driven by a continuing increase in the threats posed by the defensive and offensive capabilities of potential aggressors.

The role played by the avionic systems in a modern aircraft in enabling the crew to carry out the aircraft mission can be explained in terms of a hierarchical structure comprising layers of specific tasks and avionic system functions as shown in Fig. 1.1. This shows the prime, or 'core', functions which are mainly common to both military and civil aircraft. It must be pointed out, however, that some avionic systems have been left off this diagram for clarity. For example, the Air Traffic Control (ATC) transponder system, the ground proximity warning system (GPWS) and the collision avoidance system (TCAS), all of which are mandatory equipment for civil airliners.

Referring to Fig. 1.1, it can be seen that the main avionic subsystems have been grouped into five layers according to their role and function. These are briefly summarised below in order to provide an overall picture of the roles and functions of the avionic systems in an aircraft.

1.1.1 Systems which interface directly with the pilot

These comprise displays, communications, data entry and control and flight control.

• *Displays* The display systems provide the visual interface between the pilot and the aircraft systems and comprise head up displays (HUDs), helmet mounted displays (HMDs) and head down displays (HDDs). Most combat aircraft are now equipped with a HUD and there are a small but growing number of civil aircraft with HUDs installed. The HMD has now become an essential system in modern combat aircraft and helicopters. The prime advantages of the HUD and HMD are that they project the display information into the pilot's field of view so that the pilot can be head up and can concentrate on the outside world. The HUD now

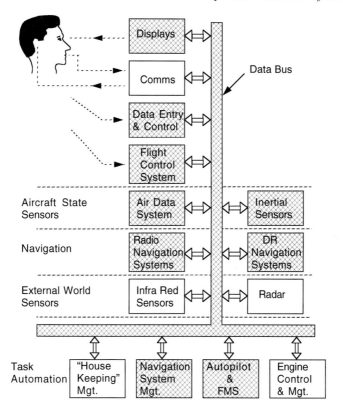

Fig. 1.1 Avionic systems.

provides the primary display for presenting the essential flight information to the pilot and in military aircraft has transformed weapon aiming accuracy. The HUD can also display a forward looking infrared (FLIR) video picture one to one with the outside world from a fixed FLIR imaging sensor installed in the aircraft. The infrared picture merges naturally with the visual scene enabling operations to be carried out at night or in conditions of poor visibility due to haze or clouds.

The helmet mounted display provides in effect a 'HUD on the helmet' with the major advantage that information can be presented to the pilot when looking in any direction as opposed to the relatively limited forward field of view of the HUD. The HMD can also form part of an indirect viewing system by servoing a gimballed infrared imaging sensor to follow the pilot's line of sight (a head position sensor derives the pilot's sight line.) Night viewing goggles (NVGs) can also be integrated into the HMD. This provides a complementary night vision capability enabling the aircraft (or helicopter) to operate at night or in conditions of poor visibility.

Colour head down displays have revolutionised the civil flightdeck with multi-function displays eliminating the inflexible and cluttered characteristics of 1960s generation flightdecks with their numerous dial type instrument displays dedicated to displaying specific information only.

The multi-function colour displays provide the primary flight displays (PFDs) of height, airspeed, Mach number, vertical speed, artificial horizon, pitch angle, bank angle and heading, and velocity vector. They provide the navigation displays, or horizontal situation indicator (HSI) displays, which show the aircraft position and track relative to the destination or waypoints together with the navigational information and distance and time to go. The weather radar display can also be superimposed on the HSI display. Engine data are presented on multi-function colour displays so that the health of the engines can easily be monitored and divergences from the norm highlighted. The aircraft systems, for example, electrical power supply system, hydraulic power supply system, cabin pressurisation system and fuel management system, can be shown in easy to understand line diagram format on the multi-function displays. The multi-function displays can also be reconfigured in the event of a failure in a particular display.

• *Communications* The vital role played by the communications radio systems and the need for reliable two way communication between the ground bases and the aircraft or between aircraft is self evident and is essential for air traffic control. A radio transmitter and receiver equipment was in fact the first avionic system to be installed in an aircraft and goes back as far as 1909 (Marconi Company). The communications radio suite on modern aircraft is a very comprehensive one and covers several operating frequency bands. Long range communication is provided by high frequency (HF) radios operating in the band 2–30 MHz. Near to medium range communication is provided in civil aircraft by very high frequency (VHF) radios operating in the band 30–100 MHz, and in military aircraft by ultra high frequency (UHF) radio operating in the band 250–400 MHz. (VHF and UHF are line of sight propagation systems) Equipment is usually at duplex level of redundancy; the VHF radios are generally at triplex level on a modern airliner. Satellite communications (SATCOM) systems are also being installed in many modern aircraft and these are able to provide very reliable world wide communication.

• *Data entry and control* Data entry and control systems are essential for the crew to interact with the avionic systems. Such systems range from keyboards and touch panels to the use of direct voice input (DVI) control exploiting speech recognition technology.

• *Flight control* Flight Control Systems exploit electronic system technology in two areas, namely auto stabilisation (or stability augmentation) systems and FBW flight control systems. Most swept wing jet aircraft exhibit a lightly damped short period oscillatory motion about the yaw and roll axes at certain height and speed conditions and require at least a yaw auto-stabiliser system to damp and suppress this motion; a roll auto-stabiliser system may also be required. The short period motion about the pitch axis can also be insufficiently damped and a pitch auto-stabiliser system is necessary. Most combat aircraft and many civil aircraft in fact require three axis auto-stabilisation systems to achieve acceptable control and handling characteristics across the flight envelope.

FBW flight control enables a lighter, higher performance aircraft to be produced compared with an equivalent conventional design by allowing the aircraft to be designed with a reduced or even negative natural aerodynamic stability. It does this by providing continuous automatic stabilisation of the aircraft by computer control of the control surfaces from appropriate motion sensors. The system can be designed to give the pilot a manoeuvre command control which provides excellent control and handling characteristics across the flight envelope. 'Care free manoeuvring' characteristics can also be achieved by automatically limiting the pilot's commands according to the aircraft's state. A very high integrity, failure survival system is of course essential for FBW flight control.

1.1.2 Aircraft state sensor systems

These comprise the air data systems and the inertial sensor systems.

• *Air Data Systems* Accurate information on the air data quantities, that is the altitude, calibrated airspeed, vertical speed, true airspeed, Mach number and airstream incidence angle is essential for the control and navigation of the aircraft. The air data computing system computes these quantities from the outputs of very accurate sensors which measure the static pressure, total pressure and the outside air temperature. The airstream incidence angle is derived from airstream incidence sensors.

• *Inertial Sensor Systems* The aircraft attitude and the direction in which it is heading are clearly essential information for the pilot in executing a manoeuvre or flying in conditions of poor visibility, flying in clouds or at night. Accurate attitude and heading information are also required by a number of avionic subsystems which are essential for the aircraft's mission – for example, the autopilot and the navigation system and weapon aiming in the case of a military aircraft.

The attitude and heading information is provided by the inertial sensor system(s). These comprise a set of gyros and accelerometers which measure the aircraft's angular and linear motion about the aircraft axes, together with a computing system which derives the aircraft's attitude and heading from the gyro and accelerometer outputs. Most modern attitude and heading reference systems (AHRS) use a strapped down (or body mounted) configuration of gyros and accelerometers as opposed to the earlier gimballed systems.

The use of very high accuracy gyros and accelerometers to measure the aircraft's motion enables an inertial navigation system (INS) to be mechanised which provides very accurate attitude and heading information together with the aircraft's velocity and position data (ground speed, track angle and latitude/longitude co-ordinates). The INS in conjunction with the air data system also provides the aircraft velocity vector information. The INS is thus a very important aircraft state sensor system – it is also completely self-contained and does not require any access to the outside world.

1.1.3 Navigation systems

Accurate navigation information, that is the aircraft's position, ground speed and track angle (direction of motion of the aircraft relative to true North) is clearly essential for the aircraft's mission, whether civil or military. Navigation systems can be divided into dead reckoning (DR) systems and position fixing systems; both types are required in the aircraft.

• *Dead reckoning navigation systems* DR navigation systems derive the vehicle's present position by estimating the distance travelled from a known position from a knowledge of the speed and direction of motion of the vehicle. They have the major advantages of being completely self contained and independent of external systems. The main types of DR navigation systems used in aircraft are:-

(a) Inertial navigation systems. These are the most accurate and widely used systems.
(b) Doppler/heading reference systems. These are widely used in helicopters.
(c) Air data/heading reference systems These systems are mainly used as a reversionary navigation system being of lower accuracy than (a) or (b).

A characteristic of all DR navigation systems is that the position error builds up with time and it is, therefore, necessary to correct the DR position error and update the system from position fixes derived from a suitable position fixing system.

• *Radio navigation systems* The position fixing systems used are now mainly radio navigation systems based on satellite or ground based transmitters. A suitable receiver in the aircraft with a supporting computer is then used to derive the aircraft's position from the signals received from the transmitters.

The prime position fixing system is without doubt GPS (global positioning system). This is a satellite navigation system of outstanding accuracy which has provided a revolutionary advance in navigation capability since the system started to come into full operation in 1989.

There are also radio navigation aids such as VOR/DME and TACAN which provide the range and bearing (R/θ) of the aircraft from ground beacon transmitters located to provide coverage of the main air routes.

Approach guidance to the airfield/airport in conditions of poor visibility is provided by the ILS (instrument landing system), or by the later MLS (microwave landing system)

A full navigation suite on an aircraft is hence a very comprehensive one and can include INS, GPS, VOR/DME, ILS, MLS. Many of these systems are at duplex level and some may be at triplex level.

1.1.4 External world sensor systems

These systems, which comprise both radar and infrared sensor, systems enable all weather and night time operation and transform the operational capability of the aircraft (or helicopter).

A very brief description of the roles of these systems is given below.

• *Radar Systems* Weather radar is installed in all civil airliners and also in many general aviation aircraft. The radar looks ahead of the aircraft and is optimised to detect water droplets and provide warning of storms, cloud turbulence and severe precipitation so that the aircraft can alter course and avoid such turbulent conditions, if possible. It should be noted that in severe turbulence, the violence of the vertical gusts can subject the aircraft structure to very high loads and stresses. These radars can also generally operate in ground mapping and terrain avoidance modes.

Modern fighter aircraft generally have a ground attack role as well as the prime interception role and carry very sophisticated multi-mode radars to enable them to fulfil these dual roles. In the airborne interception (AI) mode, the radar must be able to detect aircraft up to 100 miles away and track while scanning and keeping tabs on several aircraft simultaneously (typically at least 12 aircraft). The radar must also have a 'look down' capability and be able to track low flying aircraft below it.

In the ground attack or mapping mode, the radar system is able to generate a map type display from the radar returns from the ground, enabling specific terrain features to be identified for position fixing and target acquisition.

• *Infrared systems* These have the major advantage of being entirely passive systems. Infrared (IR) sensor systems can be used to provide a video picture of the thermal image scene of the outside world either using a fixed FLIR sensor, or alternatively, a gimballed IR imaging sensor. The thermal image picture at night looks very like the visual picture in daytime, but highlights heat sources, such as vehicle engines, enabling real targets to be discriminated from camouflaged decoys. An IR system can also be used in a search and track mode; the passive detection and tracking of targets from their IR emissions is of high operational value as it confers an all important element of surprise.

1.1.5 Task automation systems

These comprise the systems which reduce the crew workload and enable minimum crew operation by automating and managing as many tasks as appropriate so that the crew role is a supervisory management one. The tasks and roles of these are very briefly summarised below.

• *Navigation management system* Navigation management comprises the operation of all the radio navigation aid systems and the combination of the data from all the navigation sources, such as GPS and the INS systems, to provide the best possible estimate of the aircraft position, ground speed and track. The system then derives the steering commands for the autopilot so that the aircraft automatically follows the planned navigation route, including any changes in heading as particular 4waypoints are reached along the route to the destination. It should be noted that this function is carried out by the flight management system (FMS) (if installed).

• *Autopilots and flight management systems* The autopilot and FMS have been grouped together because of the very close degree of integration between these systems on modern civil aircraft. It should be noted, however, that the Autopilot is a 'stand alone' system and not all aircraft are equipped with an FMS.

The autopilot relieves the pilot of the need to fly the aircraft continually with the consequent tedium and fatigue and so enables the pilot to concentrate on other tasks associated with the mission. Apart from basic modes, such as height hold and heading hold, a suitably designed high integrity autopilot system can also provide a very precise control of the aircraft flight path for such applications as automatic landing in poor or even zero visibility conditions. In military applications, the autopilot system in conjunction with a suitable guidance system can provide automatic terrain following, or terrain avoidance. This enables the aircraft to fly automatically at high speed at very low altitudes (100 to 200 ft) so that the aircraft can take advantage of terrain screening and stay below the radar horizon of enemy radars.

Sophisticated FMS have come into wide scale use on civil aircraft since the early 1980s and have enabled two crew operation of the largest, long range civil jet airliners. The tasks carried out by the FMS include:-

- Flight planning.
- Navigation management.
- Engine control to maintain the planned speed or Mach number.
- Control of the aircraft flight path to follow the optimised planned route.
- Control of the vertical flight profile.
- Ensuring the aircraft is at the planned 3D position at the planned time slot; often referred to as 4D navigation. This is very important for air traffic control.
- Flight envelope monitoring.
- Minimising fuel consumption.

• *Engine control and management* This task covers the control and the efficient management and monitoring of the engines. The electronic equipment involved in a modern jet engine is very considerable: it forms an integral part of the engine and is essential for its operation. In many cases some of the engine control electronics is physically mounted on the engine. Many modern jet engines have a full authority digital engine control system (FADEC). This automatically controls the flow of fuel to the engine combustion chambers by the fuel control unit so as to provide a closed loop control of engine thrust in response to the throttle command. The control system ensures the engine limits in terms of temperatures, engine speeds and accelerations are not exceeded and the engine responds in an optimum manner to the throttle command. The system has what is known as full authority in terms of the control it can exercise on the engine and a high integrity failure survival control system is essential. Otherwise a failure in the system could seriously damage the engine and hazard the safety of the aircraft.

A FADEC engine control system is thus similar in many ways to a FBW flight control system.

Other very important engine avionic systems include engine health monitoring systems which measure, process and record a very wide range of parameters associated with the performance and health of the engines. These give early warning of engine performance deterioration, excessive wear, fatigue damage, high vibration levels, excessive temperature levels etc.

• *House keeping management* The term 'house keeping' management has been used to cover the automation of the background tasks which are essential for the aircraft's safe and efficient operation. Such tasks include:

- Fuel management. This embraces fuel flow and fuel quantity measurement and control of fuel transfer from the appropriate fuel tanks to minimise changes in the aircraft trim.
- Electrical power supply system management
- Hydraulic power supply system management
- Cabin/cockpit pressurisation systems
- Environmental control system
- Warning systems
- Maintenance and monitoring systems. These comprise monitoring and recording systems which are integrated into an on-board maintenance computer system. This provides the information to enable speedy diagnosis and rectification of equipment and system failures by pin-pointing faulty units and providing all the information, such as part numbers etc., for replacement units down to module level in some cases.

The above brief summaries cover the roles and importance of the avionic systems shown in Fig. 1.1. It should be pointed out, however, that there are several major systems, particularly on military aircraft, which have not been mentioned in the interest of brevity.

Space constraints limit the coverage of the avionic systems shown in Fig. 1.1 to

- Displays
- Data entry and control
- Flight control
- Inertial sensor systems
- Navigation systems
- Air data systems
- Autopilots and flight management systems
- Data buses and avionic systems integration

and it is not possible to cover communications, radar systems, infrared sensor systems, engine control and management.

The visible impact of the avionic systems on a modern aircraft can best be seen in the cockpit where the outputs of the various systems just described are displayed on the HUD and the colour HDDs.

Figure 1.2 shows the Airbus Industrie A340 long range airliner. The advanced flight deck of the A340 can be seen in Fig. 1.3 and 1.4 (this flight deck, and also the A330 flight deck, are basically the same as the A320 flight deck). An optimised layout of six, colour CRT, displays ensures that each of the two crew members can easily assimilate all relevant data, the screens being sized so that no eye scanning is necessary. In front of each pilot is a primary flight display (PFD) and a navigation display (ND); the remaining two screens display engine and systems data. The pilot's side stick controllers can be seen at the sides of the flight deck; the FBW flight control system eliminates the bulky control column between the pilots and instruments of earlier generation aircraft. This ensures an unobstructed view of the displays. There are also three multi-purpose control and display units (MCDU) which, in addition to accessing the flight management system, are also used to give systems maintenance data in the air and on the ground.

Fig. 1.2 Airbus A340 airliner (by courtesy of Airbus Industrie).

Fig. 1.3 Airbus A340 flight deck (by courtesy of Airbus Industrie).

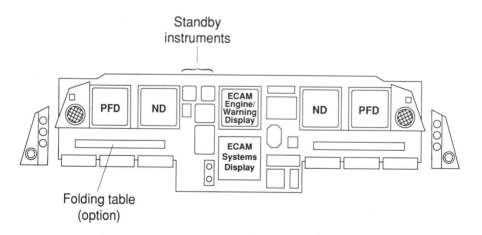

Fig. 1.4 Airbus A340 flight deck – main panel (by courtesy of Airbus Industrie).

Fig. 1.5 Eurofighter 2000 (by courtesy of British Aerospace Defence Ltd.).

Fig. 1.6 Eurofighter 2000 cockpit (by courtesy of British Aerospace Defence Ltd.).

Figure 1.5 shows the Eurofighter 2000 which is being jointly developed by the UK, Germany, Italy and Spain, and is scheduled to enter service with the respective air forces around the year 2000. EF2000 is an agile single-seater air superiority fighter with a surface attack capability. Fig. 1.6 shows the EF2000 cockpit which is designed to deliver optimum levels of tactical and functional information to the pilot without overloading him. The wide field of view holographic HUD and the HDDs, including the centrally located video colour map display, can be seen together with the small centrally mounted pilot's control stick for the FBW flight control system. The unobstructed view of the displays resulting from the use of a small control stick is apparent.

1.2 THE AVIONIC ENVIRONMENT

Avionic systems equipment is very different in many ways from ground based equipment carrying out similar functions. The reasons for these differences are briefly explained in view of their fundamental importance.

1) The importance of achieving minimum weight.
2) The adverse operating environment particularly in military aircraft in terms of operating temperature range, acceleration, shock, vibration, humidity range and electro-magnetic interference.
3) The importance of very high reliability, safety and integrity.
4) Space constraints particularly in military aircraft requiring an emphasis on miniaturisation and high packaging densities.

The effects on the design of avionic equipment to meet these requirements can result in the equipment costing up to ten times as much as equivalent ground based electronic equipment.

The aircraft environmental requirements are briefly discussed below.

1.2.1 Minimum weight

There is a gearing effect on unnecessary weight which is of the order of 10:1. For example a weight saving of 10 kg enables an increase in the payload capability of the order of 100 kg. The process of the effect of additional weight is a vicious circle. An increase in the aircraft weight due to, say, an increase in the weight of the avionics equipment, requires the aircraft structure to be increased in strength, and therefore made heavier, in order to withstand the increased loads during manoeuvres. (Assuming the same maximum normal acceleration, or 'g', and the same safety margins on maximum stress levels are maintained). This increase in aircraft weight means that more lift is required from the wings and the accompanying drag is thus increased. An increase in engine thrust is therefore required to counter the increase in drag and the fuel consumption is thus increased. For the same range it is thus necessary to carry more fuel and the payload has to be correspondingly reduced, or, if the payload is kept the same, the range is reduced. For these reasons tremendous efforts are made to reduce the equipment weight to a minimum and weight penalties can be imposed if equipment exceeds the specified weight.

1.2.2 Environmental requirements

The environment in which avionic equipment has to operate can be a very severe and adverse one in military aircraft; the civil aircraft environment is generally much more benign but is still an exacting one.

Considering just the military cockpit environment alone, such as that experienced by the HUD and HDD.   The operating temperature range is usually specified from –40°C to +70°C. Clearly, the pilot will not survive at these extremes but if the aircraft is left out in the Arctic cold or soaking in the Middle-East sun, for example, the equipment may well reach such temperatures. A typical specification can demand full performance at 20,000 ft. within two minutes of take-off at any temperature within the range.

Vibration is usually quite severe and, in particular, airframe manufacturers tend to locate the gun right under the displays. Power spectral energy levels of $0.04\ g^2$ per Hz are encountered in aircraft designed in the 1970s and levels of $0.7\ g^2$ per Hz at very low frequencies are anticipated in future installations. It is worth noting that driving over cobblestones will give about $0.001\ g^2$ per Hz.

The equipment must also operate under the maximum acceleration or 'g' to which the aircraft is subjected during manoeuvres. This can be $9\ g$ in a modern fighter aircraft and the specification for the equipment would call up at least $20\ g$.

The electromagnetic compatibility (EMC) requirements are also very demanding. The equipment must not exceed the specified emission levels for a very wide range of radio frequencies and must not be susceptible to external sources of very high levels of RF energy over a very wide frequency band

The equipment must also be able to withstand lightning strikes and the very high electromagnetic pulses (EMP) which can be encountered during such strikes. Design of electronic equipment to meet the EMC requirements is in fact a very exacting discipline and requires very careful attention to detail design.

1.2.3 Reliability

The over-riding importance of avionic equipment reliability can be appreciated in view of the essential roles of this equipment in the operation of the aircraft. It is clearly not possible to repair equipment in flight so that equipment failure can mean aborting the mission or a significant loss of performance or effectiveness in carrying out the mission. The cost of equipment failures in airline operation can be very high – interrupted schedules, loss of income during 'aircraft on the ground' situations etc. In military operations, aircraft availability is lowered and operational capability lost.

Every possible care is taken in the design of avionic equipment to achieve maximum reliability. The quality assurance (QA) aspects are very stringent during the manufacturing processes and also very frequently call for what is referred to as 'reliability shake-down testing', or RST, before the equipment is accepted for delivery. RST is intended to duplicate the most severe environmental conditions to which the equipment could be subjected, in order to try to eliminate the early failure phase of the equipment life cycle — what is generally referred to as the 'infant mortality' phase. The life cycle graph of 'mean time between failures' (MTBF) versus time has a profile which generally looks like the cross section of a bath tub, and is in fact referred to as the 'bath tub' curve. The profile of the side of the bath at the beginning of the life cycle corresponds to the relatively high number of failures before the equipment settles down to a much lower failure rate, that is the bottom of the tub, and corresponds to the in-service life phase. Eventually the effects of general wear and tear and deterioration with age cause the failure rate to go up so that the profile of MTBF versus time looks like the other side of the bath tub.

A typical RST cycle requires the equipment to operate satisfactorily through the following cycle.

- Soaking in an environmental chamber at a temperature of +70°C for a given period.
- Rapidly cooling the equipment to –55°C in 20 minutes and soaking at that temperature for a given period.
- Subjecting the equipment to vibration, for example 0.5 *g* amplitude at 50 Hz, for periods during the hot and cold soaking phases.

A typical specification would call for twenty RST cycles without a failure before acceptance of the equipment. If a failure should occur at the nth cycle, the failure must be rectified and the remaining $(20 - n)$ cycles repeated.

All failures in service (and in testing) are investigated by the QA team and remedial action taken, if necessary.

The overall cost differential in meeting all the factors imposed by the avionic equipment environment can thus be appreciated.

2

Air data and air data systems

2.1 INTRODUCTION

Air data systems provide accurate information on quantities such as, pressure altitude, vertical speed, calibrated airspeed, true airspeed, Mach number, static air temperature and air density ratio. This information is essential for the pilot to fly the aircraft safely and is also required by a number of key avionic subsystems which enable the pilot to carry out the mission. It is thus one of the key avionic systems in its own right and forms part of the essential core of avionic sub systems required in all modern aircraft, civil or military.

This chapter explains the importance of air data information and shows how the fundamental physical laws are derived and the air data quantities are generated by an air data computing system.

2.2 AIR DATA INFORMATION AND ITS USE

2.2.1 Air data measurement

The air data quantities pressure altitude, vertical speed, calibrated airspeed, true airspeed, Mach number etc. are derived from three basic measurements by sensors connected to probes which measure:

- *Total (or Pitot) pressure*
- *Static pressure*
- *Total (or indicated) air temperature*

Figure 2.1 illustrates a basic air data system.

The *total pressure*, P_T, is measured by means of an absolute pressure sensor (or transducer) connected to a Pitot tube facing the moving airstream. This measures the *impact pressure*, Q_c, that is the pressure exerted to bring the moving airstream to rest relative to the Pitot tube plus the *static pressure*, P_S, of the free airstream, ie $P_T = Q_c + P_S$.

The *static pressure* of the free airstream, P_S, is measured by an absolute pressure transducer connected to a suitable orifice located where the surface pressure is nearly the same as the pressure of the surrounding atmosphere.

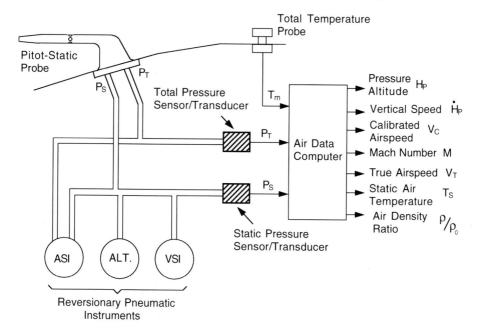

Fig. 2.1 Basic air data system.

High performance military, aircraft generally have a combined Pitot/static probe which extends out in front of the aircraft so as to be as far away as practicable from aerodynamic interference effects and shock waves generated by the aircraft structure. Some civil transport aircraft have Pitot probes with separate static pressure orifices located in the fuselage generally somewhere between the nose and the wing. The exact location of the static pressure orifices (and the Pitot tubes or probes) is determined by experience and experimentation. However, many civil transport aircraft (eg Boeing 747) have side mounted ('L') type Pitot static probes. The trend on new military aircraft is to use flush port systems for stealth reasons (protruding probes give a significant radar return). It is interesting to note that the Pitot tube, which is universally used on aircraft (and wind tunnels) for measuring airspeed because of its simplicity and effectiveness, was invented over 250 years ago by the French mathematician and scientist Henri Pitot for measuring the flow of water in rivers and canals – one invention which has stood the test of time.

From the measurements of static pressure, P_S, and total pressure, P_T, it is possible to derive the following quantities.

1) *Pressure altitude*, H_P This is derived from the static pressure, P_S, measurement by assuming a 'standard atmosphere'.
2) *Vertical speed*, $\dot{H}_P$ This is basically derived by differentiating P_S.

3) *Calibrated airspeed, V_C.* This is derived directly from the impact pressure, Q_C, which is in turn derived from the difference between the total and static pressures ($Q_C = P_T - P_S$).

4) *Mach number, M* This is the ratio of the true airspeed, V_T, to the local speed of sound, A, that is, $M = V_T/A$, and is derived directly from the ratio of the total pressure to the static pressure, P_T/P_S. (True airspeed is defined as the speed of the aircraft relative to the air.)

The third measurement, namely that of the *measured (or indicated) air temperature, T_m*, is made by means of a temperature sensor installed in a probe in the airstream. This gives a measure of the free airstream temperature, T_S, plus the kinetic rise in temperature due to the air being brought partly, or wholly, to rest relative to the temperature sensing probe. The temperature assuming the air is brought totally to rest (ie recovery ratio = 1) is known as the *total air temperature, T_T*.

The computation of the aircraft's Mach number, M, together with the known recovery ratio of the probe (which allows for the air not being brought wholly to rest) enables the correction factor for the kinetic heating effect to be derived to convert the measured (or indicated) air temperature to the free airstream or *static air temperature, T_S*. The static air temperature so derived then enables the local speed of sound, A, to be determined as this is dependent only on the air temperature. *True airspeed, V_T* can then be readily computed, viz. $V_T = MA$.

Air density ratio, $\dfrac{\rho}{\rho_0}$ can then be computed from P_S and T_S.

(ρ = air density and ρ_0 = air density at standard sea level conditions.)

2.2.2 The air data quantities and their importance

The use and importance of the air data quantities of pressure altitude, vertical speed (rate of climb/descent), calibrated airspeed, Mach number, and true airspeed by the pilot and the key avionic subsystems is discussed below.

2.2.2.1 Air data information for the pilot

The pilot is presented with displays of the above air data quantities, all of which are very important at various phases of the flight or mission. However, the two basic quantities which are fundamental for the piloting of any aircraft from a light aircraft to a supersonic fighter are the *pressure altitude*, and the *calibrated airspeed*. Pressure altitude is the height of the aircraft above sea level derived from the measurement of the static pressure assuming a standard atmosphere. Calibrated airspeed is the speed which, under standard sea level conditions, would give the same impact pressure as that measured on the aircraft. The altimeter displaying pressure altitude. and the calibrated (or indicated) airspeed display thus form part of the classic 'T' layout of vital instrument displays centred around the artificial horizon display. This layout has been largely retained for the modern electronic displays as can be seen in Fig 7.27.

The use of the air data information by the pilot is discussed below.

• *Calibrated airspeed* - The reason for the importance of calibrated airspeed information is that it provides a direct measure of the impact pressure. by definition. and the impact pressure together with the angle of incidence determine the aerodynamically generated lift and drag forces and moments acting on the aircraft. (The angle of incidence is the angle between the direction of the airflow and a datum line through the aerofoil section of the wing or control surface.) These aerodynamic forces and moments in turn determine the aircraft's ability to fly and manoeuvre. its controllability and response and its performance in terms of speed. range. operating height. etc.

The impact pressure is a function of the true airspeed. V_T and the air density. ρ. and at low airspeeds up to about 100m/sec (200 knots) where compressibility effects can be neglected is equal to $\frac{1}{2}\rho V_T^2$. (This relationship no longer holds as the speed increases and compressibility effects increase. and the impact pressure becomes a function of Mach number as well.) The air density is directly related to the altitude so that to maintain the same lift force at high altitudes as at sea level requires an increase in the true airspeed in order to produce the same impact pressure. Hence. the critical speeds which affect the aircraft's behaviour. controllability or safety are specified in terms of calibrated airspeed as this is independent of the air density variation with altitude or temperature. Such critical speeds include the *rotation speed* for take off. the *stalling speed* and the *not to exceed speed* in a dive when the aerodynamic forces and moments exerted during the pull out would approach the structural limits of the airframe or the controllability limits would be reached.

It should be noted that the quantity *indicated airspeed* is frequently used in this context. The indicated airspeed is basically the same quantity as calibrated airspeed but includes the pressure error present in the Pitot/static installation and the instrument errors present in a simple mechanical type of airspeed indicator (ASI) instrument. (Calibrated airspeed is derived by the air data computer using very much more accurate pressure sensors and the inherent pressure errors in the Pitot/static probe installation can be compensated by the computer.)

• *Pressure altitude* – Accurate measurement of the aircraft's altitude is essential for the control of the flight path in the vertical plane. For instance. to maintain adequate clearance of mountains and hills etc. under conditions of poor visibility, flying in cloud or at night. Altitude and airspeed are also vital displays during the approach and landing.

The Air Traffic Control (ATC) authorities also require very accurate measurement of the pressure altitude for air traffic control to ensure safe vertical separation in busy airways. Pressure altitude is therefore automatically reported to the ATC Ground Control by the *ATC transponder* as will be explained in the next section. The ATC authorities also require that the reported pressure altitude must be the same as that displayed on the pilot's altimeter display.

• *True airspeed* – This information is displayed to the pilot for navigation purposes.

• *Mach number* – As the aircraft speed increases and approaches the speed of sound, or exceeds it in the case of a supersonic aircraft, there is a large increase in drag, the lift characteristics change and the pitching moment characteristics change due to compressibility effects. The performance and controllability of the aircraft is dependent on the aircraft's Mach number in this high speed regimen. Accurate information on the aircraft's Mach number is thus an essential display for the pilot. It is also essential information for other aircraft subsystems which are discussed later.

• *Vertical speed or rate of climb/descent* – A display of vertical speed or rate of climb/descent is also required by the pilot and this quantity is generated within the air data computer by differentiating the static pressure. Rate of descent is particularly important during a *ground controlled approach* (GCA) where the pilot will set up a given rate of descent (and speed) in the approach to the airfield. The vertical speed indicator (VSI) display is also used during a turn to detect any tendency to lose height, the pilot applying appropriate corrective movements to the control column or 'stick' to hold a constant height turn.

• *Angle of incidence* – The importance of the angle of incidence has already been mentioned. Generally the lift force from the wings increases fairly linearly with increasing incidence angle up to near the maximum permissible incidence angle at which point the airflow starts to break away and further increase would result in the wing stalling with consequent sudden loss of lift. Airflow sensors to measure the angle of incidence are thus frequently installed so that the pilot can monitor the situation and ensure the critical value is not reached. (It should be noted that the term angle of attack is generally used in the USA for angle of incidence.)

2.2.2.2 Air data for key subsystems

The key subsystems requiring air data information and their use of this information are briefly described below.

• *Air traffic control transponder* – Pressure altitude is supplied to the air traffic control (ATC) transponder for automatic reporting to the air traffic ground control system. The ATC authorities specify the flight levels which aircraft must maintain in 'controlled airspace' in terms of pressure altitude and these are set so that there is a minimum of 1,000 ft vertical separation between aircraft flying in the vicinity of each other below 29,000 ft (FL 290) and 2,000 ft vertical separation above 29,000 ft. As stated earlier, pressure altitude is derived from the measurement of the static pressure and the assumption of a 'standard atmosphere' which enables a unique mathematical law to be derived relating altitude to the static pressure. Pressure altitude, however, can differ from the true altitude because of day to day variations from the standard atmosphere. These differences are small at low altitudes if the *ground pressure correction* is applied but can be much larger at higher altitudes.

This difference between pressure altitude and true altitude does not matter from an air traffic control standpoint providing the pressure altitude measurements are sufficiently accurate. This is because the difference is common to all the pressure altitude measurements made in the vicinity of each other. For example, suppose aircraft A is given a flight level of 33,000 ft to fly by the ATC authorities and aircraft B is given a flight level of 35,000 ft. Let the difference from the true altitude be ΔH so that ignoring any errors in the pressure altitude measurements in the two aircraft, the true altitude of aircraft A will be (33,000 + ΔH) and that of aircraft B will be (35,000 + ΔH). The required vertical separation of 2,000 ft is maintained. The cruising altitudes of jet airliners are typically in the 29,000–45,000 ft band and it will be shown in Section 2.3 that the ATC 2,000 ft separation levels place very stringent accuracy requirements on the Static Pressure Sensors.

• *Flight control systems* – Calibrated airspeed and pressure altitude information is required by the flight control system (FCS). This is to enable automatic adjustment to be made to the gains (or 'gearings') of the FCS with airspeed and height to compensate for the wide variation in control effectiveness and aircraft response over the flight envelope. This is frequently described as 'air data gain scheduling'. This is covered in more detail in Chapters 3 and 4. Adequate redundancy must be provided in the system to ensure failure survival in the case of a fly-by-wire flight control system. This may involve three or more independent air data computing systems.

• *Autopilot system* – A number of autopilot control modes require air data information, eg 'height acquire/hold', 'Mach number acquire/hold' and 'airspeed acquire/hold' (auto-throttle system). Air data gain scheduling may also be required by the autopilot control loops. Autopilots are covered in Chapter 8.

• *Navigation system* – Pressure altitude and true airspeed are required by the navigation system. Pressure altitude is required for navigation in the vertical plane. It can be combined (or mixed) with the inertially derived information from the inertial navigation system (INS) to provide vertical velocity and altitude information which is superior to either source on its own. (The technique is referred to as 'Barometric/Inertial' mixing and is covered in Chapter 6.)

The aircraft's velocity vector is derived from the vector sum of the barometric/inertial vertical velocity and the horizontal velocity (or ground speed vector) output from the INS. The velocity vector information is used for guidance and control, eg HUD, flight director system, weapon aiming system etc.

Pressure altitude is also essential information for a terrain reference navigation (TRN) system as it measures the vertical motion of the aircraft and so enables the ground profile to be derived from the radio altimeter measurements. TRN systems operate by correlating the radar altimeter measurements of the ground profile with a stored topographical map elevation data base and hence establishing the aircraft position and correcting the dead reckoning position estimate. (It is interesting to note that the standard deviation from the horizontal of an isobar (line joining points of equal pressure) over a distance of 1 nautical mile at sea level is 1 foot.)

True airspeed is required for dead reckoning (DR) navigation, DR position being computed from a knowledge of the true airspeed, aircraft heading and forecast/estimated wind velocity. True airspeed information is also used to compute the wind velocity vector using the ground speed information from the INS. Other sources of navigational information, eg GPS or Doppler, can also be used to derive wind velocity from the true airspeed information. Information on wind velocity is required by the flight management system (FMS) as will be explained. Wind velocity information is also required for weapon aiming in the case of a military aircraft.

• *Flight management system* – The flight management system requires information on all the air data quantities: pressure altitude, vertical speed, Mach number, static air temperature, true airspeed and calibrated airspeed.

Air data information is essential for the FMS to maintain the aircraft on the most fuel efficient flight path and for achieving 4D flight management (3D position and time). FMS systems are covered in Chapter 8.

• *Engine control systems* – Height and calibrated airspeed information is required by the engine control systems.

The flow of air data information from the air data computing system to the key avionic subsystems is shown in Fig. 2.2. Modern aircraft systems use a time multiplexed digital data bus system, such as MIL STD 1553B on a military aircraft,

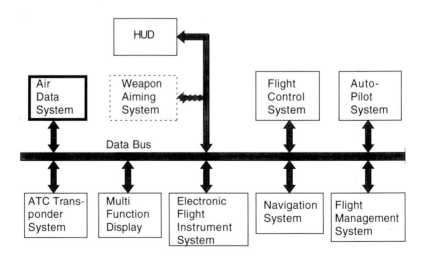

Fig. 2.2 Flow of air data to key avionic sub-systems (redundancy omitted for clarity).

or ARINC 629 in the case of a civil aircraft to transmit the data and interconnect the system. The degree of redundancy has been omitted for clarity: for instance. civil aircraft have at least two independent air data computing systems.

2.3 DERIVATION OF AIR DATA LAWS AND RELATIONSHIPS

The object of this section is to show how the mathematical laws relating altitude. air density, Mach number, calibrated airspeed. free airstream temperature and true airspeed are derived from the basic air data measurements of static pressure. total pressure and total (or indicated) air temperature.

2.3.1 Choice of units

It is appropriate at this point to explain the units which are used in this book. All quantities in this book are defined in terms of the SI system of units (Le Système Internationale d'Unités). However, some other units, notably the foot for altitude and the knot (1 nautical mile per hour) for speed are quoted in parallel for reasons which will be explained.

The SI unit of mass is the kilogram (kg), the unit of length is the metre (m) and the unit of time is the second (s). The SI unit of temperature is the Kelvin (K), zero degrees Kelvin corresponding to the absolutely lowest temperature possible at which almost all molecular translational motion theoretically stops. To convert temperature in degrees Celsius (°C) to degrees Kelvin it is necessary to add 273.15. Thus $0°C = 273.15°K$ and $30°C = 303.15°K$ and conversely $0°K = -273.15°C$.

The SI unit of force is the Newton (N), one Newton being the force required to accelerate a mass of one kilogram by 1 metre per second per second. Converting to pound, foot, second units, one Newton is approximately equal to a force of 0.22 pound (lb) weight. The apocryphal story of 'an apple falling on Newton's head triggering off his theory of gravitation' gives an appropriate twist to the unit of the Newton, as an average apple weighs about 0.22 lb! Pressure in the SI system is measured in N/m^2 or Pascals (Pa). The millibar (mb) which is equal to 10^{-3} bar is also very widely used. the bar being derived from BARometric. One bar corresponds to the atmospheric pressure under standard sea level conditions.

$$1 \text{ bar} = 10^5 \text{ N/m}^2 = 100 \text{ kPa}$$
$$1 \text{ mb} = 100 \text{ N/m}^2 = 100 \text{ Pa}$$

Inches of mercury are also widely used in the United States. The atmospheric pressure at sea level under standard temperature conditions of $288.15°K$ ($15°C$) is 101.325 kN/m^2 (kPa) or 1013.25 mb and is referred to as the *standard atmospheric pressure*. (Also equal to 29.9213 inches Hg or 760.00mm Hg.)

To return to the use of the foot for quoting altitude and the knot for speed, altitude is quoted in feet as this is the unit used by the Air Traffic Control authorities, and altimeters in the USA and the UK are calibrated in feet. Speed is also quoted in knots as this unit is widely used in navigation. One knot is one nautical mile (NM) per hour and one nautical mile is equal to the length of the arc on the Earth's surface subtended by an angle of one minute of arc measured from the Earth's centre and can be related directly to latitude and longitude. 1 NM = 6076.1155 ft (or 1852m exactly).

The conversion from knots to metres/second is given by 1 knot = 0.5144m/s and conversely 1m/s = 1.9438 knots

A useful approximate conversion (accurate to within 3%) is

$$1 \text{ knot} \approx 0.5 \text{m/s, or, } 1 \text{m/s} \approx 2 \text{ knots.}$$

2.3.2 Altitude–static pressure relationship

By making certain assumptions regarding a 'standard' atmosphere it is possible to express the altitude above sea level at any point in the Earth's atmosphere as a single valued function of the atmospheric pressure at that point. These assumptions concern the chemical constitution of the atmosphere, initial atmospheric conditions at sea level and the temperature distribution throughout the atmosphere.

These assumed conditions have been agreed internationally and enable a mathematical law to be established relating static pressure and altitude so that by measuring the pressure of the free airstream the pressure (or barometric) altitude can be derived.

These assumptions of a standard atmosphere are based on statistical data and the pressure altitude so derived can differ from the true altitude above sea level by several thousand feet. However, as explained earlier, provided the static pressure is measured accurately and the computation errors are small all aircraft flying at a particular pressure altitude will be flying at the same altitude and the fact that the pressure altitude does not coincide with the true altitude does not matter from an air traffic control aspect. A standard atmosphere also enables flight tests, wind tunnel results and general aircraft design and performance parameters to be related to a common reference.

The pressure altitude–static pressure relationship is derived as follows:
Referring to Fig. 2.3, the change in pressure, dp, of air of density, ρ, resulting from a small change in height, dH, is derived from equating the forces acting in the vertical plane on an elemental volume of air from which

$$-dp = \rho g dH \qquad (2.1)$$

where g is the gravitational constant.

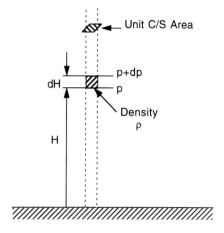

Fig. 2.3 Static pressure and altitude relationship.

From the gas law

$$p = \rho R_a T \qquad (2.2)$$

where T = air temperature (°K) and R_a = gas constant for unit mass of dry air. Combining equations (2.1) and (2.2) yields

$$-\frac{dp}{P} = \frac{g}{R_a T} \, dH \qquad (2.3)$$

T can be expressed as a function of H only, by making the following assumptions which are based on statistical data.

(1) The temperature at sea level, T_0, and the pressure at sea level, P_{S0}, are assumed to be constant.
(2) The temperature decreases linearly with increasing height until a height known as the *tropopause* is reached above which height the temperature remains constant until the *stratopause* height is reached. The region below the tropopause height is known as the *troposphere*.
    The law relating temperature, T, at altitude, H, up to the tropopause height is

$$T = T_0 - LH$$

where L = temperature lapse rate.
(3) The temperature above the tropopause height stays constant at a value $T_T{}^*$ until a height known as the *stratopause* is reached. (The asterisk * is used to distinguish the tropopause temperature $T_T{}^*$ from the total temperature T_T). The

region between the tropopause and stratopause heights is known as the stratosphere.

(4) At heights above the stratopause height the temperature starts to increase linearly with height, this region being known as the *chemosphere*.

The temperature in the chemosphere is given by

$$T = T_T{}^* + L\,(H - H_S)$$

where L = temperature rise rate and H_S = height of stratopause

The altitude–pressure law is thus made up of three parts covering the troposphere, stratosphere and chemosphere regions respectively.

The gravitational acceleration, g at an altitude, H differs from the value at the Earth's surface, g_0 and obeys an inverse square law

viz
$$g = \frac{R^2}{(R + H)^2}\,g_0 \qquad (2.4)$$

where R = radius of the Earth.

The effect of this variation in deriving the pressure altitude is very small over the normal operating altitudes and it is convenient to assume g is constant at the sea level value g_0 thereby simplifying the integration of equation (2.4). The value of altitude so derived is known as the *geopotential altitude*. This small difference does not matter from the point of view defining a pressure altitude which is a single valued function of the static pressure.

(a) Troposphere region $T = T_0 - LH$

Substituting for T and assuming g is constant and equal to g_0 in equation (2.3) and integrating both sides gives

$$-\int_{P_{S0}}^{P_S} \frac{1}{p}\,dp = \frac{g_0}{R_a} \int_0^H \frac{1}{(T_0 - LH)}\,dH$$

where P_{S0} = value of p at sea level, $H = 0$ and P_S = value of p at altitude H. From which

$$\log_e \frac{P_S}{P_{S0}} = \frac{g_0}{LR_a}\,\log_e \frac{(T_0 - LH)}{T_0}$$

Hence

$$P_S = P_{S0}\left(1 - \frac{L}{T_0}H\right)^{\frac{g_0}{LR_a}} \qquad (2.5)$$

and

$$H = \frac{T_0}{L}\left[1 - \left(\frac{P_S}{P_{S0}}\right)^{\frac{LR_a}{g_0}}\right]$$

(2.6)

(b) Stratosphere region $T = T_T^*$

Substituting T_T^* for T and g_0 for g in equation (2.3) and integrating both sides gives

$$-\int_{P_{ST}}^{P_S} \frac{1}{p}\,dp = \frac{g_0}{R_a T_T^*}\int_{H_T}^{H} dH$$

where P_{ST} = pressure at tropopause altitude, H_T

whence

$$P_S = P_{ST}\, e^{-\left(\frac{g_0}{R_a T_T^*}\right)(H - H_T)}$$

(2.7)

and

$$H = H_T + \frac{R_a T_T^*}{g_0} \log_e \frac{P_{ST}}{P_S}$$

(2.8)

(c) Chemosphere region

The static pressure vs altitude relationship can be derived by a similar process by substituting $T = T_T^* + L (H - H_S)$ and $g = g_0$ in equation (2.3) and integrating both sides between the appropriate limits.

Whence

$$P_S = P_{SS}\left[1 + \frac{L}{T_T^*}(H - H_S)\right]^{-\frac{g_0}{R_a L}}$$

(2.9)

and

$$H = H_S + \frac{T_T^*}{L}\left[\left(\frac{P_{SS}}{P_S}\right)^{\frac{R_a L}{g_0}} - 1\right]$$

(2.10)

The values of the constants used in the pressure–altitude law are set out in Table 2.1 in SI units with the alternative units in brackets.

The symbol H_P is used henceforth to denote pressure altitude.

The formulae relating P_S and H are set out below and are derived by substituting the appropriate values in the table in equations (2.5), (2.7) and (2.9) respectively, and H_P for H.

(a) *Troposphere region* – −914.4m to 11,000m (−3,000 ft to 36,089 ft)

$$P_S = 1{,}013.25 \, (1 - 2.25577 \times 10^{-5} H_P)^{\,5.255879} \text{ mb} \qquad (2.11)$$

(b) *Stratosphere region* – 11,000m to 20,000m (36,089 ft–65,617 ft)

$$P_S = 226.32 \, e^{-1.576885 \times 10^{-4} (H_P - 11{,}000)} \text{ mb} \qquad (2.12)$$

(c) *Chemosphere region* – 20,000m to 32,004m (65,617 ft–105,000 ft)

$$P_S = 54.7482 \, [1 + 4.61574 \times 10^{-6} (H_P - 20{,}000)]^{\,-34.163215} \text{ mb} \qquad (2.13)$$

H_P is expressed in metres. (The conversion factor to convert feet to metres is 1 ft = 0.3048m)

(The chemosphere region formulae have been omitted for brevity as already mentioned).

Table 2.1 Pressure–altitude law constants

Constant	Standard atmosphere value
Pressure at sea level, P_{S0}	101.325 kPa (1013.25 mb)
Temperature at sea level, T_0	288.15°K
Troposphere lapse rate, L	6.5×10^{-3} °C/m
Tropopause height, H_T	11,000m (36,089.24 ft)
Tropopause temperature, $T_T{}^*$	216.65°K (-56.5°C)
Stratopause height, H_S	20,000m (65,617 ft)
Chemosphere rise rate, L	1.0×10^{-3} °C/m
Chemosphere height limit	32,004m (105,000 ft)
g_0	9.80665m/sec²
R_a	287.0529 Joules/°K/kg
g_0/LR_a (Troposphere)	5.255879
g_0/R_aT	1.576885×10^{-4} m^{-1}
g_0/LR_a (Chemosphere)	34.163215

The static pressure–altitude relationship computed from the above formulae is plotted in Fig. 2.4. It can be shown that the difference between the pressure, or geopotential, altitude, H_P, derived from the above formulae, which assume $g = g_0$, and the geometric altitude, H_G, obtained by allowing for the variation of g with altitude is given to a close approximation over normal altitudes by $\dfrac{H_G{}^2}{R}$ where (R = radius of Earth = 6,356.8 km at latitude 45°). As mentioned earlier, the difference is small over normal altitudes, although it follows a square law. For example, the difference is 19m (62 ft) at 11,000m (36,089 ft).

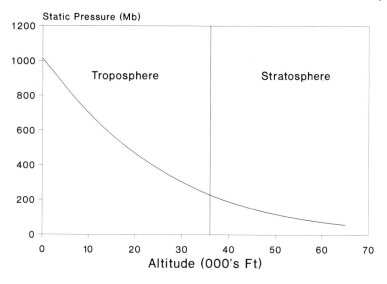

Fig. 2.4 Static pressure vs altitude.

2.3.3 Variation of ground pressure

Errors due to the variation in the ground pressure from the assumed standard value are taken out of the altimeter reading by setting a scale to a given pressure. This action affects the zero point and so alters the altimeter reading over the whole of its range by a height corresponding to the pressure set as determined by the altitude–pressure law.

The pressure that is set for any given occasion varies with the type of altitude indication to be used. If the standard ground level pressure (1013.25 mb) is set then the altimeter will read pressure altitude (or barometric altitude). If the 'QFE' system is used, the pressure at ground level (not necessarily sea level) is set and the altimeter reads height above ground at this point if the atmosphere is assumed standard. Thus if the QFE for the airfield is set, the altimeter will read zero on touch down (assuming an accurate instrument system).

If the QNH system is used the pressure at mean sea level (often computed) for the region concerned is set and the altimeter reads height above sea level, if the atmosphere is assumed standard.

The different modes of height indication are shown in Fig. 2.5.

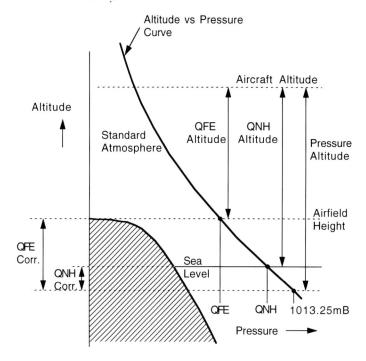

Fig. 2.5 Altimeter ground pressure adjustments.

2.3.4 Air density vs altitude relationship

The relationship between air density, ρ, and altitude, H, is derived from the equation relating P_S and H and using the gas law

$$P_S = \rho R_a T$$

to eliminate P_S.

For example, in the troposphere

$$P_S = P_{S0}\left(1 - \frac{L}{T_0}H\right)^{\frac{g_0}{LR_a}}$$

$$T = T_0 - LH$$

Hence

$$\frac{\rho}{\rho_0} = \left(1 - \frac{L}{T_0}H\right)^{(\frac{g_0}{LR_a})-1}$$ (2.14)

where ρ_0 = density at standard sea level conditions = $\dfrac{P_{S0}}{R_a T_0}$

The ratio ρ / ρ_0 is referred to as the *density reduction factor.*

Example - Compute the air density at an altitude of 10,000m (30,480 ft) given $\rho_0 = 1.225$ kg/m³ ($g_0/LR_a = 5.255879$)

$$\text{Density at 10,000m} \quad = 1.225\left(1 - \frac{6.5}{1000} \times \frac{10000}{288.15}\right)^{5.255879 - 1}$$

$$= 0.4127 \text{ kg/m}^3$$

In practice, air density ratio ρ / ρ_0 when required is normally computed from the static pressure, P_S, and the measured (or indicated) air temperature, T_m using the following relationships:

Static air temperature, $T_S = T_m /(1 + r\,0.2M^2)$ (Refer to Section 2.3.9)

and

$$\frac{\rho}{\rho_0} = \frac{P_S}{P_{S0}} \cdot \frac{T_0}{T_S}$$

from which

$$\frac{\rho}{\rho_0} = \frac{P_S}{P_{S0}} \cdot \frac{T_0\,(1 + r\,0.2\,M^2)}{T_m} \tag{2.15}$$

2.3.5 Speed of sound

The derivation of the formulae for the speed of sound, A, is set out briefly below as it is fundamental to the derivation of Mach number, calibrated airspeed, static air temperature and true airspeed from the measurements of total pressure, static pressure and indicated air temperature.

Consider first a stream tube of air of unit cross sectional area, a, through which a pressure wave is being transmitted with velocity V, travelling from right to left. Imagine now, the equivalent situation where the pressure wave is stationary and the air is moving with velocity V and travelling from left to right as shown in Fig. 2.6.

Consider a section at A where the pressure is p and air velocity V.

Let the section at B represent the adjacent portion of the stream tube where the pressure of the air increases by dp in passing through the pressure wave and let the velocity change by dV and the density by $d\rho$.

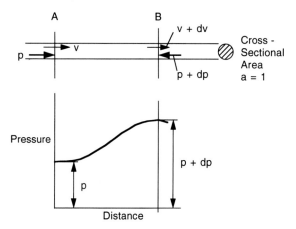

Fig. 2.6 Pressure wave.

Equating the mass flows at A and B

$$\rho a V = (\rho + d\rho)\, a\, (V + dV)$$

Neglecting second order terms

$$-V d\rho = \rho\, dV \qquad (2.16)$$

Force acting on stream tube of air between sections A and B
= Change of momentum per second
= Mass per second × change in velocity.

$$pa - (p+dp)a = \rho\, aV\, dV$$

Hence
$$dp = -\rho V\, dV \qquad (2.17)$$

Combining equations (2.16) and (2.17) yields

$$dp/d\rho = V^2 \qquad (2.18)$$

The transmission of a pressure wave closely approximates an adiabatic process because the pressure change takes place very suddenly and consequently there is no time for any appreciable interchange of heat. The law for an adiabatic process is

$$p = K\rho^{\gamma} \qquad (2.19)$$

where $\gamma = \dfrac{\text{specific heat of gas at constant pressure}}{\text{specific heat of gas at constant volume}} = \dfrac{c_p}{c_v}$

($\gamma = 1.4$ for air) and K = constant.

$$\frac{dp}{d\rho} = K\gamma \, \rho^{\gamma-1} = \frac{\gamma \, p}{\rho} \qquad (2.20)$$

Equating equations (2.18) and (2.20)

Let A = speed of sound ($= V$)

$$A = \sqrt{\frac{\gamma \, p}{\rho}} \qquad (2.21)$$

Substitute $p = \rho R_a T$ in equation (2.21)

$$A = \sqrt{\gamma R_a T} \qquad (2.22)$$

Hence, the speed of sound depends only on the temperature of the air.

The speed of sound at sea level, A_0, under standard pressure and temperature conditions is thus equal to

$$\sqrt{1.4 \times 287.0529 \times 288.15}$$

i.e. $A_0 = 340.294$m/s.

The speed of sound reduces with increasing altitude as the temperature decreases until the tropopause height is reached. Thereafter the speed stays constant within the stratosphere height band as the temperature is constant.

The variation in the troposphere region is derived from the relationship

$$A = A_0 \sqrt{1 - \frac{L}{T_0} H}$$

The variation from sea level to 65617 ft is set out in Table 2.2.

2.3.6 Pressure–speed relationships

The relationship between total pressure, P_T, static pressure, P_S, true airspeed, V_T, and the local speed of sound is explained from first principles in this section as it is fundamental to the derivation of Mach number, M, and calibrated airspeed, V_C. These are covered in Sections 2.3.7 and 2.3.8 respectively.

As stated earlier, knowledge of the Mach number enables the correction for the kinetic heating effect to be determined so that the static air temperature, T_S, can be derived from the measured air temperature, T_m, as will be shown in Section 2.2.9. True airspeed can then be derived from the Mach number and the static air temperature as will be shown in Section 2.3.10.

Table 2.2 Speed of sound variation with height

H	Speed of sound (A)	
0 ft	340.3m/s	(661.5 knots)
10,000 ft	328.4m/s	(637.4 knots)
20,000 ft	316.0m/s	(614.3 knots)
30,000 ft	303.2m/s	(589.4 knots)
36,089 ft	295.1m/s	(573.6 knots)
65,617 ft	295.1m/s	(573.6 knots)

(a) Subsonic speeds

The total and static pressure measurement system is shown schematically in Fig. 2.7. Consider first the case of low airspeeds below $M = 0.3$, where the air can be considered to be incompressible and the density therefore is constant.

The momentum equation for the air flow is

$$dp + \rho V dV = 0$$

In the free airstream $p = P_S$ and $V = V_T$
At the probe face $p = P_T$ and $V = 0$
Integrating the momentum equation between these limits

$$\int_{P_S}^{P_T} dp + \rho \int_{V_T}^{0} V dV = 0 \tag{2.23}$$

Hence

$$P_T - P_S = \tfrac{1}{2} \rho V_T^2 \tag{2.24}$$

(Equation (2.24) is Bernoulli's equation.)

$$V_T = \sqrt{\frac{2}{\rho}} \cdot \sqrt{P_T - P_S} \quad \text{(at low airspeeds)} \tag{2.25}$$

However, air is a compressible fluid and the density is not constant. The change in density due to the high impact pressures resulting from high airspeeds must

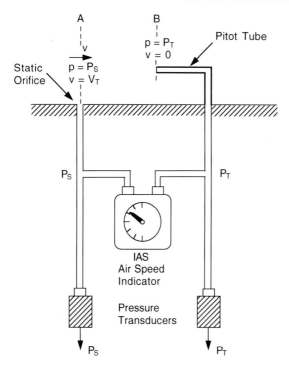

Fig. 2.7 Impact pressure measurement.

therefore be taken into account. Assuming adiabatic flow, the relationship between pressure and density is

$$P = K \rho^\gamma$$

From which
$$\rho = \frac{1}{K^{\frac{1}{\gamma}}} P^{\frac{1}{\gamma}}$$
(2.26)

Substituting for ρ in the momentum equation $dp + \rho V dV = 0$

$$dp + \frac{1}{K^{\frac{1}{\gamma}}} P^{\frac{1}{\gamma}} V dV = 0$$
(2.27)

In the free airstream $p = P_S$ and $V = V_T$
At the probe face $p = P_T$ and $V = 0$

Rearranging equation (2.27) and integrating between these limits

$$\int_{P_S}^{P_T} P^{-\frac{1}{\gamma}} \, dp + \frac{1}{K^{\frac{1}{\gamma}}} \int_{V_T}^{0} V dV = 0 \tag{2.28}$$

$$\frac{\gamma}{\gamma-1} \left[P_T^{\frac{(\gamma-1)}{\gamma}} - P_S^{\frac{(\gamma-1)}{\gamma}} \right] = \frac{1}{K^{\frac{1}{\gamma}}} \cdot \frac{V_T^{2}}{2} \tag{2.29}$$

From equation (2.26) $K^{\frac{1}{\gamma}} = \dfrac{P_S^{\frac{1}{\gamma}}}{\rho}$

Substituting for $K^{\frac{1}{\gamma}}$ in equation (2.29) and rearranging gives

$$\frac{P_T}{P_S} = \left[1 + \frac{(\gamma-1)}{2} \cdot \frac{\rho}{\gamma P_S} \cdot V_T^{2} \right]^{\frac{\gamma}{(\gamma-1)}}$$

$\dfrac{\gamma P_S}{\rho} = A^2$ and putting $\gamma = 1.4$ in the above equation gives

$$\frac{P_T}{P_S} = \left[1 + 0.2 \, \frac{V_T^{2}}{A^2} \right]^{3.5} \tag{2.30}$$

and $$Q_C = P_S \left[\left(1 + 0.2 \, \frac{V_T^{2}}{A^2} \right)^{3.5} - 1 \right] \tag{2.31}$$

(b) Supersonic speeds
The aerodynamic theory involved in deriving the relationship between P_T/P_S and Mach number (V_T/A) at supersonic speeds $(M > 1)$ is beyond the scope of this book. The formula relating pressure ratio P_T/P_S and Mach number derived by Rayleigh is therefore set out below without further explanation. This can be found in suitable aerodynamic textbooks for those readers wishing to know more.

$$\frac{P_T}{P_S} = \frac{\left[\frac{(\gamma+1)}{2} \left(\frac{V_T}{A} \right)^2 \right]^{\frac{\gamma}{(\gamma-1)}}}{\left[\frac{2\gamma}{(\gamma+1)} \left(\frac{V_T}{A} \right)^2 - \frac{(\gamma-1)}{(\gamma+1)} \right]^{\frac{1}{(\gamma-1)}}} \qquad (2.32)$$

Substituting $\gamma = 1.4$, equation (2.32) becomes

$$\frac{P_T}{P_S} = \frac{166.92 \left(\frac{V_T}{A} \right)^7}{\left[7 \left(\frac{V_T}{A} \right)^2 - 1 \right]^{2.5}} \qquad (2.33)$$

$$Q_C = P_S \left[\frac{166.92 \left(\frac{V_T}{A} \right)^7}{\left[7 \left(\frac{V_T}{A} \right)^2 - 1 \right]^{2.5}} - 1 \right] \qquad (2.34)$$

2.3.7 Mach number

Mach number can be derived from equations (2.30) and (2.33) for the subsonic and supersonic regions respectively by substituting $M = V_T/A$

(a) Subsonic speeds
From equation (2.30)

$$\frac{P_T}{P_S} = (1 + 0.2 M^2)^{3.5} \qquad (2.35)$$

(b) Supersonic speeds
From equation (2.33)

$$\frac{P_T}{P_S} = \frac{166.92 M^7}{(7M^2 - 1)^{2.5}} \qquad (2.36)$$

The pressure ratio, P_T/P_S, is plotted against Mach number, M, in Fig. 2.8 for the subsonic and supersonic regimens.

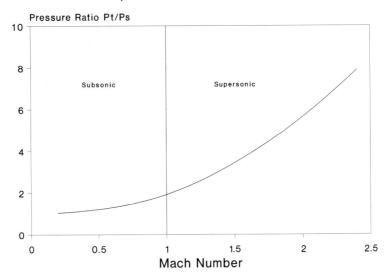

Fig. 2.8 Pressure ratio vs Mach number.

2.3.8 Calibrated airspeed

Calibrated airspeed, V_C, can be derived directly from equations (2.31) and (2.34) for the subsonic and supersonic regimens respectively by substituting for sea level conditions, that is $P_S = P_{S0}$ and $V_T = V_C$ (by definition).

(a) Subsonic speeds ($V_C \leq A_0$)
Referring to equation (2.31)

$$Q_C = P_{S0} \left\{ \left[1 + 0.2 \left(\frac{V_C}{A_0} \right)^2 \right]^{3.5} - 1 \right\} \tag{2.37}$$

A simpler approximate expression relating Q_C and V_C can be obtained by applying the binomial expansion to the inner factor $\left[1 + 0.2 \left(\frac{V_C}{A_0} \right)^2 \right]^{3.5}$ and noting

$$A_0^2 = \frac{\gamma P_{S0}}{\rho_0}$$

The expression obtained is

$$Q_c \approx \tfrac{1}{2}\rho_0 V_c^2 \left[1 + \frac{1}{4}\left(\frac{V_c}{A_0}\right)^2 \right] \tag{2.38}$$

The error in this expression being about 1.25% when $V_c = A_0$ (661.5 knots). It should be noted that the exact formula in equation (2.37) is used to compute calibrated airspeed. The approximate formula (2.38) is useful, however, for appreciating the effects of increasing airspeed and Mach number. At low airspeeds, it can be seen that $Q_c \approx \tfrac{1}{2}\rho_0 V_c^2$

(b) Supersonic speeds ($V_c > A_0$)
Referring to equation (2.34)

$$Q_c = P_{S0} \left| \frac{166.92\left(\dfrac{V_c}{A_0}\right)^7}{\left[7\left(\dfrac{V_c}{A_0}\right)^2 - 1\right]^{2.5}} - 1 \right| \tag{2.39}$$

The impact pressure, Q_c, is plotted against calibrated airspeed, V_c, in Fig. 2.9 for $V_c \leq A_0$ and $V_c > A_0$ speeds.

Note the very low impact pressure of 16.3 mb at a calibrated airspeed of 51.5m/s (100 knots) and the very high impact pressure of 1456 mb at a calibrated airspeed of 412m/s (800 knots), approx $M = 1.2$ at sea level.

2.3.9 Static air temperature

As already mentioned, the temperature sensed by a thermometer probe in the airstream is the free airstream temperature plus the kinetic rise in temperature due to the air being brought partly or wholly to rest relative to the sensing probe.

The kinetic rise in temperature can be obtained by application of Bernouilli's equation to compressible flow and assuming the pressure changes are adiabatic.

For unit mass of air

$$\frac{P_1}{\rho_1} + \tfrac{1}{2}V_1^2 + E_1 = \frac{P_2}{\rho_2} + \tfrac{1}{2}V_2^2 + E_2 \tag{2.40}$$

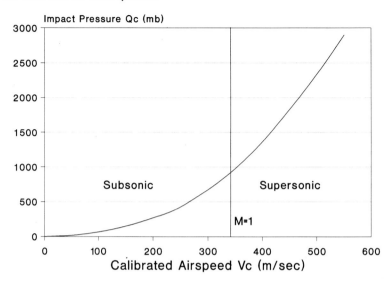

Fig. 2.9 Impact pressure vs calibrated airspeed.

where P_1, ρ_1, V_1, E_1 and P_2, ρ_2, V_2, E_2 represent pressure, density, velocity, temperature and internal energy at two points in a streamline flow, namely in the free airstream and at the probe.

$$\frac{P_1}{\rho_1} = R_a T_1 \text{ and } \frac{P_2}{\rho_2} = R_a T_2 \qquad \text{(gas law)}$$

In the free airstream, $V_1 = V_T$ and $T_1 = T_S$
At a stagnation point at the probe $V_2 = 0$ and $T_2 = T_T$
Substituting these values in equation (2.40) gives

$$\tfrac{1}{2}V_T{}^2 = (E_2 - E_1) + R_a (T_T - T_S) \qquad (2.41)$$

The change in internal energy becomes heat and is given by

$$E_2 - E_1 = Jc_v (T_T - T_S) \qquad (2.42)$$

where J = mechanical equivalent of heat (Joule's constant) and c_v = specific heat of air at constant volume. Hence equation (2.41) becomes

$$\tfrac{1}{2}V_T{}^2 = (Jc_v + R_a) (T_T - T_S) \qquad (2.43)$$

From thermodynamic theory

$$R_a = J(c_p - c_v)\qquad(2.44)$$

where c_p = specific heat of air at constant pressure
Combining equations (2.43) and (2.44) yields

$$\tfrac{1}{2}V_T^2 = \frac{R_a c_p}{(c_p - c_v)}(T_T - T_S)$$

Rearranging and substituting $\gamma = c_p/c_v$ yields

$$T_T - T_S = \frac{(\gamma - 1)}{2} \cdot \frac{1}{\gamma R_a} \cdot V_T^2\qquad(2.45)$$

$A^2 = \gamma R_a T_S$ (refer to equation (2.21))
Substituting A_2/T_S for γR_a and putting $\gamma = 1.4$ in equation (2.47) yields

$$T_S = \frac{T_T}{(1 + 0.2\,M^2)}\qquad(2.46)$$

A constant, r, known as the recovery factor, is generally introduced into the above equation, its value being dependent on the temperature probe installation,

viz
$$T_S = \frac{T_m}{(1 + r\,0.2\,M^2)}\qquad(2.47)$$

where T_m = measured (or indicated) air temperature.
At a stagnation point $r = 1.0$ and $T_m = T_T$. Although r is normally assumed to be constant, it can vary slightly due to change of heat transfer from the probe with altitude; significant variations can also occur flying through rain or cloud.

2.3.10 True airspeed

The computation of the temperature of the free airstream, T_S, enables the local speed of sound, A, to be established as $A = \sqrt{\gamma R_a T_S}$
True airspeed, V_T can then be obtained from the Mach number.

$$V_T = MA = M\sqrt{\gamma R_a T}$$

Hence
$$V_T = \sqrt{\gamma R_a} \cdot M \cdot \sqrt{\frac{T_m}{(1+r\,0.2M^2)}} \qquad (2.48)$$

$$V_T = 20.0468M \sqrt{\frac{T_m}{(1+r\,0.2M^2)}} \text{ m/sec} \qquad (2.49)$$

(This can be readily converted to knots using the conversion factor 1m/s = 1.9425 knots.)

The air data formulae are set out in Table.2.3.

2.3.11 Pressure error

The pressures measured at the Pitot head and the static orifice are not exactly equal to the true values. The error in the total pressure measured at the Pitot probe is usually very small provided the incidence angles of the probe to the airspeed vector are small. The error in the static pressure measurement, however, can be significant, and is due to disturbance of the airflow near the detecting orifices by the pressure head itself and by the aircraft structure. To minimise the latter effect, the Pitot static probe is often mounted in front of the foremost part of the aircraft in which case disturbance due to the aircraft disappears above the speed of sound. The normal error pattern for this type of installation is for the error to build up to a maximum at $M = 1$ and for it then suddenly to drop to a low value at supersonic speeds.

The pressure error for the aircraft installation is determined by extensive flight testing and in general is a function of Mach number and altitude. This function is often referred to as the *static source error correction* (SSEC) and can be a relatively complex non-linear function around $M = 1$. The SSEC is generated by the air data computer and corrections applied to the appropriate outputs.

2.4 AIR DATA SENSORS AND COMPUTING

2.4.1 Introduction

The key air data sensors, as already mentioned, comprise two pressure sensors and a temperature sensor.

The temperature sensor generally comprises a simple resistance bridge with one arm of the bridge consisting of a resistive element exposed to the airstream, the resistance of this element being a function of temperature. The siting of the probe

Table 2.3 Air data formulae

Quantity	Computational formulae
Geopotential pressure altitude H_P metres	(a) Troposphere 0–11,000m (0–36,089 ft) $$P_S = 101.325 \, (1 - 2.25577 \times 10^{-5} H_P)^{5.255879} \text{ kPa}$$ (b) Stratosphere 11,000–20,000m (36,089–65,617 ft) $$P_S = 22.632 e^{-1.576885 \cdot 10^{-4} (H_P - 11,000)} \text{ kPa}$$
Air density ratio $\dfrac{\rho}{\rho_0}$	$$\frac{\rho}{\rho_0} = \frac{P_S}{0.35164 \, T_S}$$
Mach number M	(a) Subsonic speeds ($M \leq 1$) $$\frac{P_T}{P_S} = (1 + 0.2 M^2)^{3.5}$$ (b) Supersonic speeds ($M > 1$) $$\frac{P_T}{P_S} = \frac{166.92 M^7}{\left[7 M^2 - 1\right]^{2.5}}$$
Calibrated airspeed V_C m/s	(a) $V_C \leq A_0$ $$Q_C = 101.325 \left[\left[1 + 0.2 \left(\frac{V_C}{340.294} \right)^2 \right]^{3.5} - 1 \right] \text{ kPa}$$ (b) $V_C > A_0$ $$Q_C = 101.325 \left[\frac{166.92 \left(\dfrac{V_C}{340.294} \right)^7}{\left[7 \left(\dfrac{V_C}{340.294} \right)^2 - 1 \right]^{2.5}} - 1 \right] \text{ kPa}$$
Static air temperature T_S °K	$$T_S = \frac{T_m}{1 + r \, 0.2 M^2} \, °K$$
True airspeed V_T m/s	$$V_T = 20.0468 M \sqrt{T_S} \text{ m/s}$$

and establishing the recovery factor, r (refer to section 2.3.8), are the key factors of the sensor. In terms of technology, the temperature sensor is a relatively simple and straightforward device which fairly readily meets the system accuracy requirements. The pressure sensors, however, merit considerable discussion because of their very high accuracy requirements and thus influence on the overall system accuracy, long term stability, reliability and overall cost.

A very wide variety of air data pressure sensors has been developed by different companies and organisations/technical establishments worldwide. You name an effect which is a function of pressure and some organisation has developed (or tried to develop) a pressure sensor exploiting it.

Air data pressure sensors require an extremely high accuracy (which will be explained shortly) and involve a long expensive development to establish and qualify a producible, competitive device. Like most sensors, they not only sense the quantity being measured but they can also be affected by:

Temperature changes
Vibration
Shock
Acceleration
Humidity etc.

The art of sensor design is to minimise and if possible eliminate these effects on the sensor.

2.4.2 Air data system pressure sensors

2.4.2.1 Accuracy requirements

(a) Static pressure sensor

Typical air data static pressure sensors have a full scale pressure range of 0 to 130 kPa (0–1,300 mb) and a minimum non-derangement pressure of 390 kPa (3.9 atmospheres) to cover mishandling in testing, etc. They are required to operate over a temperature range of $-60\mathrm{C}^{\circ}$ to $+90\mathrm{C}^{\circ}$ (or higher).

The effect of a small error in static pressure measurement and the resulting altitude error can be derived by rearranging equation (2.3) *viz.*

$$dH = -\frac{R_a T}{g}\frac{1}{P}dp$$

In the troposphere

$$dH = -\frac{R_a T_0}{g_0}\left(1 - \frac{L}{T_0}H\right)\frac{1}{P_S}dP_S \tag{2.50}$$

In the stratosphere

$$dH = -\frac{R_a T_T{}^{*}}{g_0}\cdot\frac{1}{P_S}dP_S \tag{2.51}$$

The effect of a 100 Pa (1 mb) error in static pressure measurement at sea level is thus equal to

$$\frac{287.0529 \times 288.15}{9.80665} \times \frac{1}{1013.25} \times 1 = 8.32\text{m} \qquad (27.3 \text{ ft})$$

However, the effect of a 100 Pa (1 mb) error in static pressure measurement at an altitude of 13,000m (42,650ft) when the static pressure is 16.5 kPa (165 mb) is equal to

$$\frac{287.0529 \times 216.65}{9.80665} \times \frac{1}{165} \times 1 = 38.43\text{m} \quad (126 \text{ ft})$$

The need for accuracy in the static pressure measurement at the high altitudes used in cruise flight can be seen bearing in mind ATC height level separations of 1,000 ft. below 29,000 ft and 2,000 ft above. A major part of the error budget in the static pressure measurement is the inherent pressure error due to the installation and location of the static pressure orifices. This means that the pressure sensors may not be measuring the true static pressure. This *pressure error*, or *position error* as it is sometimes called, has to be determined experimentally and is a function of Mach number and incidence. The uncertainty in this pressure error can be of the order of 100 to 150 Pa (1 to 1.5 mb) so that the error contribution from the pressure sensor should be as small as possible, say less than 30 Pa (0.3 mb). This would give a worst case overall error of 150 + 30 = 180 Pa (1.8 mb) at 13,000m altitude (42,650 ft) in the measurement of the static pressure which corresponds to an error in the pressure altitude of 69m (227 ft) of which the static pressure sensor contribution is less than 12m (38 ft).

The errors in static pressure measurement over the low altitude pressure range should also be less than 30 Pa (0.3 mb). This ensures the altimeter will read the true altitude on landing within 3.3m (11 ft approximately) providing the correct ground pressure has been set in and is known to an accuracy of 0.1 mb.

Modern technology air data pressure sensors generally meet or better these figures and achieve accuracies which approach those of the pressure standards used in calibration. It should be noted that an accuracy of 10 Pa (0.1 mb) in 100 kPa (1,000 mb) corresponds to an accuracy of 0.01%.

The pressure sensors must also have a very high resolution in order to be able to generate a rate of change of altitude ($\dot{H}$) output with the required dynamic range.

A typical figure aimed at for height resolution is better than 0.03m (0.1 ft) at sea level which corresponds to a pressure resolution in the region of 0.1 Pa (0.001 mb).

(b) Total pressure sensor
A typical total pressure sensor has a full scale pressure range of 0 to 260 kPa (0 to 2,600 mb). The figure of 260 kPa corresponds to a calibrated airspeed of 426m/s

(828 knots). Minimum non-derangement pressure is of the order of three times full scale pressure to cover mishandling in testing etc.

The accuracy requirements for the total pressure sensor are every bit as exacting as those required from the static pressure sensor. Using a total pressure sensor and a static pressure sensor to measure impact pressure ($P_T - P_S$) requires very high accuracy from both sensors as it involves taking the difference of two large comparable numbers at low airspeeds as the impact pressure is small.

The effect of errors in measuring the impact pressure at low airspeeds can be derived as follows.

$$Q_C = \tfrac{1}{2}\rho V_T^2$$

Hence

$$dQ_C = \rho V_T dV_T$$

Consider the requirement for the measurement of airspeed to an accuracy of 0.5m/s (approx. 1 knot) at an approach speed of 50m/s (100 knots approx.).

$$dQ_C = 1.225 \times 50 \times 0.5 = 30.6 \text{ Pa}$$

The maximum error in each sensor must therefore be less than 15 Pa (0.15 mb) to give the required accuracy at 50m/s. The accuracy requirements become even more exacting as the airspeed is reduced below 50m/s.

2.4.2.2 *Pressure sensor technology*

Two basic types of pressure sensor have now become well established in modern digital air data systems. Although there are other types of pressure sensor in service, attention has been concentrated on these two types as they account for most of the modern systems. The two main types can be divided into:

(a) Vibrating pressure sensors
(b) Solid state capsule pressure sensors.

(a) *Vibrating pressure sensors*

The basic concept of this family of sensors is to sense the input pressure by the change it produces in the natural resonant frequency of a vibrating mechanical system.

The output of the sensor is thus a frequency which is directly related to the pressure being measured. This frequency can be easily measured digitally with very high resolution and accuracy without the need for precision analogue to digital conversion circuitry. This confers a significant advantage as it enables a very simple and very accurate interface to be achieved with a micro-processor for the subsequent air data computation.

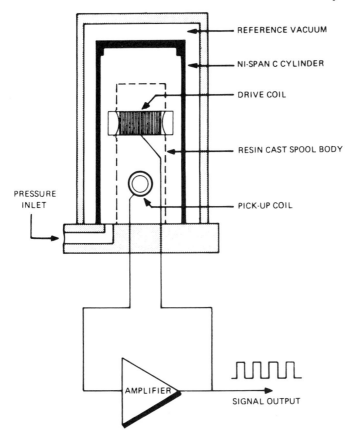

REFERENCE VACUUM

NI-SPAN C CYLINDER

DRIVE COIL

RESIN CAST SPOOL BODY

PRESSURE INLET

PICK-UP COIL

AMPLIFIER

SIGNAL OUTPUT

Fig. 2.10 Vibrating pressure sensor schematic (by courtesy of Schlumberger Industries).

The vibrating cylinder sensor is shown schematically in Fig. 2.10. The pressure sensing element consists of a thin walled cylinder with the input pressure acting on the inside of the cylinder and with the outside at zero vacuum reference pressure. The cylinder is maintained in a hoop mode of vibration by making it part of a feedback oscillator by sensing the cylinder wall displacement, processing and amplifying the signal and feeding it back to a suitable force producing device. Electro-magnetic drive and pick-off coils are used so that there is no contact with the vibrating cylinder.

The vibrating cylinder sensor is also density sensitive as the air adjacent to the cylinder wall effectively forms part of the mass of the cylinder so that a change in air density produces a small change in output frequency. This change, however, is much smaller than that due to pressure changes. The air density and pressure are directly related as shown in Section 2.2 so that this effect can be allowed for when the sensor is calibrated.

The small changes in density due to temperature variations together with the very small changes in the modulus of elasticity of the cylinder material with temperature can be compensated by measuring the sensor temperature and applying the appropriate corrections.

This type of sensor is extremely rugged, the very high Q of the cylinder conferring a very low susceptibility to vibration and shock, and the absence of moving parts enables it to withstand very high acceleration environments. The inherent simplicity and very low power consumption enables a very high reliability to be achieved together with very high accuracy and long term stability.

Acknowledgement is made to Schlumburger Industries for permission to publish the illustration in Fig. 2.11 of their Type 3088 pressure sensor. Schlumberger

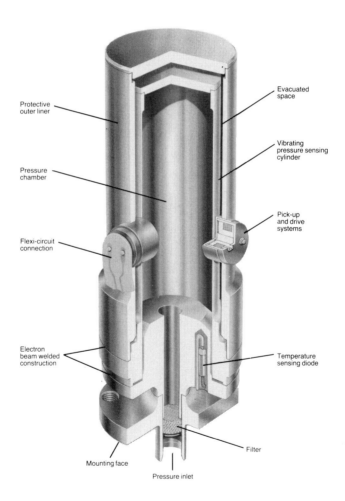

Fig. 2.11 Vibrating pressure sensor Type 3088.

Industries have developed the vibrating cylinder type of pressure sensor to an outstanding performance and reliability level. Their pressure sensors have achieved very wide scale usage in air data systems manufactured by several major avionics companies for large numbers of military and civil aircraft worldwide.

(b) Solid state capsule pressure sensors

This type of pressure sensor consists essentially of a capsule with a relatively thin diaphragm which deflects under the input pressure. They are fabricated from materials such as silicon, quartz, fused silica, or special ceramics. Figure 2.12 illustrates the basic construction. These materials exhibit virtually 'perfect mechanical properties' with negligible hysteresis and very stable and highly linear stress/strain characteristics. The modulus of elasticity variation with temperature is very stable and can be readily compensated by measuring the sensor temperature.

The deflection of the diaphragm is linear with input pressure but is also very small and a number of techniques are used to measure this deflection. These will be discussed later.

Semiconductor technology is used in the fabrication of these sensors and this together with the absence of moving parts has led to the description 'solid state'.

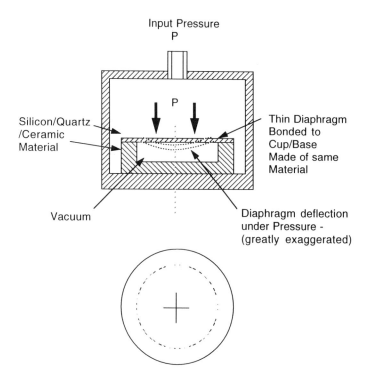

Fig. 2.12 'Solid state' capsule pressure transducer.

The technology also enables very small sensors to be fabricated with excellent repeatability because of the semiconductor processes used.

The deflection of the solid state capsule under pressure is typically only 25 to 50μm full scale, or less. The techniques which have been adopted to measure this very small deflection are briefly described below.

(i) Integral strain gauges. Piezo-resistive networks (or bridges) are ion implanted at the edge of a thin silicon diaphragm. Application of pressure causes the diaphragm to deflect thereby deforming the crystal lattice structure of the silicon which in turn causes the resistance of the piezo-resistive elements to change. The piezo-resistive elements are connected in a Wheatstone's bridge configuration as shown in Fig. 2.13 with two opposite resistive elements of the bridge located radially and the other two located tangentially on the diaphragm. The applied input pressure causes the

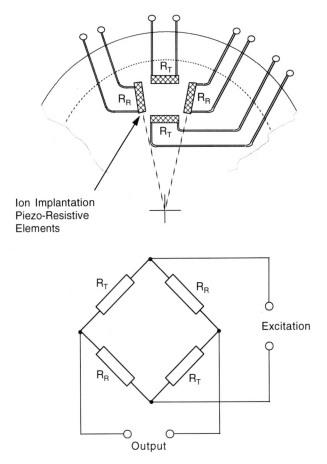

Fig. 2.13 Strain gauge configuration.

resistance of the radial elements to decrease and the tangential elements to increase, or vice versa, depending whether the pressure is increasing or decreasing and so unbalances the Wheatstone's bridge. The output of the bridge is proportional to both pressure and temperature as the modulus of elasticity of silicon is temperature dependent. A temperature sensitive resistive element is therefore incorporated into the diaphragm to measure the temperature so that the temperature dependent errors can be corrected in the subsequent processing of the sensor output. (These temperature dependent errors are very stable and repeatable.) A precision analogue to digital (A to D) conversion circuit is required to convert the sensor analogue voltage output to a digital output. The supporting electronics including the individual device characterisation, A to D conversion and digital interface to the 'outside world' can all be incorporated in the sensor with modern micro-electronic technology. This technology is sometimes referred to as 'smart sensor' technology.

(ii) Capacitive pick off. An alternative technique is to deposit a metallic film on an area at the centre of the diaphragm to form a capacitive element (or 'pick-off') whose capacitance changes as the diaphragm deflects under pressure. This forms part of a capacitance bridge network. Figure 2.14 shows the device construction. A correctly designed capacitive pick off bridge combination can have extremely high resolution and can detect the incredibly small changes in capacitance resulting from minute deflections of the diaphragm. (It is noteworthy that the highest resolution yet achieved in an angular or linear displacement sensor has been achieved with a capacitive pick off, leaving optical measuring techniques orders of magnitude behind.)

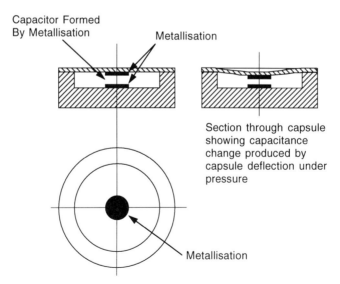

Fig. 2.14 Capacitance pick-off.

This type of pressure capsule is fabricated from quartz, fused silica, or special ceramic materials. The variation of the modulus of elasticity with temperature in these materials is less than with silicon. An integral temperature sensor is incorporated in the capsule to enable the temperature dependent errors to be compensated.

The capacitive pick off and bridge configuration must be designed so that it is not affected by changes in the stray capacitances inherent in the lead out wires and screening, etc, as these changes can be comparable or even exceed the very small changes due to the input pressure.

A precision A to D conversion is required to provide the required digital output.

It should be noted that the solid state capsule types of pressure sensor have an inherent sensitivity to acceleration (or 'g') because of the mass of the diaphragm. The sensitivity is small and can generally be ignored if the pressure sensors are mounted so that their sensitive axes are orthogonal to the normal acceleration vector.

The types of pressure sensor described in (a) and (b) are all used in modern digital air data systems. In a highly competitive market, all meet the requirements. As the saying goes – 'You pays your money and you takes your pick'.

2.4.3 Air data computation

A flow diagram of the air data computation processes is shown in Fig. 2.15. The computations are briefly described below.

Pressure Altitude The static pressure can be computed at suitable increments of altitude from −914.4m (−3,000 ft) to 32,004m (105,000 ft approx.) and the data stored in a table look-up store using the appropriate formulae relating static pressure and altitude.

(a) Troposphere −914m to 11,000m (−3,000 ft to 36,089 ft)

$$P_S = 1,013.25 \, (1\text{--}2.25577 \times 10^{-5} H_P)^{5.255879}$$

(b) Stratosphere 11,000m to 20,000m (36,089 ft to 65,617 ft)

$$P_S = 226.32 \, e^{-1.576885 \times 10^{-4} (H_P - 11,000)}$$

(c) Chemosphere 20,000m to 32,000m (65,617 ft to 105,000 ft)

$$P_S = 54.7482 \, [1+4.61574 \times 10^{-6} (H_P - 20,000)]^{-34.163215}$$

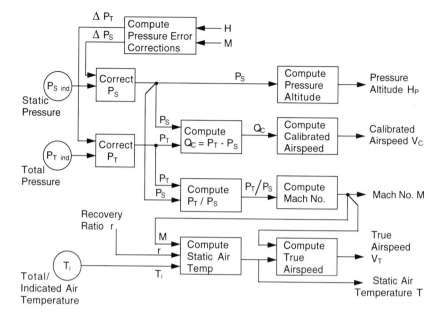

Fig. 2.15 Air data computation flow diagram.

The appropriate pressure altitude corresponding to the measured static pressure can then be derived from the table look-up store in conjunction with a suitable interpolation algorithm.

Vertical Speed, $\dot{H}_P$ The vertical speed or rate of change of altitude, $\dot{H}_P$, is derived from the rate of change of static pressure, $\dot{P}_s$, using the basic formulae relating the differentials dH and dP_s in equation (2.3). *viz.*

$$dH = -\frac{R_a T}{g_0}\frac{1}{P_s}dP_s$$

Hence

$$\dot{H}_P = -\frac{R_a T_s}{g_0}\frac{1}{P_s}\dot{P}_s$$

This can be computed from the actual measured and corrected air temperature value or from the standard atmosphere temperature. The latter is the usual case so that H_P is purely a function of P_s.

(a) Troposphere

$$\dot{H}_P = 8434.51\,(1-2.25577\times 10^{-5}\,H_P\,)\frac{1}{P_s}\dot{P}_s$$

(b) Stratosphere

$$\dot{H}_P = 6341.62 \frac{1}{P_S} \dot{P}_S$$

Deriving $\dot{H}_P$ by first differentiating P_S to generate $\dot{P}_S$ and then multiplying essentially by $1/P_S$ enables a better resolution to be obtained with fewer computational delays.

Mach number The pressure ratio, P_T/P_S, can be computed at suitable increments of Mach number and the results stored in a table look-up store using the appropriate formulae.

(a) Subsonic speeds

$$\frac{P_T}{P_S} = (1 + 0.2 M^2)^{3.5}$$

(b) Supersonic speeds

$$\frac{P_T}{P_S} = \frac{166.92 M^2}{(7 M^2 - 1)^{2.5}}$$

The actual pressure ratio P_T/P_S is then computed from the measured total pressure P_T and static pressure P_S. The appropriate Mach number corresponding to this computed pressure ratio is then derived from the table look-up store in conjunction with a suitable interpolation algorithm.

Calibrated airspeed The impact pressure, Q_C, can be computed at suitable increments over the range of calibrated airspeeds, say 25m/s (50 knots) to 400m/sec (800 knots) and the results stored in a table look-up store using the appropriate formulae.

(a) $V_C \leq 340.3$m/s (661.5 knots)

$$Q_C = 101.325 \left[1 + 0.2 \left(\frac{V_C}{340.294} \right)^2 \right]^{3.5} - 1 \right] \text{ kN / m}^2$$

(b) $V_C \geq 340.3$m/s (661.5 knots)

$$Q_C = 101.325 \left[\frac{166.92 \left(\dfrac{V_C}{340.294} \right)^7}{\left[7 \left(\dfrac{V_C}{340.294} \right)^2 - 1 \right]^{2.5}} - 1 \right]$$

Q_c is derived by subtracting the measured total and static pressures. The appropriate calibrated airspeed corresponding to this impact pressure is then derived from the table look-up store in conjunction with a suitable interpolation algorithm.

Static air temperature   The static air temperature, T_S, is derived by computing the correction factor $1/(1 + r\ 0.2M^2)$ and multiplying the measured (indicated) air temperature, T_m, by this correction factor

viz
$$T_S = \frac{T_m}{\left(1 + r\ 0.2\,M^2\right)}$$

true air speed   The true airspeed, V_T, is derived from the computed Mach number and the computed static air temperature, T_S

viz
$$V_T = 20.0468 M \sqrt{T_S}$$

The air data computations can be carried out by any modern 16 bit micro-processor, the software to carry out this task being relatively straightforward. Computations are typically carried out at an iteration rate of 20 Hz, although FBW flight control systems may require a faster iteration rate – up to 80 Hz.

A more exacting software task is the generation of the software to perform all the 'built in test' (BIT) and enable fault diagnosis to module level to be achieved.

2.4.4 Angle of incidence sensors

The importance of angle of incidence information has already been briefly explained. Knowledge of the angle of incidence can become particularly important at the extremes of the flight envelope when high values of incidence are reached and the limiting conditions approached. Many aircraft have incidence sensors installed so that the pilot can ensure the aircraft does not exceed the maximum permissible angle of incidence. Incidence sensors are also essential sensors for a Fly-by-wire (FBW) flight control system, as mentioned earlier, and a redundant, failure survival system of sensors is required. It may also be necessary to measure the sideslip incidence angle, β.

A typical incidence sensor comprises a small pivoted vane suitably located on the aircraft fuselage near the nose of the aircraft. The vane is supported in low friction bearings so that it can align itself with the incident airflow under the action of the aerodynamic forces acting on it, like a weather vane. The angle of the vane relative to the fuselage datum and hence the angle of incidence is measured by a suitable angular position pick-off such as a synchro resolver (the output of which can be readily digitised).

It should be noted that an incidence sensor measures 'indicated angle of incidence' which must be converted to 'true angle of incidence' by an appropriate correction formula which is a function of Mach number.

It is also possible to locate the Pitot probe on the vane together with the static pressure orifices to form an integrated unit which together with the appropriate sensors measures angle of incidence, total pressure and static pressure. This type of integrated Pitot probe/incidence vane is being adopted on several new high performance aircraft as it minimises incidence contamination effects on the pressure measurements at high angles of incidence as well as providing a compact integrated solution. The system is illustrated in Fig. 2.16.

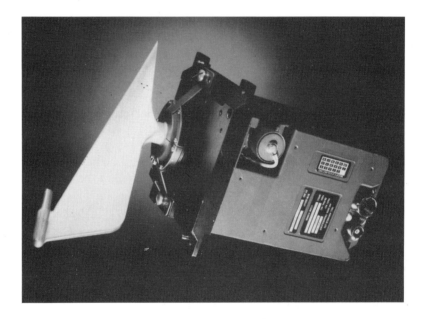

Fig. 2.16 Integrated air data transducer system (by courtesy of GEC-Marconi Avionics Ltd.).

Further reading

British Standard 2G199:1984, *Tables relating altitudes airspeed and mach numbers for use in aeronautical instrument design and calibration.*
Dommasch, D.D., Sherby, S.S., Connolly, T.F., *Airplane aerodynamics*: Pitman 1967

3

Aerodynamics and aircraft control

3.1 INTRODUCTION

The object of this chapter is to provide the reader with an introduction to aerodynamics, aircraft stability and control and the dynamic response and behaviour of an aircraft so as to be able to understand the role and tasks of automatic flight control systems. The aim is to establish the essential background to Fly-by-wire (FBW) flight control (Chapter 4) and autopilots (Chapter 8).

3.2 BASIC AERODYNAMICS

Most readers will be aware of the basic principles of aircraft flight and how the wings generate lift to support the weight of the aircraft. The aerodynamic aspects of aircraft flight are thus covered briefly, principally to recap and define the terms and aerodynamic parameters which are used.

3.2.1 Lift and drag

An aerofoil inclined at an angle to a moving air stream will experience a resultant force due to aerodynamic effects. This resultant aerodynamic force is generated by

(a) A reduction in the pressure over the upper surface of the wing as the airflow speeds up to follow the wing curvature, or camber;
(b) An increase in pressure on the undersurface of the wing due to the impact pressure of the component of the air stream at right angles to the undersurface.

About two thirds of this resultant aerodynamic force is due to the reduction in pressure over the upper surface of the wing and about one third due to the increase in pressure on the lower surface. Figure 3.1 illustrates the pressure distribution over a typical aerofoil.

The aerodynamic force exerted on an aerofoil can also be explained as being the reaction force resulting from the rate of change of momentum of the moving air stream by the action of the aerofoil in deflecting the airstream from its original direction (see Fig. 3.2).

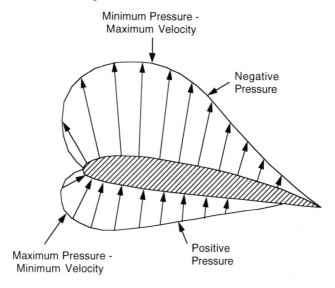

Fig. 3.1 Pressure distribution over an aerofoil.

The ability of an aerofoil to deflect a stream of air depends on the angle between the aerofoil and the airstream and on the curvature, or camber, of the aerofoil. The aerodynamic force is increased by increasing the camber or by increasing the angle between the aerofoil and the airstream. The thickness of the aerofoil also determines how efficiently the aerofoil produces a force.

This resultant aerodynamic force can be resolved into two components. The component at right angles to the velocity of the air relative to the wing (generally referred to in aerodynamic text books as *relative wind*) is called the *lift force*.

The component parallel to the air velocity is known as the *drag force* and, as its

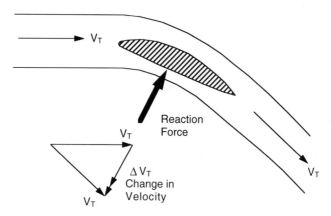

Fig. 3.2 Change in airflow momentum by aerofoil.

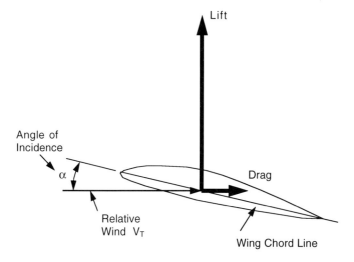

Fig. 3.3 Angle of incidence.

name implies, acts on the wing in the opposite sense to the velocity of the wing relative to the air mass (or alternatively in the direction of the relative wind). It should be noted that the drag force component of the aerodynamic force acting on the wing is known as the *wing drag* and constitutes a large part of the total drag acting on the aircraft particularly at high angles of incidence.

3.2.2 Angle of incidence/angle of attack

Referring to Fig 3.3 the angle between the direction of the air velocity relative to the wing (relative wind) and the wing chord line (a datum line through the wing section) is known both as the *angle of incidence* and the *angle of attack*. Angle of incidence is the term generally used in the UK and angle of attack in the USA. Angle of incidence is used in this book.

3.2.3 Lift coefficient and drag coefficient

Aerodynamic forces are dependent on the impact pressure which, as explained in Chapter 2, is the pressure created by the change in kinetic energy of the airstream when it impacts on the surface. Aerodynamicists use the term dynamic pressure, Q, to denote the impact pressure which would result if air were incompressible.

Dynamic pressure $\qquad\qquad Q = \frac{1}{2}\rho V_T^2$ $\qquad\qquad$ (3.1) where ρ = air density, V_T = air speed.

The aerodynamic lift force, L_W, acting on the wing is a function of the dynamic pressure, Q, the surface area of the wing, S, and a parameter which is dependent on the shape of the aerofoil section and the angle of incidence, α. This parameter is a non-dimensional coefficient, C_L, which is a function of the angle of incidence and is used to express the effectiveness of the aerofoil section in generating lift.

$$C_L = \frac{L_W}{\frac{1}{2}\rho V_T^2 S} \tag{3.2}$$

Thus
$$L_W = \frac{1}{2}\rho V_T^2 S C_L \tag{3.3}$$

The C_L versus α relationship is reasonably linear up to a certain value of angle of incidence when the airflow starts to break away from the upper surface and the lift falls off very rapidly – this is known as stalling. The slope of the C_L versus α relationship and the maximum value, C_{Lmax}, are dependent on the aerofoil section. C_{Lmax} typically ranges from about 1.2 to 1.6. Maximum values of the angle of incidence before the onset of stalling range from between 15° and 20° for conventional wing plans; for delta wing aircraft the figure can be up to 30° to 35°. The C_L versus α relationship also changes as the aircraft speed approaches the speed of sound, and changes again at supersonic speeds

Similarly, a non-dimensional characteristic called the *drag coefficient*, C_D, which is a function of the angle of incidence, α, is used to enable the drag characteristics of an aerofoil to be specified.

$$C_D = \frac{D_W}{\frac{1}{2}\rho V_T^2 S} \tag{3.4}$$

i.e. drag force
$$D_W = \frac{1}{2}\rho V_T^2 S C_D \tag{3.5}$$

The lift and drag characteristics of an aerofoil are related, the relationship being approximately
$$C_D = C_{DO} + k C_L^2 \tag{3.6}$$

where C_{DO} and k are constants for a particular aerofoil section.

Thus C_D increases rapidly at high values of C_L (and α). It also increases very rapidly indeed as the aircraft's speed approaches the speed of sound due to compressibility effects. The C_D versus α characteristics of a wing designed for supersonic operation are again very different at supersonic speeds compared with subsonic speeds.

Figure 3.4 shows the general shape for a typical subsonic aerofoil, of the relationship of C_L and C_D with α. These characteristics have been determined for a

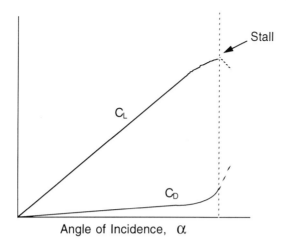

Fig. 3.4 C_L and C_D versus α.

very wide range of standardised aerofoil sections. The selection of a particular aerofoil section in the aircraft design process is determined by the optimum combination of characteristics for the aircraft mission and range of operating height and speed conditions.

Matching the wing lift requirements over the flight envelope, e.g. high lift at take-off and low drag at high speed, or increased lift for manoeuvering in combat, is achieved with the use of retractable leading edge flaps or slats and trailing edge flaps (see Fig. 3.5). These enable relatively large increases in C_L to be obtained with maximum values of C_{Lmax} between 3 to 4.

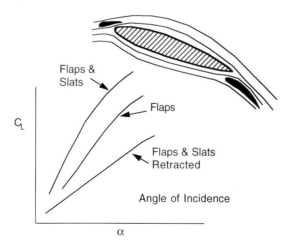

Fig. 3.5 Leading edge slats and trailing edge flaps.

3.2.4 Illustrative example on basic aerodynamics

A simple example of the application of these basic aerodynamic relationships is set out below. The object is to show how lift, incidence and manoeuvring capability can be estimated for various height and speed combinations so as to give an 'engineering feel' for the subject. The basic parameters for a hypothetical aircraft are:

Aircraft mass, m	30,000 kg
Wing area, S	75m^2
Maximum lift coeff, C_{Lmax}	1.2
Maximum angle of incidence, α_{max}	15°

C_L versus α relationship is assumed to be linear.

Air data Air density, ρ_0, at sea level and standard temperature (15°C) and pressure (1013.25 mbar) = 1.225 kg/m^3
　Air density, ρ, at 30,000 ft (9,144m) = 0.4583 kg/m^3.

Question (i) What is the wing incidence, α, when flying straight and level at a speed of 80m/s (160 knots approx) at a height of 200 ft in the approach to the airfield?

Question (ii) What is the corresponding wing incidence when flying straight and level at a speed of 200m/s (400 knots approx) at a height of 1,000 ft?
　Note: Neglect change in density 0–1,000 ft for simplicity.

Question (iii) What is the maximum normal acceleration that can be achieved at a height of 30,000 ft when flying at a speed of 225m/s (450 knots)? Neglect compressibility effects and assume adequate thrust is available to counteract the increase in drag and maintain speed of 450 knots at α_{max}.

Case 1. Low level/low speed

Dynamic pressure $Q = \frac{1}{2}\rho V_T^2 = 0.5 \times 1.225 \times 80^2 = 3920$ N/ m^2
Wing lift $= QSC_L = 3920 \times 75 \times C_L$ N
Aircraft weight $= 30,000 \times 9.81$ N ($g = 9.81$m/s^2)
Hence $3920 \times 75 \times C_L = 30,000 \times 9.81$
and required $C_L = 1.0$

C_L versus α is linear, hence $\dfrac{dC_L}{d\alpha} = \dfrac{C_{L\,max}}{\alpha_{max}} = \dfrac{1.2}{15}$ degrees^{-1}

$$C_L = \frac{dC_L}{d\alpha} \cdot \alpha$$

Hence
$$\alpha = 1.0 \times \frac{15}{1.2} = 12.5°$$

Case 2. Low level/high speed

Lift is proportional to V_T^2, hence required incidence for straight and level flight at 200m/s:

$$= \left(\frac{80}{200}\right)^2 \times 12.5 = 2°$$

Case 3. High altitude/high speed

Maximum achievable lift = ½ x 0.4583 x 225^2 x 75 x 1.2 = 1.044 x 10^3 N
Aircraft weight = 294 x 10^3 N
Hence lift available for manoeuvring = 750 x 10^3 N

$$\text{Achievable normal acceleration} = \frac{\text{Normal force}}{\text{Aircraft mass}} = \frac{750 \times 10^3}{30,000}$$

$$= 25\text{m/s}^2 = 2.5 \text{ g (approx)}$$

3.2.5 Pitching moment and aerodynamic centre

The *centre of pressure* is the point where the resultant lift and drag forces act and is the point where the moment of all the forces summed over the complete wing surface is zero. There will thus be a *pitching moment* exerted at any other point not at the centre of pressure. The centre of pressure varies with angle of incidence and for these reasons *aerodynamic centre* is now used as the reference point for defining the pitching moment acting on the wing. The aerodynamic centre of the wing is defined as the point about which the pitching moment does not change with angle of incidence (providing the velocity is constant). It should be noted that all aerofoils (except symmetrical ones) even at zero lift tend to pitch and experience a pitching moment or couple. The aerodynamic centre is generally around the quarter chord point of the wing (measured from the leading edge). At supersonic speeds it tends to move aft to the half chord point. This pitching moment or couple experienced at zero lift, M_0, is again expressed in terms of a non-dimensional coefficient, pitching moment coefficient, C_{M_0}

$$C_{M_0} = \frac{M_0}{\frac{1}{2}\rho V_T^2 Sc} \qquad (3.7)$$

where c = *mean aerodynamic chord*, equal to wing area/wing span

Thus pitching moment, $M_0 = \frac{1}{2}\rho V_T'^2 Sc\, C_{M_0}$ (3.8)

3.2.6 Tailplane contribution

The total pitching moment about the aircraft's CG and its relationship to the angle of incidence is of prime importance in determining the aircraft's stability. This will be discussed in the next section. However, it is appropriate at this stage to show how the resultant pitching moment about the CG is derived, particularly with a conventional aircraft configuration with a horizontal tailplane (refer to Fig. 3.6). The tailplane makes a major contribution to longitudinal stability and provides the necessary downward lift force to balance or trim the aircraft for straight and level flight. The moment about the CG due to the tailplane lift balances the nose down pitching moment due to the wing lift and the inherent wing pitching moment or couple, M_0. It should be noted that the trim lift exerted by the tailplane is in the opposite sense to the wing lift thereby reducing the total lift acting on the aircraft.

Resultant moment about CG, M, is given by

$$M = -L_W x - M_0 + L_t l_t$$ (3.9)

(Nose-up moments are defined as positive).

Wing lift, $L_W = \frac{1}{2}\rho V_T'^2 S C_L$

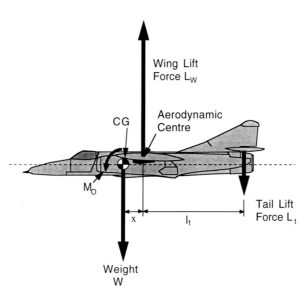

Fig. 3.6 Tailplane contribution.

Wing pitching moment, $M_0 = \frac{1}{2}\rho V_T^{'2} Sc\, C_{M_0}$ (3.10)

Tailplane lift, $L_t = k_t \frac{1}{2}\rho V_T^{'2} S_t\, C_{Lt}$ (3.11)

where S_t = tailplane area and C_{Lt} = tailplane lift coefficient

k_t = ratio of the dynamic pressure at the tailplane to the freestream dynamic pressure and is known at the *tailplane efficiency factor*. This factor takes into account the effects of downwash from the airflow over the wings on the tailplane. The tailplane efficiency factor varies from 0.65 to 0.95 and depends on several factors such as the location of the tailplane with respect to the wing wake etc.

Moment about CG due to tailplane Lift $= k_t \frac{1}{2}\rho V_T^{'2} C_{Lt}\left(S_t l_t\right)$

The term $(S_t l_t)$ is often called the *tailplane volume*.

Dividing both sides of equation (3.9) by $\frac{1}{2}\rho V_T^{'2} Sc$ yields

$$C_M = -\frac{x}{c}C_L - C_{M_0} + k_t \frac{\left(S_t l_t\right)}{Sc}C_{Lt} \qquad (3.12)$$

The equation shows the respective contributions to the total pitching moment coefficient. However, the most important characteristic from the aerodynamic stability aspect is the variation of the overall pitching moment coefficient with incidence. (This will be clarified in the next section.) Figure 3.7 shows the variation with incidence of the wing and tailplane pitching moments and the combined

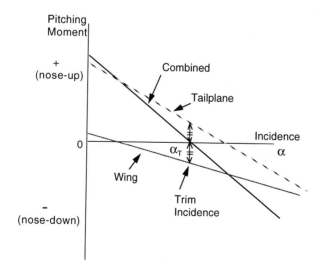

Fig. 3.7 Pitching moment vs incidence.

pitching moment. It can be seen that the effect of the tailplane is to increase the negative slope of the overall pitching moment with incidence characteristics. That is, $\dfrac{dC_M}{d\alpha}$ is more negative and this increases the aerodynamic stability.

3.3 AIRCRAFT STABILITY

A stable system is one which returns to its original state if disturbed; a neutrally stable system remains in its disturbed state and an unstable system will diverge from its state on being subjected to the slightest disturbance. This is illustrated in Fig. 3.8 which shows a ball bearing placed on concave, flat and convex surfaces respectively as an analogous example of these three stability conditions. An aircraft is said to be stable if it tends to return to its original position after being subjected to a disturbance without any control action by the pilot. Figure 3.9(a) and (b) illustrate various degrees of stability, both statically and dynamically.

3.3.1 Longitudinal stability

As already mentioned, to balance or trim the aircraft to achieve straight and level flight requires that the total pitching moment about the CG is zero and the total lift force equals the aircraft weight. However to achieve static stability it is necessary that the total pitching moment coefficient, C_M, about the CG changes with angle of incidence, α, as shown in Fig. 3.10. The value of the angle of incidence at which the pitching moment coefficient about the CG is zero is known as the *trim angle of incidence, α_T.*

For values of angle of incidence less than α_T, the pitching moment coefficient is positive so the resulting pitching moment is in the nose up sense and tends to restore the aircraft to the trim incidence angle, α_T. Conversely for angles of incidence greater than α_T, the pitching moment coefficient is negative so that the

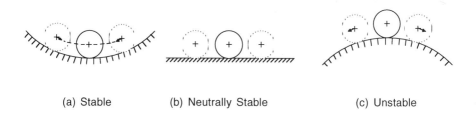

(a) Stable   (b) Neutrally Stable   (c) Unstable

Fig. 3.8 Simple example of stability conditions.

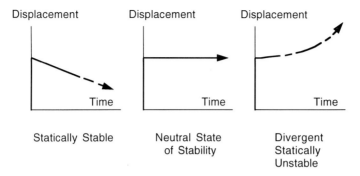

Fig. 3.9(a) Static stability.

resultant nose down moment tends to restore the aircraft to the trim incidence angle, α_T.

Thus $\dfrac{dC_M}{d\alpha}$ must be negative for longitudinal static stability

and $C_M = 0$ at $\alpha = \alpha_T$.

It should also be noted that the aircraft centre of gravity must be forward of the aerodynamic centre of the complete aircraft for static stability.

Static margin and neutral point Movement of the CG rearwards towards the aerodynamic centre decreases the static stability.

The position of the CG where $\dfrac{dC_M}{d\alpha} = 0$ is known as the *neutral point* and corresponds to the position of the aerodynamic centre of the complete aircraft. The

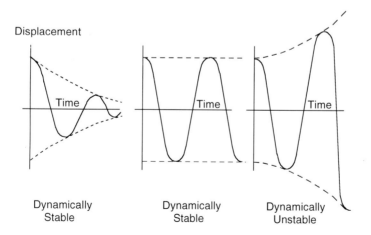

Fig. 3.9(b) Dynamic stability.

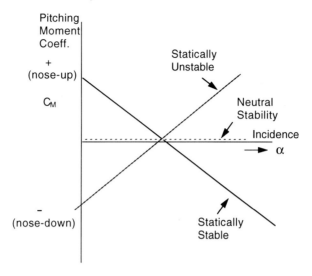

Fig. 3.10 C_M versus α.

static margin of the aircraft is defined as the distance of the CG from the neutral point divided by the mean aerodynamic chord, c. This is positive when the CG is forward of the neutral point. Thus the static margin is always positive for a stable aircraft. As an example, an aircraft with a static margin of 10% and a mean aerodynamic chord, of say, 4 metres (13.3 ft) means that a rearward shift of the CG of more than 40cm (16 inches) would result in instability.

Figure 3.11 illustrates the restoring moments when a stable aircraft is disturbed, for instance, by being subjected to an upward or downward gust.

3.3.2 Aerodynamically unstable aircraft

Figure 3.12 shows an aerodynamically unstable aircraft with the CG aft of the aerodynamic centre. The tailplane trim lift is acting in the same sense as the wing lift and so is more aerodynamically efficient. The angle of incidence required for a given lift is lower thereby reducing the induced drag; the induced drag being proportional to the square of the angle of incidence. ($C_D = C_{Do} + kC_L^2$, and C_L is approximately linear with α).

The tailplane volume ($S_t l_t$) can also be reduced further lowering the weight and drag and hence improving the performance, subject to other constraints such as rotation moment at take off.

The pilot's speed of response to correct the tendency to divergence of an unstable aircraft is much too slow and the tailplane must be controlled automatically. The time for the divergence to double its amplitude following a disturbance can be of the

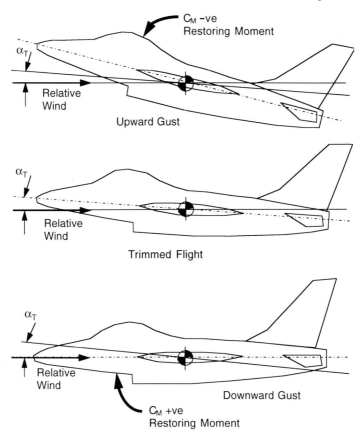

Fig. 3.11 Natural longitudinal stability.

order of 0.25 seconds or less on a modern high agility fighter which is aerodynamically unstable. The problem of flying an unstable aircraft has been compared to trying to steer a bicycle backwards (see Fig. 3.13).

Higher performance and increased agility can be obtained with an aerodynamically unstable aircraft and relying totally on an automatic stability system. The technology to implement such an automatic stability system with a manoeuvre command flight control system is now sufficiently mature to meet the very exacting safety and integrity requirements and system availability.

Such a system has become known as a 'Fly-by-wire' flight control system because of its total dependence on electrical signal transmission and electronic computing, and will be covered in the next chapter. The difference between a negative tailplane lift and positive tailplane lift to trim the aircraft is strikingly illustrated in Fig. 3.14 which shows an aerodynamically stable aircraft (Tornado) flying in formation with an aerodynamically unstable aircraft (Fly-by-wire Jaguar experimental aircraft).

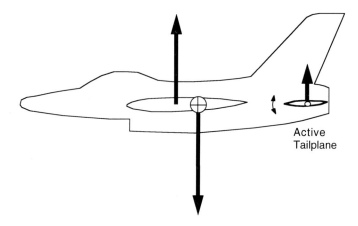

Fig. 3.12 Aerodynamically unstable aircraft.

3.3.3 Body lift contributions

It should be noted that lift forces are also generated by the fuselage. These fuselage (or body) lift forces are significant and they can also seriously affect both lateral and longitudinal aircraft stability. The engine casings in aircraft configurations with 'pod' type engine installations can also generate significant lift forces which can again affect the aircraft stability.

Fig. 3.13 Analogy of controlling aerodynamically unstable aircraft.

Fig. 3.14 Negative tailplane lift on Tornado. Positive tailplane lift on FBW Jaguar. (Photograph by Arthur Gibson provided by British Aerospace Defence Ltd.)

3.4 AIRCRAFT DYNAMICS

An aircraft has six degrees of freedom because its motion can involve both linear and angular movement with respect to three orthogonal axes. Deriving the dynamic response of the aircraft to disturbances or control surface movements first involves the generation of the differential equations which describe its motion mathematically. The solution of these equations then gives the response of the aircraft to a disturbance or control surface input. This enables the control that has to be exerted by the automatic flight control system to be examined on the basis of the aircraft's dynamic behaviour.

3.4.1 Aircraft axes – velocity and acceleration components

The aircraft motion is normally defined with respect to a set of orthogonal axes, known as body axes, which are fixed in the aircraft and move with it (see Fig. 3.15).

These axes are chosen for the following reasons:

Axis	Linear Velocity Component		Angular Velocity Component
Forward or Roll Axis, 0X	Forward Velocity	U	Roll Rate p
Slideslip or Pitch Axis, 0Y	Slideslip Velocity	V	Pitch Rate q
Vertical or Yaw Axis, 0Z	Vertical Velocity	W	Yaw Rate r

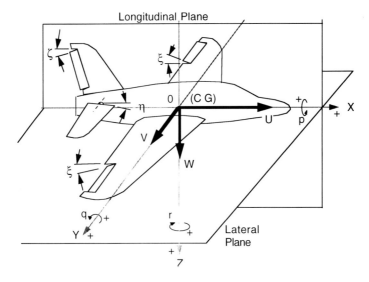

Fig. 3.15 Aircraft axes and velocity components.

(a) The equations of motion with respect to body axes are simpler.
(b) The aircraft motion can be readily measured with respect to body axes by robust motion sensors such as rate gyros and accelerometers which are body mounted (i.e. 'strapped down').
(c) Body axes are natural ones for the pilot and are the axes along which the inertia forces during manoeuvres (eg the normal acceleration during turns) are sensed.
(d) The transformation of motion data with respect to body axes to fixed space axes is not difficult with modern processors. Broadly speaking body axes are used in deriving the control dynamics and short term behaviour of the aircraft. Space axes are more suitable for the longer period guidance aspects in steering the aircraft to follow a particular flight path with respect to the Earth.

Referring to Fig. 3.15, the origin of the axes, O, is located at the centre of gravity (CG) of the aircraft with OX and OZ in the plane of symmetry of the aircraft (longitudinal plane) with OZ positive downwards and OY positive to starboard

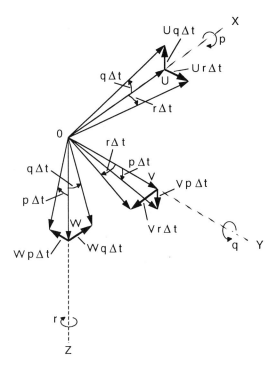

Fig. 3.16 Vector change in velocity components due to angular rotation.

(right). The figure shows the longitudinal and lateral planes of the aircraft. A fixed frame of axes is assumed to be instantaneously coincident with the moving frame and the velocity components of the aircraft CG along the OX, OY, OZ axes with respect to this fixed frame are the forward velocity, U, the sideslip velocity, V, and the vertical velocity, W, respectively. The corresponding angular rates of rotation of the body axes frame about OX, OY, OZ are the roll rate, p, the pitch rate, q, and the yaw rate, r, with respect to this fixed frame of axes. The derivation of the acceleration components when the aircraft's velocity vector is changing both in magnitude and direction is set out below as it is fundamental to deriving the equations of motion. Basically the components are made up of centrifugal (strictly speaking centripetal) acceleration terms due to the changing direction of the velocity vector as well as the components due to its changing magnitude.

In Fig. 3.16, at time $t = 0$, the axes OX, OY, OZ are as shown in dotted lines. At time $t = \Delta t$, the angular rotations of the vector components are $p\Delta t$, $q\Delta t$, $r\Delta t$ respectively. The corresponding changes in the vectors are thus $-Uq\Delta t$, $Ur\Delta t$, $Vp\Delta t$, $-Vr\Delta t$, $-Wp\Delta t$, $Wq\Delta t$. The rate of change of these vectors, that is, the centripetal acceleration components are thus: $-Uq$, Ur, Vp, $-Vr$, $-Wp$, Wq.

The changes in the velocity components due to the change in magnitude of the velocity vector are ΔU, ΔV, ΔW respectively. The rates of change of these velocity components are,

$$\frac{\Delta U}{\Delta t}, \frac{\Delta V}{\Delta t}, \frac{\Delta W}{\Delta t}, \text{ which in the limit becomes } \frac{dU}{dt}, \frac{dV}{dt}, \frac{dW}{dt},$$

The linear acceleration components are thus:

Acceleration along $\text{OX} = \dot{U} - Vr + Wq$ (3.13)

Acceleration along $\text{OY} = \dot{V} + Ur - Wp$ (3.14)

Acceleration along $\text{OZ} = \dot{W} - Uq + Vp$ (3.15)

The angular acceleration components about OX, OY and OZ are $\dot{p}, \dot{q},$ and $\dot{r}$ respectively. The dot over the symbols denotes $\dfrac{d}{dt}$ and is Newton's notation for a derivative. Thus $\dot{U} = \dfrac{dU}{dt}$ and $\dot{p} = \dfrac{dp}{dt}$ etc.

The motion of the aircraft can then be derived by solving the differential equations of motion obtained by applying Newton's second law of motion in considering the forces and moments acting along and about the OX, OY and OZ axes respectively. Namely, the rate of change of momentum is equal to the resultant force acting on the body

ie force = mass $\times$ acceleration

In the case of angular motion, this becomes:

moment (or torque) = (moment of inertia) $\times$ (angular acceleration).

3.4.2 Euler angles – definition of angles of pitch, bank and yaw

The orientation of an aircraft with respect to a fixed inertial reference frame of axes is defined by the three Euler angles. Referring to Fig. 3.17, the aircraft is imagined as being oriented parallel to the fixed reference frame of axes. A series of rotations bring it to its present orientation:

(i)   clockwise rotation in the horizontal plane, through the yaw (or heading) angle ψ, followed by

(ii) a clockwise rotation about the pitch axis, through the pitch angle θ, followed by

(iii) a clockwise rotation about the roll axis, through the bank angle Φ

The order of these rotations is very important – a different orientation would result if the rotations were made in a different order.

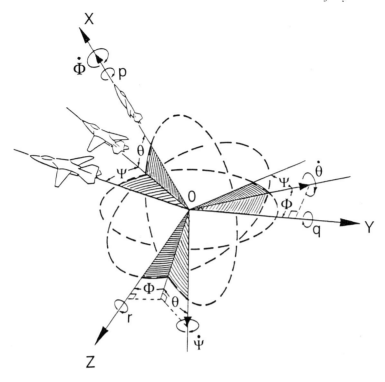

Fig. 3.17 Euler angles.

The relationship between the angular rates of roll, pitch and yaw. p, q, r (which are measured by body mounted rate gyros) and the Euler angles, ψ, θ, Φ and the Euler angle rates $\dot{\psi}$, $\dot{\theta}$, $\dot{\Phi}$ are derived as follows:

Consider Euler bank angle rate, $\dot{\Phi}$

Component of $\dot{\Phi}$ along
$$\begin{cases} OX = \dot{\Phi} \\ OY = 0 \\ OZ = 0 \end{cases}$$

Consider Euler pitch angle rate, $\dot{\theta}$

Component of $\dot{\theta}$ along
$$\begin{cases} OX = 0 \\ OY = \dot{\theta}\cos\Phi \\ OZ = -\dot{\theta}\sin\Phi \end{cases}$$

Consider Euler yaw angle rate, $\dot{\psi}$

Component of $\dot{\psi}$ along $\qquad \begin{cases} OX = -\dot{\Psi}\sin\theta \\ OY = \dot{\Psi}\cos\theta \sin\Phi \\ OZ = \dot{\Psi}\cos\theta \cos\Phi \end{cases}$

Hence

$$p = \dot{\Phi} - \dot{\Psi}\sin\theta \qquad\qquad (3.16)$$

$$q = \dot{\theta}\cos\Phi + \dot{\Psi}\cos\theta \sin\Phi \qquad\qquad (3.17)$$

$$r = \dot{\Psi}\cos\theta \cos\Phi - \dot{\theta}\sin\Phi \qquad\qquad (3.18)$$

It should be noted that these relationships will be referred to in Chapter 5, Inertial Sensors and Systems, in the derivation of attitude from strap-down rate gyros.

3.4.3 Equations of motion for small disturbances

The six equations of motion which describe an aircraft's linear and angular motion are non-linear differential equations. However, these can be linearised provided the disturbances from trimmed, straight and level flight are small. Thus in the steady state the aircraft is assumed to be moving in straight and level flight with uniform velocity and no angular rotation. That is, with no bank, yaw or side-slip and the axes OX and OZ lying in the vertical plane. The incremental changes in the velocity components of the CG along OX, OY, OZ in the disturbed motion are defined as u, v, w respectively and the incremental angular velocity components are p, q, r about OX, OY, OZ. The velocity components along OX, OY, OZ are thus $(U_0 + u)$, v, $(W_0 + w)$ in the disturbed motion where U_0 and W_0 are the velocity components along OX, OZ in steady flight ($V_0 = 0$, no side-slip). u, v, w, p, q, r are all small quantities compared with U_0 and W_0 and terms involving the products of these quantities can be neglected.

The acceleration components thus simplify to:

Acceleration along OX $= \dot{u} + W_0 q$ $\qquad\qquad (3.19)$

Acceleration along OY $= \dot{v} - W_0 p + U_0 r$ $\qquad\qquad (3.20)$

Acceleration along OZ $= \dot{w} - U_0 q$ $\qquad\qquad (3.21)$

The external forces acting on the aircraft are shown in Fig. 3.18 and comprise lift, drag, thrust and weight. These forces have components along OX, OY, OZ and can be separated into aerodynamic and propulsive force components and gravitational force components .

In the steady, straight and level trimmed flight condition, the resultant forces acting along OX, OY, OZ due to the lift, drag, thrust and weight are zero. The

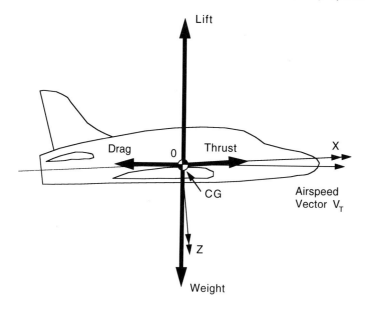

Fig. 3.18 External forces acting on aircraft.

resultant moments about OX, OY, OZ are also zero. For small disturbances it is therefore only necessary to consider the incremental changes in the forces and moments arising from the disturbance, in formulating the equations of motion (see Fig. 3.19). The incremental changes in the aerodynamic forces and moments are functions of the linear and angular velocity changes u, v, w and p, q, r respectively. The resultant incremental changes in the aerodynamic forces during the disturbance along OX, OY, OZ are denoted by X_a, Y_a, Z_a respectively.

In the disturbed motion, the incremental changes in the components of the weight along OX, OY, OZ are approximately equal to $-mg\theta$, $mg\Phi$ and zero respectively, assuming θ_0 is only a few degrees in magnitude.

From which the resultant incremental forces acting along OX, OY, OZ are:

along   OX $= X_a - mg\theta$
   OY $= Y_a + mg\Phi$
   OZ $= Z_a$

The equations of linear motion are thence derived from the relationship,

$$\text{force} = \text{mass} \times \text{acceleration} \ viz$$

$$X_a - mg\theta = m\left(\dot{u} + W_0 q\right) \tag{3.22}$$

$$Y_a + mg\Phi = m\left(\dot{v} - W_0 p + U_0 r\right) \tag{3.23}$$

$$Z_a = m\left(\dot{w} - U_0 q\right) \tag{3.24}$$

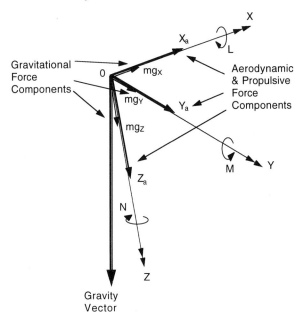

Fig. 3.19 Force components and moments.

The resultant external moments acting on the aircraft due to the aerodynamic forces about OX, OY, OZ resulting from the disturbance are denoted by L, M, N respectively.

Assuming principal axes are chosen for OX, OY, OZ so that the inertia product terms are zero.

The equations of angular motion are:

$$L = I_x \dot{p} \tag{3.25}$$
$$M = I_y \dot{q} \tag{3.26}$$
$$N = I_z \dot{r} \tag{3.27}$$

It should be noted that the equations deriving the aircraft body rates in terms of Euler angles and Euler angle rates (refer to section 3.4.2) can be simplified for small disturbances from straight and level flight. For small disturbances ψ, θ, and Φ will all be small quantities of the first order and p, q, r (and $\dot{\psi}, \dot{\theta}, \dot{\Phi}$) can be taken to be small quantities of the first order. Thus neglecting second order quantities, for small disturbances $p = \dot{\Phi}, q = \dot{\theta}$, and $r = \dot{\psi}$

3.4.4 Aerodynamic force and moment derivatives

The use of aerodynamic derivatives enables the incremental aerodynamic forces (X_a, Y_a, Z_a) and moments (L, M, N) to be expressed in terms of the force, or moment, derivatives multiplied by the appropriate incremental velocity change from the steady flight conditions. This linearises the equations of motion as the derivatives are assumed to be constant over small perturbations in the aircraft's motion. The equations of motion then become differential equations with constant coefficients and are much easier to analyse and solve. This assumption of constant derivatives is reasonable for small disturbances from a given flight condition. However it is necessary to establish the variation in these derivatives over the range of height and speed and Mach number conditions over the entire flight envelope and mission.

These variations can be large and must be allowed for in estimating the responses.

The concept of the force and moment derivatives is based on estimating the change resulting from a change in an individual variable with all the other variables held constant. The resultant change is then the sum of all the changes resulting from the individual variable changes.

Thus if a quantity, F, is a function of several independent variables x_1, x_2, x_3 ... x_n
i.e. $F = f(x_1, x_2, x_3 ... x_n)$

Then the resultant change, ΔF, is given by

$$\Delta F = \frac{\partial F}{\partial x_1} \cdot \Delta x_1 + \frac{\partial F}{\partial x_2} \cdot \Delta x_2 + \frac{\partial F}{\partial x_3} \cdot \Delta x_3 \cdots + \frac{\partial F}{\partial x_n} \cdot \Delta x_n$$

$\dfrac{\partial F}{\partial x_1}, \dfrac{\partial F}{\partial x_2}, \dfrac{\partial F}{\partial x_3} \cdots \dfrac{\partial F}{\partial x_n}$ are the partial derivatives of F with respect to x_1, x_2, x_3 ... x_n

The notation adopted for derivatives is to indicate the partial derivative variable by means of a suffix

Thus $\qquad \dfrac{\partial F}{\partial x_1} = F_{x_1}, \quad \dfrac{\partial F}{\partial x_2} = F_{x_2}, \quad \dfrac{\partial F}{\partial x_3} = F_{x_3}, \cdots \dfrac{\partial F}{\partial x_n} = F_{x_n}$

hence $\qquad \Delta F = F_{x_1} \Delta x_1 + F_{x_2} \Delta x_2 + F_{x_3} \Delta x_3 \cdots + F_{x_n} \Delta x_n$

The aerodynamic forces and moments acting on the aircraft are functions of the linear and angular velocity components u, v, w, p, q, r of the aircraft (all small quantities of the first order).

The forces acting along OX, OY, OZ are X, Y, Z respectively and the corresponding moments about OX, OY, OZ are L, M and N. The derivatives of these forces and moments due to a change in a particular variable are thus denoted by the appropriate suffix eg X_u, X_w, Y_v, Z_w, L_p, M_q, N_r, N_v etc., corresponding forces moments being $X_u u$, $X_w w$, $Y_v v$, $Z_w w$, $L_p p$, $M_q q$, $N_r r$, $N_v v$ etc. It should be noted that some text books and reference books use a 'dressing' (°) over the top of the symbol to denote a derivative, the (°) indicating the value is expressed in ordinary SI units, eg $\overset{o}{X}_u$ $\overset{o}{Y}_v$ $\overset{o}{M}_q$.

There are also forces and moments exerted due to the movement of the control surfaces from their trimmed position. Thus in the case of controlling the aircraft's motion in the longitudinal plane by movement of the tailplane (or elevator) from its trimmed position by an amount, η , the corresponding force exerted along the OZ axis would be $Z_\eta \eta$ and the corresponding moment exerted about the OY axis would be $M_\eta \eta$.

Because the disturbances are small there is no cross coupling between longitudinal motion and lateral motion. For example a small change in forward speed or angle of pitch does not produce any side force or rolling or yawing moment. Similarly a disturbance such as side-slip, rate of roll and rate of yaw produces only second order forces or moments in the longitudinal plane. Thus the six equations of motion can be split into two groups of three equations.

(a) Longitudinal equations of motion involving linear motion along the OX and OZ axes and angular motion about the OY axis.
(b) Lateral equations of motion involving linear motion along the OY axis and angular motion about the OX and OZ axes.

Thus each group of three equations can be solved separately without having to deal with the full set of six equations.

3.4.4.1 *Longitudinal motion derivatives*

These comprise the forward and vertical force derivatives and pitching moment derivatives arising from the changes in forward velocity, u, vertical velocity, w, rate of pitch, q, and rate of change of vertical velocity $\dot{w}$.

The main longitudinal derivatives due to u, v, q and w are shown below.

Forward velocity, u

$$\begin{cases} \text{Forward force derivative } X_u \\ \text{Vertical force derivative } Z_u \end{cases}$$

Vertical velocity, w

$$\begin{cases} \text{Forward force derivative } X_w \\ \text{Vertical force derivative } Z_w \\ \text{Pitching moment derivative } M_w \end{cases}$$

Rate of pitch, q

$$\begin{cases} \text{Vertical force derivative } Z_q \\ \text{Pitching moment derivative } M_q \end{cases}$$

Rate of change of vertical velocity, $\dot{w}$

$$\begin{cases} \text{Vertical force derivative } Z_{\dot{w}} \\ \text{Pitching moment derivative } M_{\dot{w}} \end{cases}$$

Force derivatives due to forward velocity X_u, Z_u and vertical velocity X_w, Z_w. The changes in the forward velocity, u, and vertical velocity, w, occurring during the disturbance from steady flight result in changes in the incidence angle, α, and air speed, V_T. These changes in incidence and air speed result in changes in the lift and drag forces. The derivatives depend upon the aircraft's lift and drag coefficients and the rate of change of these coefficients with incidence and speed.

The forward force derivative X_u due to the change in forward velocity, u, and the forward force derivative, X_w, due to the change in vertical velocity, arise from the change in drag resulting from the changes in airspeed and incidence respectively. The resulting forward force components are $X_u u$ and $X_w w$. The *vertical force derivative Z_u*, due to the forward velocity change, u, and the *vertical force derivative Z_w*, due to the vertical velocity change, w, arise from the change in lift due to the changes in airspeed and incidence respectively. The resulting vertical force components are equal to $Z_u u$ and $Z_w w$.

Pitching moment derivative due to vertical velocity, M_w. The change in incidence resulting from the change in the vertical velocity, w, causes the pitching moment about the CG to change accordingly. The resulting pitching moment is equal to $M_w w$ where M_w is the pitching moment derivative due to the change in vertical velocity, w. This derivative is very important from the aspect of the aircraft's longitudinal stability.

It should be noted that there is also a pitching moment derivative M_u due to the change in forward velocity, u, but this derivative is generally small compared with M_w.

Force and moment derivatives due to rate of pitch Z_q, M_q. These derivatives arise from the effective change in the tailplane angle of incidence due to the normal velocity component of the tailplane. This results from the aircraft's angular rate of pitch, q, about the CG and the distance of the tailplane from the CG, l_t (see Fig. 3.20). This normal velocity component is equal to $l_t q$ and the effective change in the tailplane angle of incidence is equal to $l_t q / V_T$. This incidence change results in a lift force acting on the tailplane which is multiplied by the tail moment arm, l_t, to give a significant damping moment to oppose the rate of rotation in pitch.

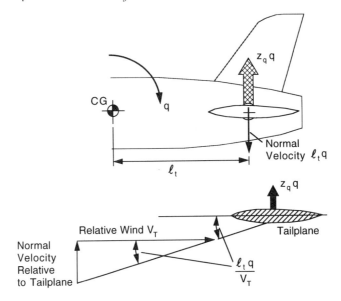

Fig. 3.20 Damping moment due to pitch rate.

The vertical force (acting on the tailplane) is equal to $Z_q q$ where Z_q is the vertical force derivative due to rate of pitch, q. The resultant pitching moment about the CG is equal to $M_q q$ where M_q is the pitching moment derivative due to the rate of pitch, q. The M_q derivative has a direct effect on the damping of the aircraft's response to a disturbance or control input and is a very important derivative from this aspect. The use of auto-stabilisation systems to artificially augment this derivative will be covered later in this chapter.

Force and moment derivatives due to the rate of change of vertical velocity $Z_{\dot{w}}$, $M_{\dot{w}}$. These force and moment derivatives result from the lag in the downwash from the wing acting on the tailplane and are proportional to the rate of change of downwash angle with wing incidence. The resulting vertical force component is equal to $Z_{\dot{w}} \dot{w}$ and the resulting pitching moment is equal to $M_{\dot{w}} \dot{w}$ where $Z_{\dot{w}}$ is the vertical force derivative due to rate of change of vertical velocity, $\dot{w}$, and $M_{\dot{w}}$ is the pitching moment derivative due to the rate of change of vertical velocity, $\dot{w}$.

Control derivatives Z_η, M_η. Control of the aircraft in the longitudinal plane is achieved by the angular movement of the tailplane (or elevator) by the pilot. The lift force acting on the tailplane as a result of the change in tailplane incidence creates a pitching moment about the aircraft's CG because of the tail moment arm. The pitching moment is proportional to the tailplane (or elevator) angular movement. The vertical force resulting from the tailplane (or elevator) movement is equal to

$Z_\eta \eta$ where Z_η is the vertical force derivative due to the angular movement of the tailplane, η, from its steady trimmed flight position. The pitching moment is equal to $M_\eta \eta$ when M_η is the pitching moment derivative due to the change in tailplane angle, η.

Longitudinal forces and moments
The main forces and moments acting along and about the OX, OY, OZ axes affecting longitudinal motion are set out below

$$X_a = X_u u + X_w w \tag{3.28}$$

$$Z_a = Z_u u + Z_w w + Z_{\dot{w}} \dot{w} + Z_q q + Z_\eta \eta \tag{3.29}$$

$$M = M_w w + M_q q + M_{\dot{w}} \dot{w} + M_\eta \eta \tag{3.30}$$

3.4.4.2 *Lateral motion derivatives*

Changes in the sideslip velocity, v, rate of roll, p, and rate of yaw, r, following a disturbance from steady flight produces both rolling and yawing moments. This causes motion about the roll axis to cross couple into the yaw axis and vice versa. The sideslip velocity also results in a side force being generated. The main lateral motion derivatives due to v, p and r are shown below.

Sideslip velocity v

$$\begin{cases} \text{Side force derivative } Y_v \\ \text{Yawing moment derivative } N_v \\ \text{Rolling moment derivative } L_v \end{cases}$$

Rate of roll, p

$$\begin{cases} \text{Rolling moment derivative } L_p \\ \text{Yawing moment derivative } N_p \end{cases}$$

Rate of yaw, r

$$\begin{cases} \text{Yawing moment derivative } N_r \\ \text{Rolling moment derivative } L_r \end{cases}$$

Side force derivative due to sideslip velocity Y_v. The change in sideslip velocity, v, during a disturbance changes the incidence angle, β, of the aircraft's velocity vector, V_T, (or relative wind) to the vertical surfaces of the aircraft comprising the fin and fuselage sides (see Fig. 3.21). The change in incidence angle v/V_T results in a sideways lifting force being generated by these surfaces. The net side force from

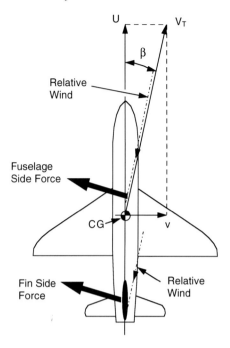

Fig. 3.21 Lateral forces.

the fuselage and fin combined is equal to $Y_v v$ where Y_v is the sideforce derivative due to the sideslip velocity.

Yawing moment derivative due to sideslip velocity N_v. The side force on the fin due to the incidence, β, resulting from the sideslip velocity, v, creates a yawing moment about the CG which tends to align the aircraft with the relative wind in a similar manner to a weathercock (refer to Fig. 3.21).

The main function of the fin is to provide this directional stability (often referred to as weathercock stability). This yawing moment is proportional to the sideslip velocity and is dependent on the dynamic pressure, fin area, fin lift coefficient and the fin moment arm, the latter being the distance between the aerodynamic centre of the fin and the yaw axis through the CG. However, the aerodynamic lateral forces acting on the fuselage during side-slipping also produce a yawing moment which opposes the yawing moment due to the fin and so is destabilising. The net yawing moment due to sideslip is thus dependent on the combined contribution of the fin and fuselage. The fin area and moment arm, known as the *fin volume*, is thus sized to provide good directional stability under all conditions and subject to other constraints such as engine failure in the case of a multi-engined aircraft.

The yawing moment due to sideslip velocity is equal to $N_v v$ where N_v is the yawing moment derivative due to sideslip velocity.

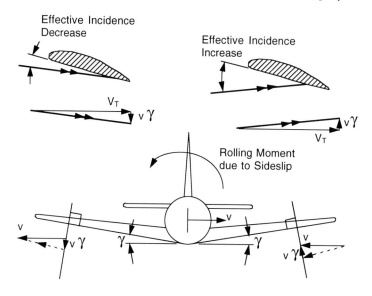

Fig. 3.22 Effect of dihedral and side-slip.

Rolling moment derivative due to sideslip velocity L_v. When the aircraft experiences a sideslip velocity, the effect of wing dihedral is for this sideslip velocity to increase the incidence on one wing and reduce it on the other (see Fig. 3.22). Thus if one wing tends to drop whilst sideslipping, there is a rolling moment created which tends to level the wings.

The effect of wing sweepback is also to give a differential incidence change on the wings and a rolling moment is generated, even if the dihedral is zero. There is also a contribution to the rolling moment due to sideslip from the fin. This is because of the resulting lift force acting on the fin and the height of the aerodynamic centre of the fin above the roll axis. There is also a contribution from the fuselage due to flow effects round the fuselage which affect the local wing incidence. These are beneficial in the case of a high wing and detrimental in the case of low wing. It is in fact necessary to incorporate considerably greater dihedral for a low wing than that required for a high wing location. The rolling moment derivative due to sideslip is denoted by L_v and the rolling moment due to sideslip is equal to $L_v v$.

Rolling moment derivative due to rate of roll L_p. When the aircraft is rolling, the angular velocity causes each section of wing across the span to experience a tangential velocity component which is directly proportional to its distance from the centre. Referring to Fig. 3.23, it can be seen that one section of wing experiences an increase in incidence whilst the corresponding section on the other wing experiences a decrease. The lift force exerted on one wing is thus increased whilst

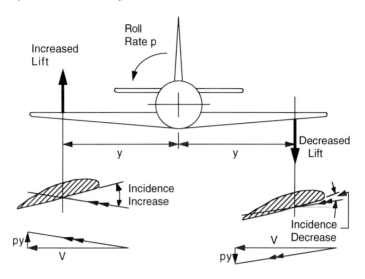

Fig. 3.23 Rolling moment due to rate of roll.

that on the other wing is decreased and a rolling moment is thus generated. The rolling moment due to the rate of roll, p, acts in the opposite sense to the direction of rolling and is equal to $L_p p$ where L_p is the rolling moment derivative due to rate of roll.

Yawing moment derivative due to rate of roll N_p. The rate of roll which increases the lift on the outer part of one wing and reduces it on the other also creates a differential drag effect. The increase in lift is accompanied by an increase in drag in the forward direction and the decrease in lift on the other wing by a corresponding reduction in drag. A yawing moment is thus produced by the rate of roll, p, which is equal to $N_p p$ where N_p is the yawing moment derivative due to rate of roll.

Yawing moment derivative due to rate of yaw N_r. The rate of yaw, r, produces a tangential velocity component equal to $l_f r$ where l_f is the distance between the aerodynamic centre of the fin and the yaw axis through the CG. The resulting change in the effective fin incidence angle, $l_f r / V_T$, produces a lift force which exerts a damping moment about the CG opposing the rate of yaw. The yawing moment due to the rate of yaw is equal to $N_r r$ where N_r is the yawing moment derivative due to rate of yaw.

Rolling moment derivative due to rate of yaw L_r. When the aircraft yaws, the angular velocity causes one wing to experience an increase in velocity relative to the airstream and the other wing a decrease. The lift on the leading wing is thus increased and the trailing wing decreased thereby producing a rolling moment. The rolling moment derivative due to rate of yaw is denoted by L_r and the rolling moment due to rate of yaw is equal to $L_r r$.

Lateral control derivatives due to ailerons and rudder. The ailerons and rudder are illustrated in Fig. 3.15. The angle through which the ailerons are deflected differentially from their position in steady trimmed flight is denoted by ξ and the angle the rudder is deflected from the position in steady trimmed flight is denoted by ζ.

Differential movement of the ailerons provides the prime means of lateral control by exerting a controlled rolling moment to bank the aircraft to turn; this will be covered later in the chapter.

Rolling moment derivative due to aileron deflection L_ξ. The effect of the differential deflection of the ailerons from their steady trimmed flight position, ξ, is to increase the lift on one wing and reduce it on the other thereby creating a rolling moment. The rolling moment due to the aileron deflection is equal to $L_\xi \xi$ where L_ξ is the rolling moment derivative due to aileron deflection.

Yawing moment derivative due to aileron deflection N_ξ. The differential lift referred to above is also accompanied by a differential drag on the wings which results in a yawing moment being exerted. The yawing moment due to the aileron deflection is equal to $N_\xi \xi$ where N_ξ is the yawing moment derivative due to aileron deflection.

Deflection of the rudder creates a lateral force and yawing moment to counteract sideslipping motion and the yawing moment due to the movement of the ailerons. (The control function of the rudder will be covered in more detail later in the chapter).

Side force derivative due to rudder deflection Y_ζ. The sideways lift force acting on the fin due to the deflection of the rudder, ζ, from its steady trimmed flight position is equal to $Y_\zeta \zeta$ where Y_ζ is the side force derivative due to rudder deflection.

Yawing moment derivative due to rudder deflection N_ζ. The yawing moment exerted by the rudder is equal to $N_\zeta \zeta$ where N_ζ is the yawing moment derivative due to the deflection of the rudder from its steady, trimmed flight position.

Lateral forces and moments
The main sideways force, Y_a, the rolling moment, L, and the yawing moment, N, resulting from changes in the sideslip velocity, v, rate of roll, p, and rate of yaw, r, from the steady flight condition are set out below.

$$Y_a = Y_v v + Y_\zeta \zeta \qquad (3.31)$$

$$L = L_v v + L_p p + L_r r + L_\xi \xi + L_\zeta \zeta \tag{3.32}$$

$$N = N_v v + N_p p + N_r r + N_\xi \xi + N_\zeta \zeta \tag{3.33}$$

Normalisation of derivatives

It should be noted that the derivatives are sometimes normalised and non-dimensionalised by dividing by the appropriate quantities – air density, airspeed, mean aerodynamic chord, wing span etc. The appropriate quantities depend on whether it is a force/velocity derivative or a moment/angular rate derivative etc. The equations of motion are also sometimes non-dimensionalised by dividing by the aircraft mass, moments of inertia etc.

This enables aircraft responses to be compared independently of speed, air density, size, mass, inertia etc. This stage has been omitted in this book for simplicity.

Readers wishing to find out more about non-dimensionalised derivatives and equations are referred to Further Reading at the end of the chapter.

3.4.5 Equations of longitudinal and lateral motion

Equations (3.28), (3.29) and (3.30) for X_a, Z_a and M can be substituted in the equations of motion (3.22), (3.24) and (3.26) respectively.

The equations of longitudinal motion for small disturbances then become:

$$X_u u + X_w w - mg\theta = m(\dot{u} + W_0 q) \tag{3.34}$$

$$Z_u u + Z_w w + Z_{\dot{w}} \dot{w} + Z_q q + Z_\eta \eta = m(\dot{w} - U_0 q) \tag{3.35}$$

$$M_w w + M_q q + M_{\dot{w}} \dot{w} + M_\eta \eta = I_y \dot{q} \tag{3.36}$$

Equations (3.31), (3.32) and (3.33) for Y_a, L and N can similarly be substituted in the equations of motion (3.23), (3.25) and (3.27) respectively. The equations of lateral motion for small disturbances then become:

$$Y_v v + Y_\zeta \zeta + mg\Phi = m(\dot{v} - W_0 p + U_0 r) \tag{3.37}$$

$$L_v v + L_p p + L_r r + L_\zeta \zeta + L_\xi \xi = I_x \dot{p} \tag{3.38}$$

$$N_v v + N_p p + N_r r + N_\zeta \zeta + N_\xi \xi = I_z \dot{r} \tag{3.39}$$

These equations can be rearranged as a set of first order differential equations to express the first derivative of each variable as a linear equation relating all the variables in the set. The longitudinal equations have been simplified for clarity by omitting the $Z_q q$, $Z_{\dot{w}} \dot{w}$ and $M_{\dot{w}} \dot{w}$ terms. Hence from equations (3.34). (3.35) and (3.36) and noting that $q = \dot{\theta}$ for small disturbances

$$
\left.
\begin{aligned}
\dot{u} &= \frac{X_u}{m} u + \frac{X_w}{m} w - W_0 q - g\theta \\[2ex]
\dot{w} &= \frac{Z_u}{m} u + \frac{Z_w}{m} w + U_0 q & + \frac{Z_\eta}{m}\eta \\[2ex]
\dot{q} &= \frac{M_w}{I_y} w + \frac{M_q}{I_y} q & + \frac{M_\eta}{I_y}\eta \\[2ex]
\dot{\theta} &= \qquad\qquad\quad q
\end{aligned}
\right\} \tag{3.40}
$$

Similarly from equations (3.37), (3.38) and (3.39) and noting that $p = \dot{\Phi}$ for small disturbances

$$
\left.
\begin{aligned}
\dot{v} &= \frac{Y_v}{m} v + W_0 p - U_0 r + g\Phi & + \frac{Y_\zeta}{m}\zeta \\[2ex]
\dot{p} &= \frac{L_v}{I_x} v + \frac{L_p}{I_x} p + \frac{L_r}{I_x} r & + \frac{L_\zeta}{I_x}\zeta + \frac{L_\xi}{I_x}\xi \\[2ex]
\dot{r} &= \frac{N_v}{I_z} v + \frac{N_p}{I_z} p + \frac{N_r}{I_z} r & + \frac{N_\zeta}{I_z}\zeta + \frac{N_\xi}{I_z}\xi \\[2ex]
\dot{\Phi} &= \qquad\qquad\quad p
\end{aligned}
\right\} \tag{3.41}
$$

These equations can be solved in a step-by-step, or iterative, manner by continuously computing the first derivative at each increment of time and using it to derive the change in the variables over the time increment and hence update the value at the next time increment.

As a simplified example, suppose the first order equations relating three variables x, y, z are

$$
\begin{aligned}
\dot{x} &= a_1 x + b_1 y + c_1 z \\
\dot{y} &= a_2 x + b_2 y + c_2 z \\
\dot{z} &= a_3 x + b_3 y + c_3 z
\end{aligned}
$$

Let Δt = time increment, value of x at time $n\Delta t$ be denoted by x_n and value of $\dot{x}$ at time $n\Delta t$ be denoted by $\dot{x}_n$

Value of x at time $(n+1)$ Δt , that is $x_{(n+1)}$, is given by

$$x_{(n+1)} = x_n + \dot{x}_n \cdot \Delta t$$

(Assuming Δt is chosen as a suitably small time increment).

Hence

$$x_{(n+1)} = x_n + (a_1 x_n + b_1 v_n + c_1 z_n) \Delta t$$

a form suitable for implementation in a digital computer.

These equations can be expressed in the matrix format

$$\begin{bmatrix} \dot{x} \\ \dot{y} \\ \dot{z} \end{bmatrix} = \begin{bmatrix} a_1 & b_1 & c_1 \\ a_2 & b_2 & c_2 \\ a_3 & b_3 & c_3 \end{bmatrix} \begin{bmatrix} x \\ y \\ z \end{bmatrix}$$

Similarly equations (3.40) and (3.41) can be expressed:

$$\begin{bmatrix} \dot{u} \\ \dot{w} \\ \dot{q} \\ \dot{\theta} \end{bmatrix} = \begin{bmatrix} \dfrac{X_u}{m} & \dfrac{X_w}{m} & -W_0 & -g \\ \dfrac{Z_u}{m} & \dfrac{Z_w}{m} & U_0 & 0 \\ \dfrac{M_w}{I_y} & \dfrac{M_q}{I_y} & 0 & 0 \\ 0 & 0 & 1 & 0 \end{bmatrix} \begin{bmatrix} u \\ w \\ q \\ \theta \end{bmatrix} + \begin{bmatrix} 0 \\ \dfrac{Z_\eta}{m} \\ \dfrac{M_\eta}{I_y} \\ 0 \end{bmatrix} [\eta] \qquad (3.42)$$

$$\begin{bmatrix} \dot{v} \\ \dot{p} \\ \dot{r} \\ \dot{\Phi} \end{bmatrix} = \begin{bmatrix} \dfrac{Y_v}{m} & W_0 & -U_0 & g \\ \dfrac{L_v}{I_x} & \dfrac{L_p}{I_x} & \dfrac{L_r}{I_x} & 0 \\ \dfrac{N_v}{I_z} & \dfrac{N_p}{I_z} & \dfrac{N_r}{I_z} & 0 \\ 0 & 1 & 0 & 0 \end{bmatrix} \begin{bmatrix} v \\ p \\ r \\ \Phi \end{bmatrix} + \begin{bmatrix} \dfrac{Y_\zeta}{m} & 0 \\ \dfrac{L_\zeta}{I_x} & \dfrac{L_\xi}{I_x} \\ \dfrac{N_\zeta}{I_z} & \dfrac{N_\xi}{I_z} \\ 0 & 0 \end{bmatrix} \begin{bmatrix} \zeta \\ \xi \end{bmatrix} \qquad (3.43)$$

The use of a matrix format for the equations of motion gives much more compact expressions and enables the equations to be manipulated using matrix algebra.

The equations can be expressed in the general form

$$\dot{X} = AX + BU$$

The bold letters indicate that **X, A, B, U** are matrices.

X is the state vector matrix, the elements comprising the state variables
A is the state coefficient matrix
B is the driving matrix
U is the control input vector matrix, the elements comprising the control input variables

For example, in equation (3.43)
The state variables v, p, r, Φ form the state vector

$$X = \begin{bmatrix} v \\ p \\ r \\ \Phi \end{bmatrix}$$

The state coefficient matrix

$$A = \begin{bmatrix} \dfrac{Y_v}{m} & W_0 & -U_0 & g \\[2mm] \dfrac{L_v}{I_x} & \dfrac{L_p}{I_x} & \dfrac{L_r}{I_x} & 0 \\[2mm] \dfrac{N_v}{I_z} & \dfrac{N_p}{I_z} & \dfrac{N_r}{I_z} & 0 \\[2mm] 0 & 1 & 0 & 0 \end{bmatrix}$$

The driving matrix

$$B = \begin{bmatrix} \dfrac{Y_\varsigma}{m} & 0 \\[2mm] \dfrac{L_\varsigma}{I_x} & \dfrac{L_\varepsilon}{I_x} \\[2mm] \dfrac{N_\varsigma}{I_z} & \dfrac{N_\varepsilon}{I_z} \\[2mm] 0 & 0 \end{bmatrix}$$

The control input vector

$$\mathbf{U} = \begin{bmatrix} \varsigma \\ \xi \end{bmatrix}$$

Further treatment of state variable matrix equations is beyond the scope of this chapter, the main objective being to introduce the reader who is not familiar with the methods and terminology to their use. Appropriate text books are listed in the Further Reading at the end of the chapter.

3.5 LONGITUDINAL CONTROL AND RESPONSE

3.5.1 Longitudinal control

In conventional (i.e. non Fly-by-wire) aircraft, the pilot controls the angular movement of the tailplane/elevators directly from the control column, or 'stick', which is mechanically coupled by rods and linkages to the tailplane/elevator servo actuator. (Fully powered controls are assumed). To manoeuvre in the longitudinal (or pitch) plane, the pilot controls the tailplane/elevator angle and hence the pitching moment exerted about the CG by the tailplane lift. This enables the pilot to rotate the aircraft about its CG to change the wing incidence angle and hence control the wing lift to provide the necessary normal, or centripetal, force to change the direction of the aircraft's flight path (see Fig. 3.24).

The initial response of the aircraft on application of a steady tailplane/elevator angular movement from the trim position is as follows.

The resulting pitching moment accelerates the aircraft's inertia about the pitch axis causing the aircraft to rotate about its CG so that the wing incidence angle increases. The wing lift increases accordingly and causes the aircraft to turn in the pitch plane and the rate of pitch to build up. This rotation about the CG is opposed by the pitching moment due to incidence which increases as the incidence increases and by the pitch rate damping moment exerted by the tailplane. A steady condition is reached and the aircraft settles down to a new steady wing incidence angle and a steady rate of pitch which is proportional to the tailplane/elevator angular movement from the trim position.

The aircraft's inertia about the pitch axis generally results in some transient overshoot before a steady wing incidence is achieved, the amount of overshoot being mainly dependent on the pitch rate damping generated by the tailplane.

The normal acceleration is equal to the product of the forward speed and the rate of pitch and is directly proportional to the increase in wing lift resulting from the increase in wing incidence angle. For a given constant forward speed, the rate of pitch is thus proportional to the tailplane/elevator angular movement from the trim position which in turn is proportional to the stick deflection. The rate of pitch is

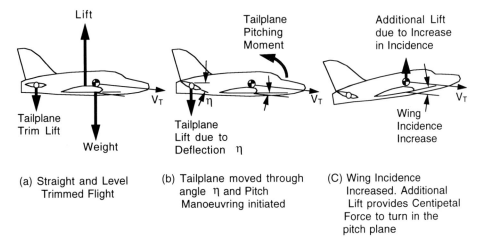

Fig. 3.24 Manoeuvring in the pitch plane.

reduced to zero by returning, that is 'centralising', the stick to its trimmed flight position.

3.5.2 Stick force/g

The wing lift is, however, proportional to both the wing incidence and the dynamic pressure, $\frac{1}{2}\rho V_T'^2$, so that the required wing incidence for a given normal acceleration will vary with height and speed over the flight envelope. Only a small change in wing incidence is required at the high dynamic pressures resulting from high speed low level flight, and vice versa.

The tailplane/elevator angular movement required per g normal acceleration, and hence stick displacement/g, will thus vary with height and speed: the variation can be as high as 40:1 over the flight envelope for a high performance aircraft.

The aerodynamic forces and moments are proportional to the dynamic pressure and with manually operated flying controls (ie no hydraulically powered controls) the pilot has a direct feedback from the stick of the forces being exerted as a result of moving the elevator. Thus, at higher speeds and dynamic pressures only a relatively small elevator movement is required per g normal acceleration. The high dynamic pressure experienced, however, requires a relatively high stick force although the stick movement is small. At lower speeds and dynamic pressures a larger elevator angle is required per g and so the stick force is still relatively high. The stick force/g thus tends to remain constant for a well designed aeroplane operating over its normal flight envelope.

However, with the fully powered flying controls required for high speed aircraft there is no direct feedback at the stick of the control moments being applied by moving the tailplane/elevators.

Stick displacement is an insensitive control for the pilot to apply, stick force being a much more natural and effective control and matches with the pilot's experience and training flying aircraft with manually operated controls.

The stick force is made proportional to stick deflection by either a simple spring loading system or by an 'artifical feel unit' connected directly to the stick mechanism. The artificial feel unit varies the stick stiffness (or 'feel') as a function of dynamic pressure $\frac{1}{2}\rho V_T^2$ so as to provide the required stick force/g characteristics. The rate of pitch at a given forward speed (i.e. normal acceleration) is thus made proportional to the stick force applied by the pilot. However, as already mentioned, the effectiveness of the tailplane/elevator can vary over a wide range over the height and speed envelope. Automatic control may be required to improve the aircraft response and damping over the flight envelope and give acceptable stick force per g characteristics.

3.5.3 Pitch rate response to tailplane/elevator angle

The key to analysing the response of the aircraft and the modifying control required from the automatic flight control system is to determine the aircraft's basic control transfer function. This relates the aircraft's pitch rate, q, to the tailplane/elevator angular movement, η, from the trimmed position.

This can be derived from equations (3.34), (3.35) and (3.36) together with the relationship $q = d\theta/dt$ by eliminating u and w to obtain an equation in q and η only. This results in a transfer function of the form below:

$$\frac{q}{\eta} = \frac{K(D^3 + b_2 D^2 + b_1 D + b_0)}{(D^4 + a_3 D^3 + a_2 D^2 + a_1 D + a_0)}$$

where $D = d/dt$.

K, b_2, b_1, b_0, a_3, a_2, a_1, a_0 are constant coefficients comprising the various derivatives. The transient response (and hence the stability) is determined by the solution of the differential equation $(D^4 + a_3 D^3 + a_2 D^2 + a_1 D + a_0)\, q = 0$

$q = Ce^{\lambda t}$, where C and λ are constants, is a solution of this type of linear differential equation with constant coefficients. Substituting $Ce^{\lambda t}$ for q yields

$$(\lambda^4 + a_3\lambda^3 + a_2\lambda^2 + a_1\lambda + a_0)\, Ce^{\lambda t} = 0$$

i.e.
$$(\lambda^4 + a_3\lambda^3 + a_2\lambda^2 + a_1\lambda + a_0) = 0$$

This equation is referred to as the characteristic equation. This quartic equation can be split into two quadratic factors which can be further factorised into pairs of conjugate complex factors as shown below

$$(\lambda + \alpha_1 + j\omega_1)(\lambda + \alpha_1 - j\omega_1)(\lambda + \alpha_2 + j\omega_2)(\lambda + \alpha_2 - j\omega_2) = 0$$

The solution is

$$q = \underbrace{A_1 e^{-\alpha_1 t} \sin(\omega_1 t + \phi_1)}_{\text{Short period motion}} + \underbrace{A_2 e^{-\alpha_2 t} \sin(\omega_2 t + \phi_2)}_{\text{Long period motion}}$$

A_1, ϕ_1, A_2, ϕ_2 are constants determined by the initial conditions
i.e. value of $\dddot{q}$, $\ddot{q}$, $\dot{q}$, q at time $t = 0$

For the system to be stable, the exponents must be negative so that the exponential terms decay to zero with time. (Positive exponents result in terms which diverge exponentially with time, ie an unstable response.)

The solution thus comprises the sum of two exponentially damped sinusoids corresponding to the short period and long period responses respectively.

The initial basic response of the aircraft is thus a damped oscillatory response known as the short period response. The period of this motion is in the region of 1 to 10 seconds, depending on the type of aircraft and its forward speed, being inversely proportional to the forward speed, e.g. typical fighter aircraft would be about 1 second period whereas a large transport aircraft would be about 5 to 10 seconds period. This short period response is generally fairly well damped for a well behaved (stable) aircraft but may need augmenting with an auto-stabilisation system over parts of the flight envelope of height and speed combinations.

The second stage comprises a slow lightly damped oscillation with a period ranging from 40 seconds to minutes and is known as the long period/motion. It is basically similar to the *phugoid* motion and is again inversely proportional to forward speed. The phugoid motion consists of a lightly damped oscillation in height and airspeed whilst the angle of incidence remains virtually unchanged and is due to the interchange of kinetic energy and potential energy as the aircraft's height and speed change. The damping of the phugoid motion is basically a task for the autopilot, the long period making it very difficult for the pilot to control.

Figure 3.25 illustrates the two types of motion in the response.

A simpler method of obtaining a good approximation to the q/n transfer function which accurately represents the aircraft's short period response can be obtained by assuming the forward speed remains constant. This is a reasonable assumption as the change in forward speed is slow compared with the other variables.

This simpler method is set out in the next section as it gives a good 'picture' in control engineering terms of the behaviour of the basic aircraft.

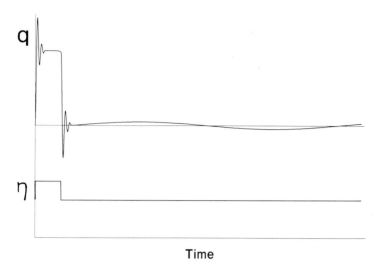

Fig. 3.25 Pitch response.

3.5.4 Pitch response assuming constant forward speed

The transfer functions relating pitch rate and wing incidence to tailplane (or elevator) angle, that is q/η and α/η, are derived below from first principles making the assumption of constant forward speed and small perturbations from steady straight and level, trimmed flight.

The derivative terms due to the rate of change of vertical velocity, $\dot{w}$, that is $Z_{\dot{w}}\dot{w}$ and $M_{\dot{w}}\dot{w}$ are neglected and also the $Z_{q}q$ term.

Referring to Fig. 3.26, an orthogonal set of axes OX, OY, OZ moving with the aircraft with the centre O at the aircraft's centre of gravity is used to define the aircraft motion. OX is aligned with the flight path vector, OY with the aircraft's pitch axis and OZ normal to the aircraft's flight path vector (positive direction downwards). These axes are often referred to as 'stability axes' and enable some simplification to be achieved in the equations of motion.

Velocity along OX axis $= U$ (constant)
Velocity increment along OZ axis $= w$
Velocity along OY axis $= 0$
Rate of rotation about pitch axis, OY $= q$
Change in angle of incidence from trim value $= \alpha$
Aircraft mass $= m$
Aircraft moment of inertia about pitch axis, OY $= I_y$

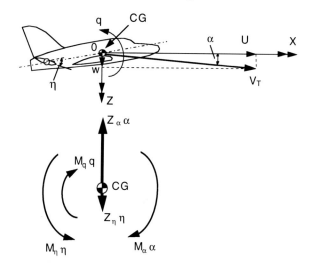

Fig. 3.26 Forces and moments – longitudinal plane (stability axes).

Considering forces acting in OZ direction
(1) Change in lift force acting on wing due to change in angle of incidence, α from trim value = $Z_\alpha\alpha$
(2) Change in lift force acting on tailplane due to change in tailplane angle, η, from trim value = $Z_\eta\eta$

Normal acceleration along OZ axis = $\dot{w} - Uq$
where $\dot{w} = dw/dt$ (Refer to section 3.4.3 equation (3.21)).

Equation of motion along OZ axis
$$Z_\alpha\alpha + Z_\eta\eta = m(\dot{w} - Uq) \tag{3.44}$$
Considering moments acting about CG

(1) Pitching moment due to change in tailplane angle, η, from trim value = $M_\eta\eta$
(2) Pitching moment due to change in angle of incidence, α, from trim value = $M_\alpha\alpha$
(3) Pitching moment due to angular rate of rotation, q, about the pitch axis = $M_q q$

Equation of angular motion about pitch axis, OY

$$M_\eta\eta + M_\alpha\alpha + M_q q = I_y\dot{q} \tag{3.45}$$
where $\dot{q} = dq/dt$ = angular acceleration
change in angle of incidence $\alpha = w/U$ $\qquad$ (3.46)
  These simultaneous differential equations can be combined to give equations in terms of q and η only by eliminating w, or, α and η only by eliminating q. However, it is considered more instructive to carry out this process using block

diagram algebra as this gives a better physical picture of the aircraft's dynamic behaviour and the inherent feedback mechanisms relating q and α.

Equation (3.45) can be written

$$q = \frac{1}{I_y} \int \left(M_\alpha \alpha + M_q q + M_\eta \eta \right) dt \tag{3.47}$$

substituting $U\dot{\alpha}$ for w in equation (3.44) and re-arranging yields

$$\alpha = \int \left(q + \frac{Z_\alpha}{mU} \alpha + \frac{Z_\eta}{mU} \eta \right) dt \tag{3.48}$$

Figure 3.27 represents equation (3.47) in block diagram form. It should be noted that the signs of the derivatives have been indicated at the summation points in the block diagram. M_q is negative and an aerodynamically stable aircraft is assumed so that M_α is also negative. The $M_q q$ and $M_\alpha \alpha$ terms are thus negative feedback terms. (Positive M_α can be allowed for in the subsequently derived transfer function, by the appropriate sign change.) The overall block diagram, Fig. 3.28, is then obtained by adding the block diagram representation of equation (3.48) to Fig. 3.27.

The q and α subsidiary feedback loops can be simplified using the relationship between output, θ_o, and input, θ_i, in a generalised negative feedback process of the type shown in Fig. 3.29.

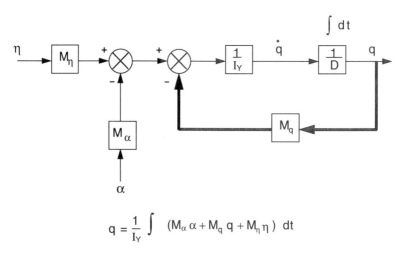

$$q = \frac{1}{I_Y} \int \ (M_\alpha\, \alpha + M_q\, q + M_\eta \eta) \ dt$$

Note   Signs of Derivatives indicated at Summation Points
        i.e. M_q and M_α are both negative

Fig. 3.27 Block diagram representation of equation.

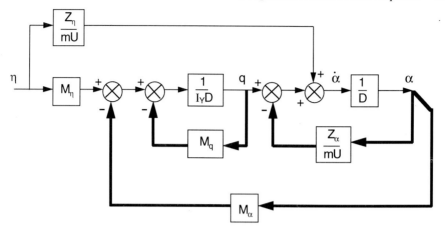

Fig. 3.28 Block diagram representation of longitudinal dynamics.

That is

$$\frac{\theta_o}{\theta_i} = \frac{KG(D)}{1 + KG(D).K_FG_F(D)}$$

(3.49)

where $KG(D)$ and $K_FG_F(D)$ are the transfer functions of the forward path and feedback path respectively.

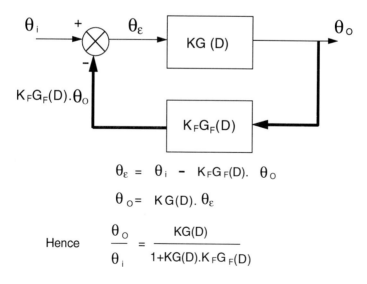

$$\theta_\varepsilon = \theta_i - K_FG_F(D).\theta_o$$

$$\theta_o = KG(D).\theta_\varepsilon$$

Hence

$$\frac{\theta_o}{\theta_i} = \frac{KG(D)}{1 + KG(D).K_FG_F(D)}$$

Fig. 3.29 Generalised negative feedback system.

Thus, considering the $M_q q$ feedback loop

$$\frac{q}{(M_\eta \eta - M_\alpha \alpha)} = \frac{\dfrac{1}{I_y D}}{1 + \dfrac{1}{I_y D} \cdot M_q}$$

as $KG(D) = 1/I_y D$ and $K_F G_F(D) = M_q$

$$\frac{q}{(M_\eta \eta - M_\alpha \alpha)} = \frac{1}{M_q} \frac{1}{(1 + T_1 D)} \tag{3.50}$$

where $T_1 = I_y / M_q$ $\hspace{3cm}$ (3.51)
This represents a simple first órder lag transfer function.
Similarly the α/q inner loop can be simplified to

$$\frac{\alpha}{q} = T_2 \cdot \frac{1}{(1 + T_2 D)} \tag{3.52}$$

where $T_2 = mU/Z_\alpha$ $\hspace{3cm}$ (3.53)
This represents another first order lag transfer function. The overall block diagram thus simplifies to Fig. 3.30.
It should be noted that the $Z_\eta \eta / mU$ input term has been omitted for simplicity as it is small in comparison with the other terms. Also, because it is an external input to the loop, it does not affect the feedback loop stability and transient response to a disturbance.
Referring to Fig. 3.30 and applying equation (3.49) yields

$$\frac{q}{\eta} = M_\eta \cdot \frac{\dfrac{\frac{1}{M_q}}{(1 + T_1 D)}}{1 + \dfrac{\frac{1}{M_q}}{(1 + T_1 D)} \cdot \dfrac{T_2}{(1 + T_2 D)} M_\alpha} \tag{3.54}$$

This simplifies to

$$\frac{q}{\eta} = \frac{M_\eta}{I_y T_2} \cdot \frac{(1 + T_2 D)}{D^2 + \left(\dfrac{1}{T_1} + \dfrac{1}{T_2}\right) D + \dfrac{M_\alpha}{I_y} + \dfrac{1}{T_1 T_2}} \tag{3.55}$$

i.e. $\hspace{2cm}$ $$\frac{q}{\eta} = \frac{K(1 + T_2 D)}{(D^2 + A_1 D + A_2)} \tag{3.56}$$

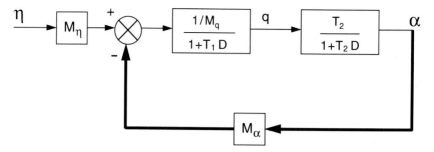

Fig. 3.30 Simplified overall block diagram.

where

$$K = \frac{M_\eta}{I_y T_2} = \frac{M_\eta Z_\alpha}{I_y m U} \tag{3.57}$$

$$A_1 = \frac{1}{T_1} + \frac{1}{T_2} = \frac{M_q}{I_y} + \frac{Z_\alpha}{mU} \tag{3.58}$$

$$A_2 = \frac{M_\alpha}{I_y} + \frac{1}{T_1 T_2} = \frac{M_\alpha}{I_y} + \frac{M_q Z_\alpha}{I_y m U} \tag{3.59}$$

Substituting $\alpha = \dfrac{T_2}{1+T_2 D}.q$ in equation (3.56) yields

$$\frac{\alpha}{\eta} = \frac{KT_2}{(D^2 + A_1 D + A_2)} \tag{3.60}$$

The resulting overall block diagram is shown in Fig. 3.31.
From equation (3.56)

$$(D^2 + A_1 D + A_2)q = K(1 + T_2 D)\eta \tag{3.61}$$

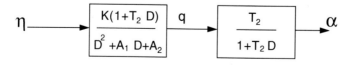

Fig. 3.31 Further simplified overall block diagram.

The transient response is given by the solution of the differential equation

$$(D^2 + A_1 D + A_2)q = 0 \qquad (3.62)$$

The solution is determined by the roots of the characteristic equation

$$(\lambda^2 + A_1\lambda + A_2) = 0$$

which are $\lambda = -\tfrac{1}{2}A_1 \pm \sqrt{A_1^2 - 4A_2}\,/2$

In a conventional aircraft these are complex roots of the form $-\alpha_1 \pm j\omega$ where

$\alpha_1 = A_1/2$ and $\omega = \sqrt{4A_2 - A_1^2}\,/2$

The solution in this case is an exponentially damped sinusoid

$$q = Ae^{-\alpha_1 t} \sin(\omega t + \phi)$$

when A and ϕ are constants determined from the initial conditions (value of $\dot{q}$ and q at time $t = 0$). The coefficient A_1 determines the value of $\alpha_1 (= A_1/2)$ and this determines the degree of damping and overshoot in the transient response to a disturbance. The coefficient A_2 largely determines the frequency, ω, of the damped oscillation $\omega = \sqrt{4A_2 - A_1^2}\,/2$ and hence the speed of the response.

In order for the system to be stable the roots of the characteristic equation must be negative or have a negative real part, so that the exponents are negative and exponential terms decay to zero with time. This condition is met provided A_1 and A_2 are both positive.

Conversely, an aerodynamically unstable aircraft with a positive M_α would result in A_2 being negative thereby giving a positive exponent and a solution which diverges exponentially with time.

It is useful to express the quadratic factor $(D^2 + A_1 D + A_2)$ in terms of two generalised parameters ω_0 and ζ, thus $(D^2 + 2\zeta\omega_0 D + \omega_0^2)$

Where ω_0 = undamped natural frequency = $\sqrt{A_2}$

$$\zeta = \text{damping ratio} = \frac{\text{coefficient of } D}{\text{coefficient of } D \text{ for critical damping}}$$

$$\text{i.e. } \zeta = \frac{A_1}{2\sqrt{A_2}} \qquad (3.63)$$

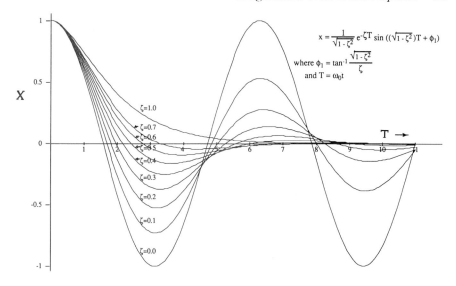

Fig. 3.32 Generalised second order system transient response.

(Critical damping results in a non oscillatory response and is the condition for equal roots which occurs when $A_1^2 = 4A_2$.)

The transient response to an initial disturbance to a general second order system of the type $(D^2 + 2\zeta\omega_0 D + \omega_0^2) x = 0$ when x is any variable is plotted out in Fig. 3.32 for a range of values of damping ratio, ζ, and non-dimensional time $T = \omega_0 t$. These graphs can be used for any second order system.

The steady state values on application of a given tailplane (or elevator) angle can be obtained for both pitch rate and incidence by equating all the terms involving rates of change to zero. Thus in equation (3.61), $\ddot{q}$, $\dot{q}$ and $\dot{\eta}$ are all zero in the steady state. The suffix ss is used to denote steady state values.

Whence $A_2 q_{ss} = K\eta_{ss}$

$$q_{ss} = \frac{K}{A_2}\eta_{ss} \qquad (3.64)$$

The steady state change in wing incidence, α_{ss}, can be obtained in a similar manner from equation (3.60) and is given by

$$\alpha_{ss} = \frac{KT_2}{A_2}\eta_{ss} \qquad (3.65)$$

It can be seen that changes in the aerodynamic derivatives over the flight envelope will alter both the speed and damping of the transient response to a tailplane movement or a disturbance. The steady state wing incidence and pitch rate for a given tailplane (or elevator) movement will also change and hence the *Stick Force per g* will change over the flight envelope.

3.5.5 Worked example on q/η transfer function and pitch response

The aerodynamic data below are very broadly representative of a conventional combat aircraft of transonic performance. These data have been derived by a 'reverse engineering' process from a knowledge of typical control 'gearings' in flight control systems (e.g. control surface movement/g) and actual aircraft short period responses in terms of undamped natural frequency and damping ratio.

The objective of the worked example is to try to bring together the basic aerodynamic and aircraft control theory presented so far and show how this can be applied to give an appreciation of the aircraft's pitch response and also the need for autostabilisation.

The data and calculations are rounded off to match the accuracy of the estimates, and the aerodynamic behaviour has been greatly simplified and compressibility effects etc ignored.

In practice, the necessary aerodynamic derivative data would normally be available. The worked example below has been structured to bring out some of the aerodynamic relationships and principles.

Aircraft Data

Aircraft mass, m = 25,000 kg (55,000 lb)
Forward speed, U = 250m/s (500 knots approx)
Moment of inertia about pitch axis, $I_y = 6 \times 10^5$ kg m^2
Pitching moment due to incidence, $M_\alpha = 2 \times 10^7$ Nm/radian
Tailplane moment arm, $l_t = 8$m
Wing incidence/g:- 1.6° increase in wing incidence at 500 knots produces 1g normal acceleration
Tailplane angle/g:- 1° Tailplane angular movement produces 1g normal acceleration at 500 knots

Questions

(a) Derive an approximate q/n transfer function, assuming forward speed is constant.
(b) What is the undamped natural frequency and damping ratio of the aircraft's pitch response?

(c) What is the the percentage overshoot of the transient response to a disturbance or control input?

$$\frac{q}{\eta} = \frac{K(1+T_2 D)}{D^2 + A_1 D + A_2}$$

The problem is to derive K, T_2, A_1 and A_2 from the data given.

The undamped natural frequency, $\omega_0 = \sqrt{A_2}$ and damping ratio $\zeta = \dfrac{A_1}{2\sqrt{A_2}}$ can then be calculated knowing A_1 and A_2.

• *Derivation of T_2*

Increase in wing lift $= Z_\alpha \alpha = m$ (ng)
 Where (ng) = normal acceleration
 $1.6°$ change in wing incidence produces 1g normal acceleration at 500 knots
 Hence Z_α 1.6/60 $= m$ 10
 (Taking $1° \approx 1/60$ radian and $g \approx 10\text{m/s}^2$)
 i.e. $Z_\alpha = 375$m
 $T_2 = mU/Z_\alpha = m$ 250/375m
 i.e. $T_2 = 0.67$ s

• *Relationship between M_q and M_η*

Assuming an all moving tailplane
 Pitching moment due to tailplane angle change, $\eta = M_\eta \eta$
 Change in tailplane incidence due to pitch rate, $q = l_t q/U$
 Pitching moment due to pitch rate $= M_\eta \, l_t q/U = M_q q$
 Hence for an all moving tailplane, $M_q = l_t/U$. M_η
 $l_t = 8$m, $U = 250$m/s
 $M_q = 8/250$. $M_\eta = 0.032 \, M_\eta$

• *Derivation of M_η and M_q*

1g normal acceleration at 250m/s forward speed corresponds to a rate of pitch equal to 10/250 rad/s, that is 2.4°/s approx (taking $g = 10\text{m/s}^2$)
1.6° Change in wing incidence produces 1g normal acceleration and requires 1° tailplane angular movement.
Thus in the steady state (denoted by suffix ss), $M_\eta \eta_{ss} = M_\alpha \alpha_{ss} + M_q q_{ss}$
Working in degrees, $M_\eta.1 = (2 \times 10^7 \times 1.6) + (0.032 \, M_\eta \times 2.4)$
Hence, $M_\eta = 3.5 \times 10^7$ Nm/rad
and $M_q = 0.032 \times 3.5 \times 10^7$ Nm/rad per s
i.e. $M_q = 1.1 \times 10^6$ Nm/rad per s

• *Derivation of T_1*

$$T_1 = \frac{6 \times 10^5}{1.1 \times 10^6}$$

ie $T_1 = 0.55$ s

• *Derivation of coefficient A_1*

$$A_1 = \frac{1}{T_1} + \frac{1}{T_2} = \frac{1}{0.55} + \frac{1}{0.67}$$

ie $A_1 = 2.3$

• *Derivation of coefficient A_2*

$$A_2 = \frac{M_\alpha}{I_y} + \frac{1}{T_1 T_2} = \frac{2 \times 10^7}{6 \times 10^5} + \frac{1}{0.55 \times 0.67}$$

ie $A_2 = 36$

• *Derivation of K*

$$K = \frac{M_\eta}{I_y T_2} = \frac{3.5 \times 10^7}{6 \times 10^5 \times 0.67}$$

ie $K = 87$

Hence

$$\frac{q}{\eta} = \frac{87(1 + 0.67D)}{(D^2 + 2.3D + 36)}$$

By inspection $\omega_0{}^2 = 36$

• *ie* $\omega_0 = 6$ rad/s (0.95 Hz)
  $\zeta = 2.3/(2 \times 6)$

• *ie* $\zeta = 0.2$ approx
  From examination of second order system responses in Fig. 3.32
  Percentage overshoot of response to a disturbance or control input is just over 50%.

• *A candidate for auto-stabilisation*
These data will be used in a worked example on auto-stabilisation in Section 3.6.

3.6 LATERAL CONTROL

3.6.1 Aileron control and bank to turn

The primary means of control of the aircraft in the lateral plane are the ailerons (see Fig. 3.15). These are moved differentially to increase the lift on one wing and reduce it on the other thereby creating a rolling moment so that the aircraft can be banked to turn.

Aircraft bank to turn so that a component of the wing lift can provide the essential centripetal force towards the centre of the turn in order to change the aircraft's flight path (see Fig. 3.33). The resulting centripetal acceleration is equal to (aircraft velocity) × (rate of turn) i.e. $V_T \dot{\Psi}$. This centripetal acceleration causes an inertia force to be experienced in the reverse direction i.e. a centrifugal force.

In a steady banked turn with no sideslip the resultant force the pilot experiences is the vector sum of the gravitational force and the centrifugal force and is in the direction normal to the wings. The pilot thus experiences no lateral forces.

Banking to turn is thus fundamental to effective control and manoeuvring in the lateral plane because of the large lift forces that can be generated by the wings. The lift force can be up to nine times the aircraft weight in the case of a modern fighter aircraft. The resulting large centripetal component thus enables small radius, high g turns to be executed.

Referring to Fig. 3.33:

Horizontal component of the lift force = $Z \sin\Phi$

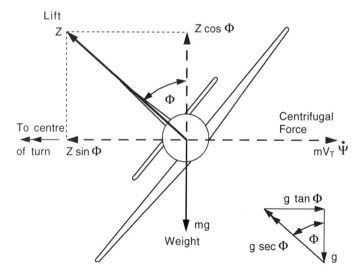

Fig. 3.33 Forces acting in a turn.

Equating this to the centrifugal acceleration gives

$$Z \sin\Phi = mV_T\dot{\Psi}$$

Vertical component of the lift force = $Z \cos\Phi$
Equating this to the aircraft weight gives

$$Z \cos\Phi = mg$$

from which

$$\tan\Phi = \frac{V_T\dot{\Psi}}{g} \tag{3.66}$$

Thus the acceleration towards the centre of the turn = $g \tan\Phi$

Referring to the inset vector diagram in Fig. 3.33, the normal acceleration component is thus equal to $g \sec\Phi$. Thus a 60° banked turn produces a centripetal acceleration of 1.73 g and a normal acceleration of 2 g. At a forward speed of 100m/s (200 knots approximately) the corresponding rate of turn would be 10.4°/s.

The lift required from the wings increases with the normal acceleration and the accompanying increase in drag requires additional engine thrust if the forward speed is to be maintained in the turn. The ability to execute a high g turn thus requires a high engine thrust/aircraft weight ratio.

To execute a coordinated turn with no sideslip requires the operation of all three sets of control surfaces, that is the ailerons and the tailplane (or elevator) and to a lesser extent the rudder. It is also necessary to operate the engine throttle(s) to control the engine thrust. The pilot first pushes the stick sideways to move the ailerons so that the aircraft rolls, the rate of roll being dependent on the stick movement. The rate of roll is arrested by centralising the stick when the desired bank angle for the rate of turn has been achieved. The pilot also pulls back gently on the stick to pitch the aircraft up to increase the wing incidence and hence the wing lift to stop loss of height and provide the necessary centripetal force to turn the aircraft. A gentle pressure is also applied to the rudder pedals as needed to counteract the yawing moment created by the differential drag of the ailerons. The throttles are also moved as necessary to increase the engine thrust to counteract the increase in drag resulting from the increase in lift and hence maintain airspeed.

The dynamics of the aircraft response to roll commands are complex and are covered later in Section 3.6.4. However, an initial appreciation can be gained by making the simplifying assumption of pure rolling motion, as follows.

Angular movement of the ailerons from the trim position, ξ, produces a rolling moment equal to $L_\xi\xi$ where L_ξ is the rolling moment derivative due to aileron angle movement. This is opposed by a rolling moment due to the rate of roll (refer to Section 3.4.4.3 and Fig. 3.23) which is equal to $L_p p$

The equation of motion is thus

$$L_\xi \xi + L_p p = I_x \dot{p} \qquad (3.67)$$

where I_x = moment of inertia of the aircraft about the roll axis.
    This can be expressed in the form

$$(1 + T_R D)p = \frac{L_\xi}{L_p} \cdot \xi \qquad (3.68)$$

where T_R = roll response time constant = $I_x / -L_p$ (note L_p is negative).
    This is a classic first order system, the transient solution being $p = Ae^{-t/T_R}$, where A is a constant determined by the initial conditions.
    For a step input of aileron angle, ξ_i the response is given by

$$p = \frac{L_\xi}{L_p}(1 - e^{-t/T_R}) \, \xi_i \qquad (3.69)$$

This is illustrated in Fig. 3.34.
    It can be seen that variations in the derivatives L_p and L_ξ over the flight envelope will affect both the speed of response and the steady state rate of roll for a given aileron movement. It should be noted that aerodynamicists also refer to the roll

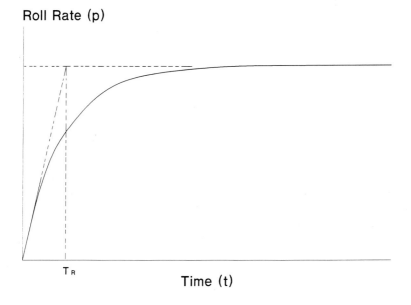

Fig. 3.34 Roll rate response.

response as a subsidence as this describes the exponential decay following a roll rate disturbance.

3.6.2 Rudder control

Movement of the rudder creates both a lateral force and a yawing moment. The control action exerted by the rudder is thus used to:

(a) Counteract the yawing moment due to movement of the ailerons to bank the aircraft to turn as already explained.
(b) Counteract sideslipping motion.
(c) Counteract asymmetrical yawing moments resulting, say, from loss of engine power in the case of multi-engined aircraft or carrying asymmetrical stores/weapons in the case of a combat aircraft (or both).
(d) Deliberate execution of a sideslipping manoeuvre. The yawing moment created by the rudder movement will produce a sideslip incidence angle, β, and this will result in a side force from the fuselage and fin (see Fig. 3.21). This side force enables a flat sideslipping turn to be made. However, the sideways accelerations experienced in flat turns are not comfortable for the pilot (or crew and passengers for that matter). In any case the amount of sideways lift that can be generated by the fin and fuselage body will only enable very wide flat turns to be made in general.
(e) To 'kick-off' the drift angle just prior to touch down when carrying out a cross wind landing. This is to arrest any sideways motion relative to the runway and so avoid side forces acting on the undercarriage at touch down.

Directional stability in the lateral plane is provided by the fin and rudder. Referring to Fig. 3.21, a sideslip velocity, v, with the aircraft's fore and aft axis at an incidence angle, β, to the aircraft velocity vector, V_T, (or relative wind) results in the fin developing a side force. As explained earlier this side force tends to align the aircraft with the relative wind in a similar manner to a weather cock.

3.6.3 Short period yawing motion

This weathercock action can result in a lightly damped oscillatory motion under certain flight conditions because of the aircraft's inertia, if the yaw rate damping moment is small. The analysis of this motion can be simplified by assuming pure yawing motion and neglecting the cross coupling of yawing motion with rolling motion (and vice versa). The direction of the aircraft's velocity vector is also assumed to be unchanged.

(It should be noted that the lateral response allowing for the effects of the roll/yaw cross coupling is covered later in Section 3.6.4.) Referring to Fig. 3.21, it can be seen that:

Sideslip velocity, $v = V_T \sin \beta = V_T \beta$ (as β is a small angle) $= V_T \psi$

as $\beta = \psi$, the incremental change in heading (or yaw) angle.

Thus yawing moment due to sideslip $= N_v V_T \psi$

It should be noted that in general the yaw angle ψ is not the same as the sideslip incidence angle β, the yaw angle being the angle in the lateral plane between the aircraft's fore and aft axis and the original direction of motion. The two angles are only the same when the direction of the aircraft's velocity vector is unchanged.

The moments acting about the CG are:

1) Yawing moment due to sideslip $= N_v V_T \psi$
2) Yawing moment due to yaw rate $= N_r r = N_r \dot{\psi}$

The equation of motion is :

$$N_v V_T \psi + N_r \dot{\psi} = I_z \ddot{\psi} \tag{3.70}$$

where I_z = moment of inertia about yaw axis

That is
$$\left(D^2 + \frac{-N_r}{I_z} D + \frac{-N_v}{I_z} V_T \right) \psi = 0 \tag{3.71}$$

Note that derivatives N_r and N_v are both negative.

This is a second order system with an undamped natural frequency $\omega_0 = \sqrt{\dfrac{N_v V_T}{I_z}}$

and damping ratio $\zeta = \dfrac{1}{2} \dfrac{N_r}{\sqrt{N_v V_T I_z}}$.

The period of the oscillation $(2\pi/\omega_0)$ and the damping ratio are both inversely proportional to the square root of the forward speed, i.e. $\sqrt{V_T}$. Thus increasing the speed shortens the period but also reduces the damping. The period of the oscillation is of the order of 3 to 10 seconds.

This short period yawing motion can be effectively damped out automatically by a yaw axis auto-stabilisation system (or yaw damper) and will be covered in Section 3.7.

In practice there can be a significant cross coupling between yawing and rolling motion and the yawing oscillatory motion can induce a corresponding oscillatory motion about the roll axis 90° out of phase known as 'Dutch roll'. (Reputedly after the rolling-yawing gait of a drunken Dutchman). The ratio of the amplitude of the rolling motion to that of the yawing motion is known as the 'Dutch roll ratio' and can exceed unity, particularly with aircraft with large wing sweepback angles.

3.6.4 Combined roll–yaw–sideslip motion

The cross coupling and interactions between rolling, yawing and sideslip motion and the resulting moments and forces have already been described in section 3.4.4.2. The interactions between rudder and aileron controls have also been mentioned.

The equations of motion in the lateral plane following small disturbances from straight and level flight are set out in Section 3.4.5, equations (3.37), (3.38) and (3.39) respectively.

The solution to this simplified lateral set of equations in terms of the transient response to a disturbance and resulting stability is a fifth order differential equation. This can usually be factorised into two complex roots and three real roots as shown below.

$$(D^2 + 2\zeta\omega_0 D + \omega_0{}^2)\left(D + \frac{1}{T_R}\right)\left(D + \frac{1}{T_Y}\right)\left(D + \frac{1}{T_{Sp}}\right)x = 0$$

where x denotes any of the variables v, p, r.

The transient solution is of the form

$$x = A_1 e^{-\zeta\omega_0 t}\sin(\sqrt{1 - \zeta^2}.\omega_0 t + \phi) + A_2 e^{-t/T_R} + A_3 e^{-t/T_Y} + A_4 e^{-t/T_{Sp}}$$

where A_1, ϕ, A_2, A_3 and A_4 are constants determined by the initial conditions.

The quadratic term describes the Dutch roll motion with undamped natural frequency ω_0 and damping ratio ζ. T_R describes the roll motion subsidence and T_Y the yaw motion subsidence.

T_{Sp} is usually negative and is the time constant of a slow spiral divergence. It is a measure of the speed with which the aircraft, if perturbed, would roll and yaw into a spiral dive because of the rolling moment created by the rate of yaw. The time constant of this spiral divergence is of the order of 0.5 to 1 minute and is easily corrected by the pilot (or autopilot).

3.7 AUTO-STABILISATION SYSTEMS

3.7.1 Limited authority auto-stabilisation systems

The possible need for improved damping and stability about all three axes has been referred to in the preceding sections. This can be achieved by an auto-stabilisation system, or, as it is sometimes referred to, a stability augmentation system.

Yaw auto-stabilisation systems are required in most jet aircraft to suppress the lightly damped short period yawing motion and the accompanying oscillatory roll motion due to yaw/roll cross coupling known as Dutch roll motion which can occur over parts of the flight envelope (refer to Section 3.6.4). In the case of military aircraft, the yaw damper system may be essential to give a steady weapon aiming platform as the pilot is generally unable to control the short period yawing motion and can in fact get out of phase and make the situation worse.

A yaw damper system is an essential system in most civil jet aircraft as the undamped short period motion could cause considerable passenger discomfort.

As mentioned earlier a yaw damper system may be insufficient with some aircraft with large wing sweepback to suppress the effects of the yaw/roll cross coupling and a roll damper (or roll auto-stabilisation) system may also be necessary. The possible low damping of the short period pitch response (refer to Section 3.5) can also require the installation of a pitch damper (or pitch auto-stabilisation) system. Hence, three axis auto-stabilisation systems are installed in most high performance military jet aircraft and very many civil jet aircraft.

A single channel, limited authority auto-stabilisation system is used in many aircraft which have an acceptable though degraded and somewhat lightly damped response over parts of the flight envelope without auto-stabilisation. 'Single channel' means that there is no back-up or parallel system to give a failure survival capability.

The degree of control, or authority, exerted by the auto-stabiliser is limited so that in the event of a failure in the autostabilisation system, the pilot can over-ride and disengage it. Typical authority of the auto-stabiliser system is limited to a control surface movement corresponding to $\pm 0.5g$.

A simple auto-stabiliser system is shown schematically in Fig. 3.35. It comprises a rate gyroscope which senses the angular rate of rotation of the aircraft about the auto-stabilised axis and which is coupled to an electronic unit which controls the auto-stabiliser actuator. The auto-stabiliser actuator controls the main servo actuator (driving the control surface) in parallel with the pilot's control by means of a differential mechanism – typically a lever linkage differential.

The control surface angular movement is thus the sum of the pilot's stick input and the auto-stabiliser output. The autostabiliser system produces a control surface movement which is proportional to the aircraft's rate of rotation about the axis being stabilised and hence a damping moment proportional to the angular rate of rotation of the aircraft about that axis. In effect a synthetic $N_r r$ or $L_p p$ or $M_q q$ term. In practice, the auto-stabiliser damping term is more complicated than a simple (constant) $\times$ (aircraft rate of rotation) term because of factors such as auto-stabiliser servo performance and the need to avoid exciting structural resonances. i.e. 'tail wagging the dog' effect. This latter effect is due to the force exerted by the control surface actuator reacting back on the aircraft structure which has a finite structural stiffness.

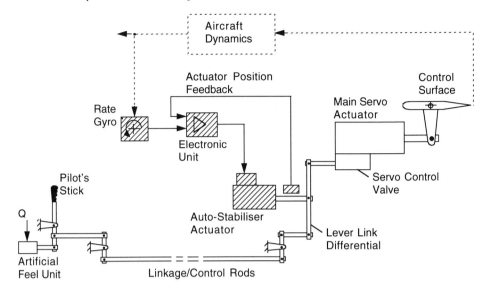

Fig. 3.35 Simple auto-stabiliser system.

The resulting bending of the aircraft structure may be sensed by the auto-stabiliser rate gyro sensor and fed back to the actuator and so could excite a structural mode of oscillation, if the gain at the structural mode frequencies was sufficiently high. In the case of auto-stabilisation systems with a relatively low auto-stabiliser gearing, it is generally sufficient to attenuate the high frequency response of the auto-stabiliser with a high frequency cut-off filter to avoid exciting the structural modes. The term auto-stabiliser gearing is defined as the control surface angle/degree per second rate of rotation of the aircraft.

With higher auto-stabiliser gearings, it may be necessary to incorporate *notch filters* which provide high attenuation at the specific body mode flexural frequencies. (This is briefly covered again in Chapter 4, Fly-by-wire flight control.)

Most simple, limited authority yaw dampers feed the angular rate sensor signal to the auto-stabiliser actuator through a 'band pass filter' with a high frequency cut-off (see Fig. 3.36). The band pass filter attenuation increases as the frequency decreases and becomes infinite at zero frequency (ie dc) and so prevents the auto-stabiliser actuator from opposing the pilot's rudder commands during a steady rate of turn. (This characteristic is often referred to as 'dc blocking'.) The yawing motion frequency is in the centre of the band pass range of frequencies. The angular rate sensor output signal resulting from the short period yawing motion thus passes through the filter without attenuation or phase shift and so provides the required signal to the auto-stabiliser actuator to damp the oscillatory yawing motion. (It should be noted that this type of filter in flight control systems is often referred to as a 'wash-out filter', particularly in the United States, as it effectively 'washes out' the dc response.)

The variation in the aircraft response and control effectiveness over the flight envelope may require the auto-stabiliser gearings to be switched as a function of height and airspeed to possibly two or three different values.

3.7.1.1 'Worked example' of simple pitch auto-stabiliser

An example of a typical representative simple pitch auto-stabiliser is given below to show how initial calculations can be made to obtain an order of magnitude estimate of the auto-stabiliser 'gearing' required using the transfer functions derived in Section 3.5 (auto-stabiliser 'gearing' being the tailplane angle in degrees per degree/s rate of pitch). It should be pointed out that much more complex computer modelling and simulation is necessary to establish the optimum values and allow for a more complex transfer function. For example:

(a) Certain derivatives have been ignored.
(b) Lags in the actuator response have been neglected together with any non-linear behaviour which might be present.

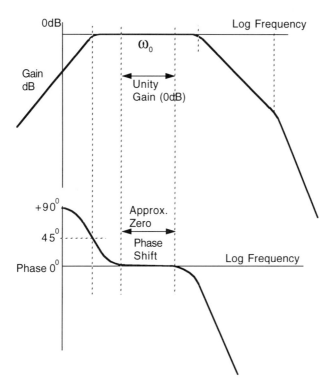

Fig. 3.36 DC blocking or (wash-out) filter with high frequency cut off.

(c) The transfer functions of any filters required to attenuate rate gyro noise and avoid excitation of structural resonances have been omitted.

Referring to Section 3.5.4, pitch rate, q, to tailplane angle, η, transfer function is:

$$\frac{q}{\eta} = \frac{K(1+T_2 D)}{(D^2 + 2\zeta\omega_0 D + \omega_0{}^2)}$$

where $T_2 = mU/Z_\alpha$ and $K = M_\eta /I_y T_2$

Referring to Fig. 3.37 (ignoring lags in auto-stab actuator, main actuator, gyro and filters etc):

Auto-stabiliser output $= G_q q$

$$\text{Tailplane angle, } \eta = \delta_i - G_q q \qquad (3.72)$$

where G_q = auto-stabiliser gearing and δ_i = pilot's stick input

whence

$$q = \frac{K(1+T_2 D)}{(D^2 + 2\zeta\omega_0 D + \omega_0{}^2)} . (\delta_i - G_q q)$$

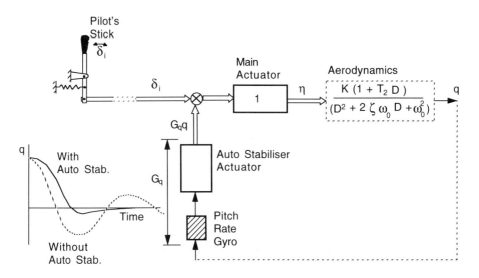

Note: Main Actuator, Auto-Stabiliser Actuator, Rate Gyro & Filters
assumed to have scalar transfer functions (i.e. no lags)

Fig. 3.37 Pitch auto-stabiliser loop.

from which

$$[D^2 + (2\zeta\omega_0 + KT_2G_q)D + (\omega_0^2 + KG_q)]q = K(1+T_2D)\delta_i$$

The worked example in 3.5.2 provides the data for the basic aircraft response with:

$K = 87$
$T_2 = 0.67$
$\omega_0 = 6$ rad/s
$\zeta = 0.2$

The required value of G_q to improve the damping of the aircraft's pitch response to correspond to a damping ratio of 0.7 is derived below.

Let ω_{0stab} and ζ_{stab} denote the undamped natural frequency and damping ratio of the quadratic factor in the transfer function q/δ_i of the aircraft with pitch auto-stabiliser.

Hence

$$\omega_{0stab}^2 = \omega_0^2 + KG_q$$
$$2\zeta_{stab}\omega_{0stab} = 2\zeta\omega_0 + KT_2G_q$$
$$\zeta_{stab} = 0.7$$

$\therefore 2 \times 0.7 \times \omega_{0stab} = (2 \times 0.2 \times 6) + (87 \times 0.67 \times G_q)$
whence $\omega_{0stab} = 1.71 + 41.6G_q$
viz $1734\ G_q^2 + 55\ G_q - 33 = 0$
from which $G_q = 0.12°$ tailplane/degree per second rate of pitch.
whence $\zeta_{stab} = 0.7$ and $\omega_{0stab} = 6.8$ rad/s (1.1 Hz) approx.
The q/δ_i transfer function with auto-stabilisation is thus

$$\frac{q}{\delta_i} = \frac{87(1+0.67D)}{D^2 + 9.5D + 46}$$

The improvement in damping is shown inset in Fig. 3.37.

3.7.2 Full authority auto-stabilisation systems

In many cases, aerodynamic performance and manoeuvrability do not always match with stability and damping over the whole of the flight envelope and can lead to restricting compromises. There may thus be a need for full authority auto-stabilisation over part of the flight envelope (as opposed to the limited authority systems just described). Because the system has full authority, a failure could be catastrophic and so a failure survival system of triplex level redundancy (or higher) is essential.

Triplex level redundancy means three totally independent parallel systems of sensors, electronic units (or controllers) and actuation units with independent electrical and hydraulic supplies.

Basically the triplex system operates on a majority voting of the 'odd man out' principle. Because the three parallel systems or 'channels' are totally independent, the probability of two systems failing together at the same instant in time is negligibly low (of the order typically of 10^{-6}/hour failure probability). A failure in a channel is thus detected by cross comparison with the two 'good' channels and the failed channel can be disconnected leaving the system at duplex level redundancy. Failure survival redundant configurations are discussed in some detail in the following Chapter 4, 'Fly-by-wire flight control', and at this stage it is sufficient to appreciate the basic triplex concept.

A second failure results in a stalemate situation with the failed auto-stabiliser actuator being counteracted by the 'good' channel the result is a passive failure situation and the control surface does not move.

The operational philosophy with such a system after the first failure would generally be to minimise the time at risk at duplex level by, say, decelerating and possibly descending until a speed and height condition is reached where auto-stabilisation is not essential.

The relatively high auto-stabiliser gearings that may be required together with the failure survival system requirements lead to what is known as a manoeuvre command 'Fly-by-wire' flight control system. This is discussed in Chapter 4.

Further reading

Babister, A.W., *Aircraft dynamic stability and response*: Pergammon Press.

Bertin, J.J., and Smith, M.L., *Aerodynamics for engineers*: Prentice Hall.

Blakelock, J.H., *Automatic control of aircraft and missiles*: Wiley

Dommasch, D.D., Sherby, S.S., Connolly, T.F., *Airplane aerodynamics*: Pitman

Etkin, B., *Dynamics of atmospheric flight*: Wiley.

McLean, D., *Automatic flight control systems*: Prentice Hall.

McRuer, D., Ashkenas, I., and Graham, D., *Aircraft dynamics and automatic control*: Princeton U.P.

Nelson, R.C., *Flight stability and automatic control*: McGraw-Hill

4

Fly-by-wire flight control

4.1 INTRODUCTION

Fly-by-wire (FBW) flight control has come into use in production aircraft (as opposed to experimental prototypes) since the late 1970s and can enable a major increase in flight control capability to be achieved. One of the unique benefits of a FBW flight control system is the ability to fly aircraft configurations which have minimal stability margins, or which may be aerodynamically unstable over part or even the whole of the flight envelope. Such aircraft configurations and shapes are chosen for aerodynamic efficiency or for achieving 'stealth', that is minimisation of detection by radar. Higher agility can be achieved with an aerodynamically unstable aircraft compared with a conventional stable configuration, agility being defined as the ability to change the direction of the aircraft's velocity vector. Increases in turn rate of 35% are being claimed for the new generation FBW fighters with aerodynamically unstable configurations over conventional designs – a major edge in air combat.

For civil aircraft, FBW enables very significant reductions in the weight, cost and complexity of the flight control system to be achieved by the elimination of the bulky mechanical control system and the integration of the autostabilisation system.

The high integrity full authority FBW system can also provide the automatic stabilisation required to enable a civil aircraft to be designed with reduced stability margins in certain loading configurations and flight conditions. This enables a lighter aircraft to be produced with an increased performance and payload capability compared with a conventional design.

The flight control computing system can also automatically limit the control power available to the pilot to allow the pilot to have safe and 'carefree' handling throughout the flight envelope.

The FBW flight control system can be designed so that the pilot is able to exercise a closed loop manoeuvre command control of the aircraft. This is achieved by the pilot controlling the aircraft through the FBW computers with feedback of the aircraft motion from the appropriate sensors, and will be covered in the next section. A well designed and integrated combination of aircraft and FBW manoeuvre command flight control system can achieve very good pilot handling qualities with a fast, well damped response which is substantially constant over a wide flight envelope and range of loading configurations.

These reasons have led to the adoption of a manoeuvre command FBW control system in high performance aircraft which are aerodynamically stable and which also require a high performance from the autopilot system for such applications as automatic terrain following (for example, the Tornado aircraft).

As already mentioned, the use of electrical signalling has also enabled the mechanical control system of rods and linkages connecting the pilot's control stick to the main control surface actuators to be eliminated. Significant weight saving is achieved as well as the elimination of problems such as friction, backlash etc. A small control stick can also be used instead of the large, bulky control column which is necessary to exert the forces required with mechanically signalled controls.

As with many systems, the basic concepts of FBW manoeuvre command control are not new. It has been a question of the essential enabling technologies becoming sufficiently mature to enable a high integrity system to be implemented economically with the required safety levels, reliability and availability. A major factor has been the development of failure survival digital flight control systems and their implementation in VLSI (very large scale integration) micro-electronic circuits. It should be noted that there have also been several other key technologies whose development has been essential for FBW control such as failure survival actuation systems.

4.2 FLY-BY-WIRE FLIGHT CONTROL

4.2.1 FBW system basic concepts

The essential features of a FBW system can be best appreciated by first examining the operation of a conventional mechanically signalled flight control system and its inherent limitations. Figure 3.35, which shows a simple autostabiliser system, also illustrates the essential features of a conventional flight control system where the pilot directly controls the movement of the control surfaces. The variations in the aerodynamic derivatives across the flight envelope can affect the aircraft response although the auto-stabilisation gearing can be adjusted to compensate for these variations to a considerable extent. The problems can really arise at the 'corners' of the flight envelope, at certain height and speed conditions and high angles of incidence where the 'safe' control available to the pilot is limited. The aircraft loading configuration may also be near the safe limit as regards CG position relative to the aerodynamic centre and this further compounds the flight control problems. The carriage of a large load of external 'stores' on a military strike aircraft generally exerts a destabilising effect and markedly degrades the aircraft handling. Particular problems can also arise when a high performance is required from the autopilot system to minimise the aircraft's excursions from the desired flight path. Any lags in the aircraft response to the steering commands from the autopilot affect the stability of the autopilot loop and limit the loop gain which can be used.

Lowering the autopilot loop gain inevitably increases the magnitude of the excursions from the commanded flight path and so degrades the autopilot performance.

Figure 4.1 illustrates the essential features of a FBW flight control system. It can be seen that the pilot controls the movement of the control surfaces through the flight control computers. The computers are also supplied with the information on the aircraft's motion from appropriate sensors with failure survival redundancy. The aircraft motion sensors comprise:

(a) Body mounted rate gyroscopes to measure the aircraft rates of rotation about the roll, pitch and yaw axes.
(b) Linear accelerometers to measure the normal and lateral acceleration components.
(c) Air data sensors to measure height and airspeed.
(d) Airstream sensors to measure the airstream incidence angles in the longitudinal and lateral planes.

The essential features are:

- *Electrical signal transmission*

Pilot's stick sensor signals, aircraft motion sensor signals and control surface actuator position signals are all transmitted electrically to the flight control computers. The commanded control surface angles derived by the flight control computers are also transmitted electrically to the control surface servo actuators.

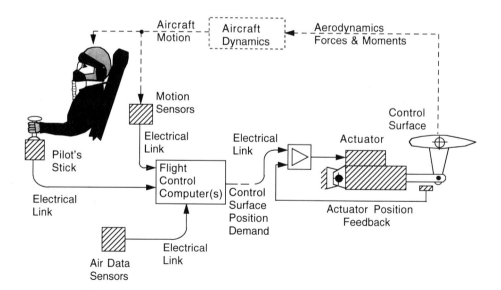

Fig. 4.1 Fly-by-wire flight control system (redundancy omitted for clarity).

It should be noted that modern digital FBW systems use a very high integrity digital data bus system for transmission of these signals using time division multiplexing of the data. Independent triplex, or even quadruplex, data buses are used to meet the failure survival requirements.

- *Motion sensor feedback*

The aircraft 'state' information is supplied to the flight control computers from a failure survival, redundant configuration of aircraft motion sensors – e.g. quadruplex pitch, roll, yaw rate gyros etc.

- *High integrity, failure survival computing system*

The flight control computing system must be of very high integrity and have the failure survival capability to meet the flight safety requirements. The tasks carried out by the computing system comprise:

a) Failure detection.
b) Fault isolation and system reconfiguration in the event of a failure.
c) Computation of the required control surface angles.
d) Monitoring.
e) Built in test.

- *High integrity, failure survival actuation system*

The actuation system operating the control surfaces must be of very high integrity and have the failure survival capability to meet the aircraft safety requirements.

The servo actuation systems driving the control surfaces comprise a two stage servo system with the FBW servo actuators driving the duplex control valves of the main power control actuator. The system is illustrated schematically in Fig. 4.2. The normal mechanical feedback from the output of the main power control actuator to this control valve is removed and replaced with electrical feedback from position pick-offs on the main power control actuator to the servo amplifiers controlling the FBW servo actuators. This enables a faster and more precise response to be obtained from the control surface actuation system because the servo loop gain is increased. It is essential that the lags in the control surface servo actuators are very small over the operating range of control frequencies as these lags have a destabilising effect on the FBW control loop. Typical performance requirements for a modern agile fighter correspond to a phase lag of less than $12°$ at 1 Hz input frequency.

- *Very high overall system integrity*

The overall system integrity must be as high as the mechanical control system it replaces. The probability of a catastrophic failure must not exceed 10^{-9}/hour for a civil aircraft and 10^{-7}/hour for a military aircraft.

It should be noted that the FBW system can be configured to be a direct electrical implementation of the mechanical system shown in Fig. 3.35. The advantages being sought in such an application are:

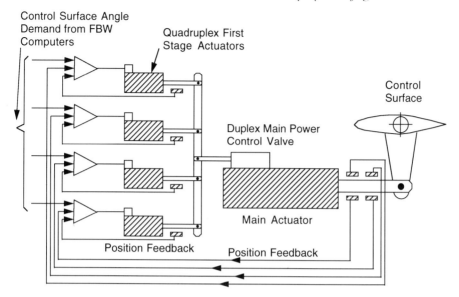

Note - Measures to alleviate force fight between actuators due to tolerance
        variations between actuator loops omitted for clarity
        - e.g. Differential Pressure Feedback

Fig. 4.2 Failure survival control surface actuation system schematic.

- Aircraft control system is 'transparent' to the pilot, the aircraft responding in the same way as a mechanically controlled aircraft. This is a considerable advantage in civil airline operation where pilots may be interchanging with existing aircraft with mechanically operated control systems.
- Ability to limit the control power normally available to the pilot. The system can however be designed so that the pilot, if he wishes, can break through or over-ride the bank angle and stall limits if he is concerned about the aircraft's behaviour in extreme conditions.
- Integration of the autostabilisation system.
- Ease of integration of the autopilot, electrical interfaces only being involved.
- Flexibility. Control gearings are implemented through software.
- Lower weight and elimination of mechanical control problems such as friction, backlash etc.

A closed loop manoeuvre command control is achieved by increasing the gain of the motion sensor feedback loops. The control surface actuators are thus controlled by the difference, or error, between the pilot's command signals and the measured aircraft motion from the appropriate sensors, for example pitch rate in the case of a pitch rate command system and roll rate in the case of a roll rate command system. Other control terms may also be included such as airstream incidence angles and possibly normal and lateral accelerations. The flight control computer derives the required control surface movements for the aircraft to follow the pilot's commands in a fast, well damped manner. It should be stressed that the achievement of such a response requires extensive design and testing and a well integrated combination of aircraft and FBW control system. For instance, adequate control power must be available from the control surfaces and non linear behaviour taken into account (e.g. rate limiting in the actuation system).

A well designed closed loop control system offers the following advantages over an open loop control system.

(i)   The steady state output to input relationship is substantially independent of changes in the loop gain provided this remains sufficiently high.
(ii) The system bandwidth is improved and the phase lag when following a dynamically varying input is reduced.
(iii) A fast, well damped response can generally be achieved by suitable design of the control loop, this response being little affected by normal changes in the loop gain.

The improvement brought by closed loop control can be illustrated by considering the roll response of an aircraft with a roll rate command system by making certain simplifying assumptions. The roll rate/aileron angle response is derived in the previous Chapter, Section 3.6.1.

Assuming pure rolling motion from equation (3.68)

$$\frac{p}{\xi} = \frac{L_\xi}{L_p} \frac{1}{1 + T_R D} \tag{4.1}$$

where p = roll rate
ξ = aileron angle
L_ξ = rolling moment derivative due to aileron angle, ξ
L_p = rolling moment derivative due to roll rate, p
I_x = moment of inertia about roll axis
T_R = roll time constant = I_x/L_p

Referring to Fig. 4.3, and ignoring the lags in the aileron servo actuators and any filters which may be required

$$\frac{p}{p_D} = \frac{\dfrac{K_c L_\xi / L_p}{(1 + T_R D)}}{1 + \dfrac{K_c L_\xi / L_p}{(1 + T_R D)}} \tag{4.2}$$

where p_D = demanded roll rate
K_c = controller gain = aileron angle in degrees/degree per second roll rate error.

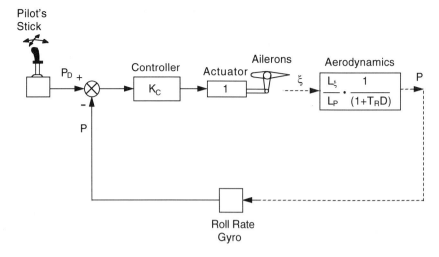

Fig. 4.3 Simplified FBW roll rate command system.

From this

$$\frac{p}{p_D} = \frac{K}{(1+K)} \cdot \frac{1}{1 + \dfrac{T_R}{(1+K)}D} \qquad (4.3)$$

where $K = \dfrac{K_c L_\xi}{L_p}$ = open loop gain

$$\text{If } K \gg 1, \quad \frac{p}{p_D} \approx \frac{1}{1 + \dfrac{T_R}{K}D} \qquad (4.4)$$

The steady state value of the roll rate for a given stick input is thus substantially constant over the flight envelope providing K is sufficiently large. The time constant of the roll response is also greatly reduced, being divided by the open loop gain K.

The difference in performance between a conventional control system response and a roll rate command response is illustrated in Fig. 4.4. It should be noted that high agility requires a fast roll rate response to bank the aircraft rapidly so as to be able to turn in the lateral plane. (The incidence response must also be fast and this can be achieved by a pitch rate command system.)

It should be stressed again that the above treatment has been greatly simplified by assuming pure rolling motion and no roll/yaw/sideslip cross coupling and also ignoring lags in the actuator and filters etc. The object is to explain basic principles without resorting to a mathematically complex model.

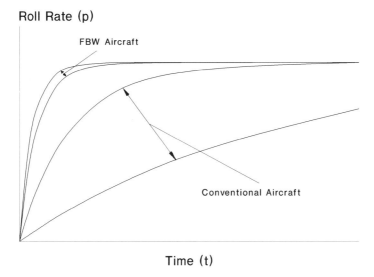

Fig. 4.4 Roll rate response.

Consider now FBW manoeuvre command control systems in general. The flight control computers can generate additional terms such as derivatives (suitably filtered) or integrals of particular quantities to 'shape' and improve the loop response and performance and ensure adequate stability margins. Reference has already been made in Chapter 3, Section 3.5.2, to the very wide range of control effectiveness, or stick force/g, that can occur over the flight envelope. For instance, a high performance aircraft can have a variation in stick force/g over the flight envelope as high as 40:1. It is not possible to achieve a stable closed loop control system with such a wide variation in the open loop gain. It is therefore necessary for the flight control computer to adjust the gains or gearings of the various control terms as a function of height and airspeed so that the overall variation in the open loop gain is acceptable. This is sometimes referred to as air data gain scheduling.

A 'hands off' automatic stabilisation of the aircraft is effected in the case of an aerodynamically unstable aircraft by the FBW manoeuvre command system. Any disturbance, such as a gust, which tends to disturb the aircraft from its flight path is sensed by the motion sensors and causes the control surfaces to move automatically to restore the aircraft to its flight path.

A further feature of the closed loop FBW control system with its relatively high bandwidth can be the ability to filter the pilot's commands through a 'model filter' which represents the 'ideal aircraft' response, as far as the pilot is concerned. The closed loop FBW control with its higher bandwidth ensures that the aircraft follows the filtered command as closely as possible and thus approaches the pilot's 'ideal aircraft' response.

For example, statistical analysis of military pilot opinion as to what constitutes a satisfactory longitudinal response has been found to correspond to a second order system response with an undamped natural frequency of 0.5 Hz and damping ratio 0.7.

ie
$$\frac{\theta_o}{\theta_i} = \frac{10}{D^2 + 4.4D + 10}$$
(4.5)

where θ_o = aircraft response parameter and θ_i = pilot's input command.

However, it should be noted that not all FBW manoeuvre command control systems use such a technique and opinions are divided on its merits.

4.2.2 Advantages of FBW control

The main advantages of a well designed FBW flight control system are summarised below.

- Good consistent handling which is sensibly constant over a wide flight envelope and range of load conditions.
- Automatic stabilisation.
- Autopilot integration – the electrical interface and the manoeuvre command control of the FBW system greatly ease the autopilot integration task. The autopilot provides steering commands as pitch rate or roll rate commands to the FBW system. The relatively high bandwidth manoeuvre command 'inner loop' FBW system ensures that response to the outer loop autopilot commands is fast and well damped, ensuring good control of the aircraft flight path in the autopilot modes.
- Carefree manoeuvring – the FBW computers monitor the aircraft's state continually and assess how close it is to its manoeuvre boundaries and control limits and automatically limit the pilot's command inputs. This ensures that the aircraft does not enter an unacceptable attitude or approach too near the limiting angle of attack or carry out manoeuvres which would exceed the structural limits of the airframe. This facility increases the safety of the aircraft as the accident statistics show that a number of aircraft are lost each year due to flying too close to their manoeuvre limits and the very high pilot work load in the event of a subsequent emergency.
- Ability to integrate additional controls automatically. In addition to the primary controls (tailplane (or elevator), ailerons, rudder) there may be other controls such as:
- Leading edge slats/flaps and trailing edge flaps which can be automatically controlled to generate additional lift for manoeuvring and not just landing and take off.
- Variable wing sweep.
- Thrust vector control nozzles and engine thrust .

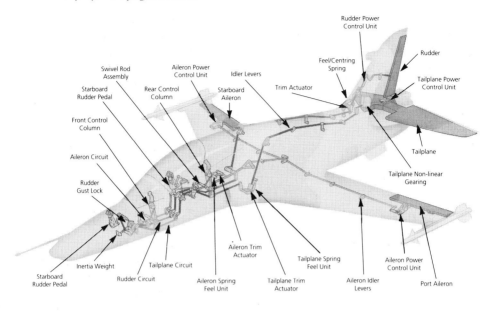

Fig. 4.5 Mechanically signalled controls – BAe Hawk (by courtesy of British Aerospace Defence Ltd.).

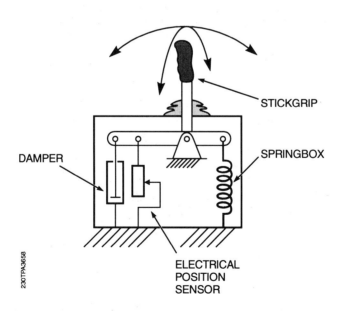

Fig. 4.6 FBW control stick schematic.

- Elimination of mechanical control runs – the bulk and weight and mechanical complexity of the control rods and linkages coupling the pilot's control column to the control valves of the actuators which move the control surfaces in conventional systems are very significant.

The problems of friction, backlash and flexure of the airframe structure can also be very considerable, particularly with large aircraft. Figure 4.5 illustrates a typical mechanical control system. The adoption of electrical signalling eliminates these problems and enables significant weight savings to be made.

- Small control 'stick' – electrical signalling enables a small control stick to be used. The location of the FBW control stick can be either a 'sidearm' or central mounting, the respective merits being fairly evenly divided. Figures 4.6 and 4.7 illustrate a FBW control stick. Such sticks can have a carefully tailored combination of displacement and force characteristics (see Fig. 4.8). The small control stick also has additional advantages in terms of the greater freedom it provides in optimising the cockpit layout of controls and displays.

- Ability to exploit aerodynamically unstable aircraft configurations – overall stability is achieved by the high integrity FBW control system enabling the conflicting design requirements of aerodynamic performance and stability to be decoupled. For instance, at supersonic speeds the aerodynamic centre moves

Fig. 4.7 FBW control sticks – left, side arm mounted, right, centrally mounted (by courtesy of GEC-Marconi Avionics Ltd.).

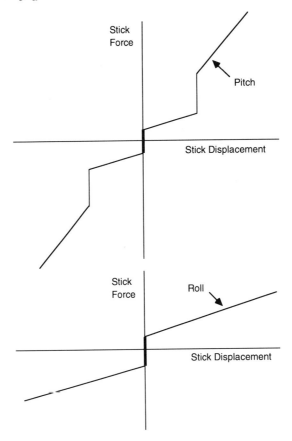

Fig. 4.8 Stick force/displacement characteristics.

aft, increasing the longitudinal static stability so that larger pitch control moments are required. A highly stable aircraft resists changing its incidence whereas manoeuvrability requires rapid changes in incidence. As already covered in Section 3.3.2, an aerodynamically unstable aircraft can have a lower weight and drag and increased performance over a comparable aircraft which is inherently stable. The increase in agility in the case of a fighter is a deciding factor with an increase in turn rate of 35% being claimed – a major edge in air combat in the ability to evade a missile or manoeuvre into a firing opportunity.

- Aerodynamics versus 'stealth' – the concept of reducing the radar cross section of an aircraft so that it is virtually undetectable (except at very close range) has been given the name 'stealth' in the USA. Radar reflection returns are minimised by faceted surfaces which reflect radar energy away from the direction of the source, engine intake design and the extensive use of radar

Fig. 4.9 Lockheed F-117A Stealth fighter. 'STEALTH + AERODYNAMICS = FBW' (by courtesy of Lockheed Advanced Development Company).

energy absorbing materials in the structure. An example of a stealth aircraft is the Lockheed F117A shown in Fig. 4.9.

Stealth considerations and requirements can conflict with aerodynamics requirements and FBW flight control is essential to give acceptable, safe handling across the flight envelope.

4.3 CONTROL LAWS

The term 'control laws' is used to define the algorithms relating the control surface demand to the pilot's stick command and the various motion sensor signals and the aircraft height, speed and Mach number.

As an example, a very simple basic pitch rate command law would be :

$$\eta_D = K(\theta_i - G_q q) \tag{4.6}$$

where η_D = tailplane demand angle
θ_i = pilot's input command
q = pitch rate

K = forward loop gain
G_q = pitch rate gearing

In practice additional control terms from other sensors (e.g. incidence and normal acceleration may be required. The value of K and the sensor gearings would also almost certainly need to be varied with height and airspeed by an air data gain scheduling system, as already mentioned.

Control terms proportional to the derivative or rate of change of error and integral of error are also used to shape and improve the closed loop response. A brief explanation of the action of these control terms is given below in view of their importance in closed loop flight control systems.

Proportional plus derivative of error control provides a phase advance characteristic to compensate for the lags in the system, for instance actuator response, and hence improve the loop stability. This increases the damping of the aircraft response and reduces the overshoot to a minimum when responding to an input or disturbance.

The control term is filtered to limit the increase in gain at high frequencies and smooth the differentiation process which amplifies any noise present in the error signal. Figures 4.10 and 4.11 illustrate the gain and phase versus frequency characteristics of a simple filtered 'one plus derivative' error control. (usually known as phase advance control) with a transfer function $\dfrac{(1+TD)}{(1+\frac{T}{n}D)}$, the

characteristics being plotted for n = 3, 5, 7, 10. The amount of phase advance is determined by the value of n and is generally limited by the resulting increase in

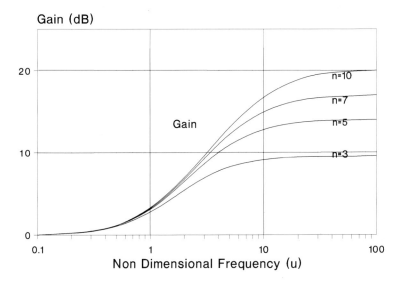

Fig. 4.10 Phase advance function frequency response.

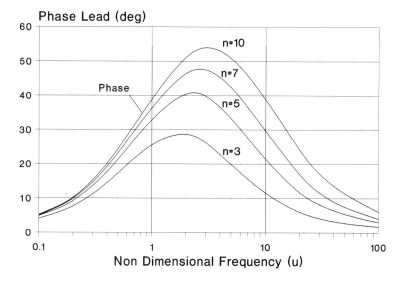

Fig. 4.11 Phase advance function frequency response.

gain at high frequencies and avoiding exciting structural resonance modes. Typical values used for *n* are between 4 and 5 giving a maximum phase lead of about 40°.

A more physical explanation of the damping action of a phase advance element is shown in Fig. 4.12. This shows how a control moment which is proportional to error plus rate of change of error (suitably smoothed) changes sign before the error reaches zero and hence applies a retarding moment to decelerate and progressively

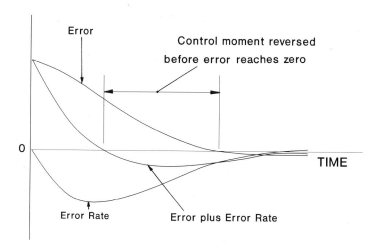

Fig. 4.12 Damping action – 'one plus derivative' of error.

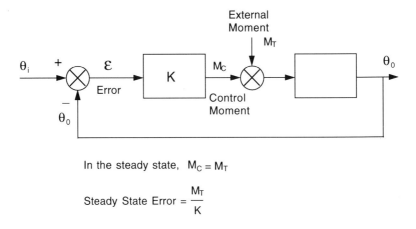

In the steady state, $M_C = M_T$

$$\text{Steady State Error} = \frac{M_T}{K}$$

Fig. 4.13 Simple closed loop control system subjected to external moment.

reduce the velocity before the error reaches zero and hence minimises the overshoot.

Proportional plus integral of error control eliminates steady state errors and reduces the following lag. The integral of error term increases the loop gain at low frequencies up to theoretically infinite gain at dc so that there are zero steady state errors due to out of trim external moments or forces acting on the aircraft. The error when following a low frequency input command is also reduced and is zero for a constant input rate. The need for an integral term control can be seen by examining the simple system shown in Fig. 4.13. In the steady state if θ_i is constant, θ_o must also be constant and thus the resultant moment acting on the aircraft must also be zero and the control surface moment is equal and opposite to the out of trim moment. However to generate a control surface moment requires an error signal so that θ_o cannot equal θ_i exactly. The magnitude of the steady state error is dependent on the loop gain, K which in turn is limited by the loop stability.

The effect of 'proportional plus integral' error control can be seen in Fig. 4.14. The steady state error is zero, however the integral of the error is not zero and reaches a value sufficient to generate the necessary control moment to balance the trim moment. The value of an integral term control can thus be seen in achieving automatic trimming of the control surfaces.

Figures 4.15 and 4.16 illustrate the gain and phase versus frequency characteristics for proportional plus integral control. The transfer function being

$$\left(1 + \frac{1}{TD}\right), \text{ that is } \frac{(1 + TD)}{TD}$$

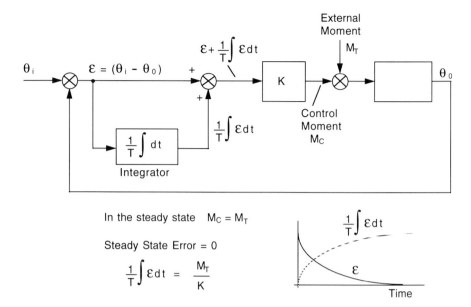

Fig. 4.14 'One plus integral' of error control.

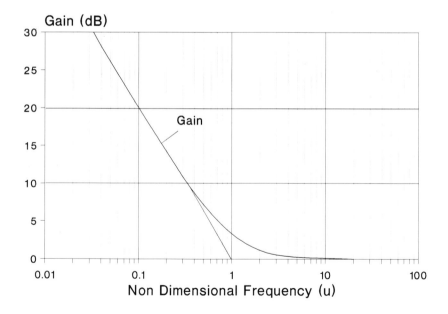

Fig. 4.15 'One plus integral' function frequency response.

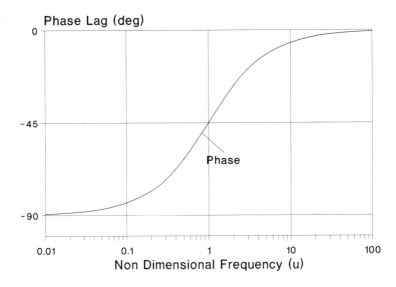

Fig. 4.16 'One plus integral' function frequency response.

It can be seen that the gain approaches unity (0 dB) at high frequencies, with zero phase shift so that the loop stability is not degraded, providing the integral time constant, T, is suitably chosen.

It should be noted that integral terms need to be initially synchronised to the initial conditions defined at start up to avoid large transients on switching on.

Mention has already been made of the 'carefree manoeuvring' characteristic that can be incorporated in a FBW system. This is achieved by continually monitoring the aircraft's state and automatically limiting the authority of the pilot's command input by means of a suitable control law which takes into account the aircraft's manoeuvre boundaries and control limits. This prevents the pilot from attempting to manoeuvre the aircraft into an unacceptable attitude, or approach too near to the limiting angle of incidence, or exceed the structural limits of the aircraft in a manoeuvre.

4.3.1 Pitch rate command control

A block diagram of a pitch rate command FBW flight control system is shown in Fig. 4.17 (the essential redundancy has been omitted for clarity).

The primary feedback of pitch rate for the pitch rate command loop is provided by the pitch rate gyro(s).

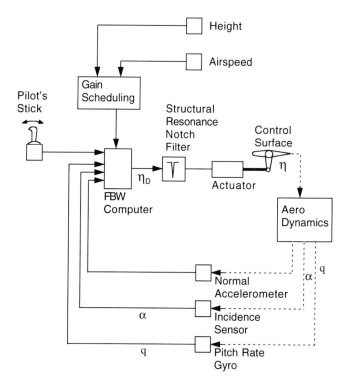

Fig. 4.17 Pitch rate command FBW loop (redundancy omitted for clarity).

In the case of an aerodynamically unstable aircraft, an integral of pitch rate term is also used to provide a 'quasi incidence term' to counteract a positive (unstable) M_α.

The 'quasi incidence' term provided by $\int qdt$ can be explained as follows.

The transfer function relating incidence, α, and pitch rate, q, is derived in Chapter 3, section 3.5.4. From equation (3.52)

$$\alpha = T_2 \frac{1}{1+T_2 D} q \qquad (4.7)$$

Compared with

$$\int qdt = \frac{1}{D} q \qquad (4.8)$$

Suppose q varies sinusoidally with time, that is $q = q_{max} \sin \omega t$. In the steady state, α and $\int qdt$ will also vary sinusoidally and can be obtained by substituting $j\omega$ for D in (4.7) and (4.8).

Hence
$$\alpha(j\omega) = T_2 \frac{1}{1+j\omega T_2} q(j\omega)$$

$$|\alpha| = \frac{T_2}{\sqrt{1+\omega^2 T_2^{\;2}}} q_{max}$$

$$\angle\alpha = -\tan^{-1}\omega T_2$$

For values of ω such that $\omega T_2 \gg 1$

$$|\alpha| \rightarrow \frac{1}{\omega} q_{max}$$

$$\angle\alpha \rightarrow -\frac{\pi}{2}$$

Now $\displaystyle\int qdt = \frac{1}{D} q$

$$\left| \int qdt \right| = \frac{1}{\omega} q_{max}$$

$$\angle \int qdt = -\frac{\pi}{2}$$

Hence $\int qdt \approx \alpha$ for values of ω such that $\omega T_2 \gg 1$

The proportional plus integral control is generally applied to the pitch rate error, $q_E = (q_D - q)$, where q_D = demanded pitch rate.

This has the advantages of eliminating steady state following errors and enabling the pilot to change the pitch attitude easily. (It should be noted that as far as loop stability is concerned, there is no difference between placing the stabilising transfer function in the pitch rate error path as opposed to the pitch rate feedback path.)

In the case of an aerodynamically unstable aircraft, a well designed *one plus integral of pitch rate error* control can provide an acceptable stable loop, although not an optimum one, without any other sensor terms such as airstream incidence or normal acceleration. This has the advantage that the key motion feedback sensors, namely the pitch rate gyros are very rugged and reliable devices.

The advent of solid state gyros (e.g. Fibre Optic Gyros) offers even higher reliability which should approach airframe life (50,000 to 100,000 hours) – 'fit and forget' devices.

It should be noted that while airstream incidence sensors can also provide the stabilising term to counter an unstable M_α, they have two inherent shortcomings, as far as a 'core' stabilising function is concerned.

(a) Vulnerability to damage on the ground, or bird strikes in flight, as the sensors have to be mounted externally in the airstream.
(b) Problems in matching due to the differences in the local airflow at different locations, eg two sensors located on the port side of the fuselage and two sensors on the starboard side.

The incidence terms from the airstream sensors can be blended with the pitch rate error terms and the 'gearings' (or gain coefficients) of these terms adjusted as higher angles of attack are approached. This enables the angle of incidence to be kept within the safe limits whilst manoeuvring at high angles of incidence, ie angle of incidence limiting.

Normal acceleration terms from the normal accelerometers are also blended with the other control terms and the gearings of the terms adjusted as high normal accelerations are demanded so as to give normal acceleration limiting.

The blending of normal acceleration and pitch rate is sometimes referred to as C_N* ('C star') control. The blending of pitch rate and normal acceleration to meet pilot handling criteria is, however, a more complex subject and beyond the scope of this chapter. Sufficient to say that optimum handling requires a blend of control terms the mixing of which needs to be adjusted over the flight envelope.

Air data gain scheduling is necessary to adjust the gearings of the control terms to cope with the changes in control effectiveness over the flight envelope as mentioned earlier.

The aircraft structural resonance modes require the use of 'notch filters' in the FBW loop to attenuate the loop gain at the structural resonance frequencies, as shown in Fig. 4.18, so that the resonance modes are not excited. Figure 4.19 illustrates the first body bending mode in the aircraft's longitudinal plane. The frequency of this first vibrational mode is typically between 5 and 12 Hz for a fighter/strike aircraft and 2–4 Hz for a large transport type aircraft. There is also a corresponding second body mode vibration at 2 × (first body mode frequency) and possibly significant even higher modes. The location of the motion sensors with respect to the nodes and antinodes of these flexural modes is also important. Ideally the pitch rate gyros would be located on an antinode (direction of the gyro input axis is unchanged) and the normal accelerometers on a node (zero linear movement; see Fig. 4.19). This may not be practical and acceptable compromises have to be made on the sensor locations.

4.3.1.1 Agile fighter pitch rate command loop

A worked example for a pitch rate command loop for a hypothetical FBW agile fighter with a 12% negative (unstable) static margin is set out below. It is stressed

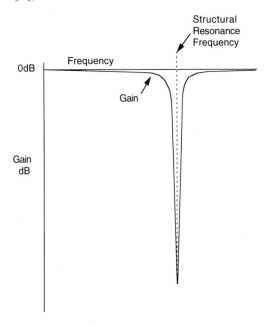

Fig. 4.18 Structural resonance 'notch filter' frequency response.

that the aircraft is a hypothetical one based on published data on several modern fighter aircraft and order of magnitude estimates.

Aircraft data

Mass, $m = 16,000$ kg
Overall length = 14.5m
Wing span = 11m
Wing area $S = 50m^2$
Moment of inertia about pitch axis, $I_y = 2.5 \times 10^5$ kg m^2
Airspeed $V_T = 300$m/s (600 knots approximately)

- Wing incidence/g at 600 knots = 2/3 degree/g
- Static margin = 12% negative
- Pitching moment derivative due to control surface deflection,
    $M_\eta = 5 \times 10^6$ Nm/radian
- Pitching moment derivative due to pitch rate,
    $M_q = 5 \times 10^5$ Nm/radian per s.

Problem
From the data above derive a suitable pitch control law, neglecting lags in the actuator response, non linear effects, structural resonance notch filters etc.

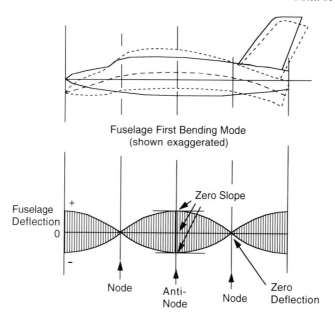

Fuselage First Bending Mode
(shown exaggerated)

Fig. 4.19 First body-bending mode.

The stages in the proposed solution are set out below.

Stage 1 Derivation of q/η transfer function of basic aircraft

• Derivation of T_1 and T_2

$$T_1 = \frac{I_Y}{M_q} = \frac{2.5 \times 10^5}{5 \times 10^5}$$

ie $T_1 = 0.5$ s.
Given 2/3 degree wing incidence/g at 600 knots

$$Z_\alpha \frac{2}{3} \times \frac{1}{60} = 16,000 \times 10$$

($1° = 1/60$ rad and $g = 10\text{m/s}^2$ approximately)
Hence $Z_\alpha = 1.44 \times 10^7$ N/rad

$$T_2 = \frac{mU}{Z_\alpha} = \frac{16,000 \times 300}{1.44 \times 10^7}$$

$T_2 = 0.33$ s

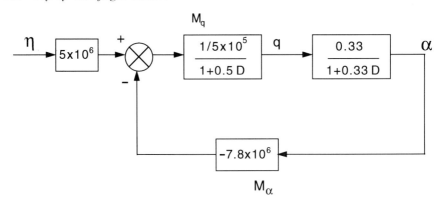

Fig. 4.20 Block diagram of basic aircraft pitch transfer function.

- Derivation of M_α

Aerodynamic mean chord, $c = \dfrac{\text{wing area}}{\text{wing span}} = \dfrac{50}{11}$

i.e. $c = 4.5\text{m}$

Static margin $= \dfrac{\text{distance between CG and aerodynamic centre}}{\text{aerodynamic mean chord}} = 12\%$

Distance between CG and aerodynamic centre $= -0.12 \times 4.5 = -0.54\text{m}$
$M_\alpha = Z_\alpha \times$ (Distance CG to aerodynamic centre) $= 1.44 \times 10^7 \times 0.54$
i.e. $M_\alpha = 7.8 \times 10^6$ Nm/radian

The block diagram representation of the basic aircraft dynamics is shown in Fig. 4.20 with the numerical values of M_η, M_q, M_α, T_1, T_2 inserted (refer to Fig. 3.30 in Section 3.5.4, Chapter 3).

The q/η is obtained using block diagram algebra:

$$\frac{q}{\eta} = 5 \times 10^6 \cdot \frac{\dfrac{1/5 \times 10^5}{(1+0.5D)}}{1 + \dfrac{1/5 \times 10^5}{(1+0.5D)} \cdot \dfrac{0.33}{(1+0.33D)} \cdot (-7.8 \times 10^6)}$$

This simplifies to

$$\frac{q}{\eta} = \frac{60\,(1+0.33D)}{(D^2 + 5D - 25)}$$

$$\frac{q}{\eta} = \frac{60\,(1+0.33D)}{(D+8.1)\,(D-3.1)}$$

The response of the basic aircraft to a disturbance or a control input is divergent with a divergent exponential component with a time constant of 1/3.1 s, ie 0.32 s. ie $e^{t/0.32}$

Now $e^{0.7} \approx 2$, so that the time to double amplitude t_{x2} is given by

$$\frac{t_{x2}}{0.32} = 0.7$$

Hence time to double amplitude = 0.22 s.

Stage 2 Derivation of stabilising transfer function
The pitch rate demand loop can be stabilised by a controller with a transfer function

$$G_q = \left(1+\frac{1}{TD}\right) q_E$$

Where G_q = controller gearing = degrees control surface movement per degree per second pitch rate error
T = integrator time constant
q_E = pitch rate error
The overall loop is shown in block diagram form in Fig. 4.21.
    The open loop transfer function is

$$\frac{q}{q_E} = \frac{60\,G_q(1+TD)\,(1+0.33D)}{TD\,(D+8.1)\,(D-3.1)} = KG(D)$$

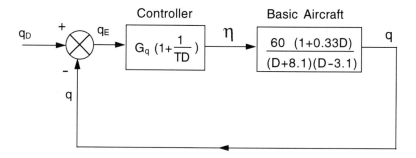

Fig. 4.21 Block diagram of overall loop.

The closed loop transfer function is

$$\frac{q}{q_D} = \frac{KG(D)}{1+KG(D)}$$

The stability is determined by the roots of the equation

$$(1+KG(D))\, q = 0$$

which must have a negative real part for stability.

$$\left[1 + \frac{60\, G_q\, (1+TD)\, (1+0.33D)}{TD\, (D+8.1)\, (D-3.1)}\right] q = 0$$

This simplifies to

$$\left[D^3 + (5+20\,G_q)\, D^2 + \left(60\,G_q + 20\frac{G_q}{T} - 25\right) D + 60\frac{G_q}{T}\right] q = 0 \qquad (4.9)$$

This can be factorised

$$(D+a)\, (D^2 + 2\zeta\,\omega_0 D + \omega_0{}^2)\, q = 0 \qquad (4.10)$$

The desired value of ω_0 is about 6.3 rad/s (1 Hz) which is well below the likely first body flexure mode frequency of, say, about 12 Hz (high g capability and aeroelastic deformation effects at high values of dynamic pressure require high stiffness structures). An ω_0 of about 1 Hz also corresponds to that of many existing conventional fighters with good handling qualities. The value of damping ratio, ζ, should be about 0.6 to achieve a well damped response and allow for the erosion of the loop phase margin due to lags in the actuator and structural resonance 'notch' filters etc.

The desired quadratic factor is thus $(D^2 + 7.56D + 39.7)$, the roots of which are $(-3.8 + j5)$ and $(-3.8 - j5)$

Having decided on these roots, the values of G_q and T can be obtained from the open loop response using the root locus method. This method is considered to be beyond the scope of an introductory book. It is, however, well covered in most standard text books on control engineering and readers wishing to find out more about the method are referred to appropriate references at the end of the chapter.

It is, however, possible to derive the required values of G_q and T by equating coefficients of the two equations, although this is not a route which would normally be used in practice. Nevertheless, it does convey the concept of manipulating the system gains and time constants to achieve the desired roots for the closed loop response.

Expanding equation (4.10) yields

$$\left[D^3 + (a + 2\zeta\omega_0)D^2 + (\omega_0^2 + 2\zeta\omega_0 a)D + a\omega_0^2\right]q = 0 \qquad (4.11)$$

Equating the coefficients of D^2, D and the constant terms and substituting $\zeta = 0.6$ and $\omega_0 = 6.3$ yields:

$$5 + 20G_q = a + 7.56 \qquad (4.12)$$

$$60G_q + 20\frac{G_q}{T} - 25 = 39.7 + 7.56a \qquad (4.13)$$

$$60\frac{G_q}{T} = 39.7a \qquad (4.14)$$

Solving these three simultaneous equations for the three unknowns G_q, T and a gives:

$G_q = 0.46$ degrees control surface/degree per s pitch rate error
$T = 0.105$ s
$a = 6.58$ s^{-1}

The transient response to a disturbance or control input is thus made up of a subsidence with a time constant of $1/6.58$ s, that is 0.15 s, and a 0.6 critically damped sinusoid with an undamped natural frequency of 1 Hz (damped natural frequency is 5 rad/s or 0.8 Hz) – that is a fast well damped response.

The required stabilising transfer function is:

$$0.46\left[1 + \frac{1}{0.105D}\right]$$

4.3.2 Roll rate command control

A roll rate command system is shown in Fig. 4.22, the primary feedback term being roll rate from the roll rate gyros. Other sensors are also used to control the rudder automatically to counteract and suppress the lateral cross coupling effects described earlier in Chapter 3 (Section 3.5). A yaw rate, r, feedback term derived from the yaw rate gyros provides the yaw autostabilisation function. Other terms can include

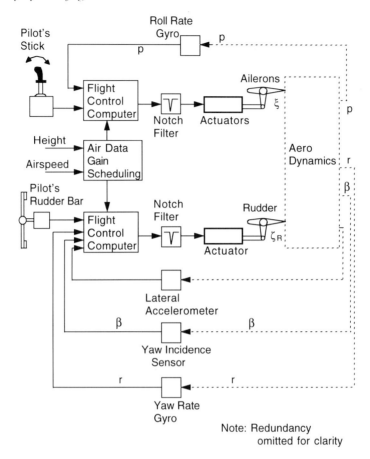

Note: Redundancy
omitted for clarity

Fig. 4.22 Lateral FBW control loops – roll rate command.

yaw incidence angle, β, from yaw incidence sensors and lateral acceleration from lateral accelerometers.

As in the pitch rate command loop, air data gain scheduling is used to adjust the gearings (or gains) of the control terms. Derivative and integral terms may also be used. Notch filters are also used to attenuate the loop gain at the structural resonance frequencies.

4.3.3 Modern control theory

The treatment of closed loop flight control systems in this book has used the classic linear control theory approach. The aircraft dynamics have been simplified where possible to enable a good basic understanding to be obtained – for example, simplifying the pitch dynamics to a second order system transfer function by assuming the forward speed is constant over the short period motion, deriving the roll rate response to aileron movement by assuming pure rolling motion, and neglecting roll/yaw cross coupling effects.

Similarly, the short period yawing motion is first derived by assuming pure yawing motion and the yaw/roll cross coupling effects neglected. A more accurate representation of the dynamic behaviour has then been derived and the representation of the aircraft dynamics as a matrix array of first order state equations has been explained in Chapter 3 (Section 3.4.5). The representation of the system dynamics and the control processes in the form of matrix arrays of first order equations is a key feature of what is now referred to as modern control theory.

Flight control systems come into the category of multi-input/multi-output (MIMO) closed loop control systems as control is exerted about 3 axes and there are six degrees of freedom. The classic control theory approach is very suitable for single input/single output (SISO) closed loop control systems and some single input/multi-output (SIMO) systems (eg two outputs).

However, modern control theory has been specifically developed to deal with MIMO systems and is now widely used in the design of modern flight control systems. The techniques are very powerful and can deal with non-linear systems. The methods, however, require a good understanding of matrix algebra and not all readers are familiar with this subject. For this reason, the classic control theory approach has been considered more suitable for this book as it is intended to act as an introduction to what is a large multi-discipline subject. It should also be stressed that classic control theory methods are extensively used in parallel with modern control theory methods because of their robustness and visibility in particular areas.

Appropriate references on modern control theory methods are given at the end of the chapter for those readers wishing to know more about the subject.

4.4 REDUNDANCY AND FAILURE SURVIVAL

4.4.1 Safety and integrity

Clearly the FBW flight control system must be no less safe than the simple mechanical control systems which it replaces. The safety levels required are specified in terms of the probability of a catastrophic failure occurring in the system from any cause whatsoever which could result in loss of control of the aircraft. It is generally specified that the probability of a catastrophic failure in the flight control system must not exceed 1×10^{-7} /hour for a military aircraft or 1×10^{-9} /hour for a civil aircraft

These very low probability figures are difficult to appreciate and also impossible to verify statistically. To give some idea of their magnitude, a failure probability figure of 1×10^{-9} /hour means that a fleet of 3,000 aircraft flying an average of 3,000 hours per annum would experience one catastrophic failure of the FBW system in 100 years! Military aircraft utilisation per annum is much lower than civil transport aircraft so that a figure of 1×10^{-7} /hour for the FBW system is acceptable.

It should be noted that the statistical level of safety currently being achieved with civil aircraft transport corresponds to a figure of around 1×10^{-6} /hour. This figure is derived from the total number of civil aircraft crashes occurring in a year from all causes divided by the total number of aircraft flying and their annual operating hours.

The mean time between failures, or MTBF, of a single channel FBW system is in the region of 3,000 hours. The FBW system must thus possess redundancy with multiple parallel channels so that it is able to survive at least two failures, if these very low failure probability figures are to be met.

There is clearly an economic limit to the number of parallel redundant channels. Apart from the cost of the additional channels, the overall system MTBF is reduced and hence the system availability is lowered. While it may be acceptable to fly with one failed channel, assuming sufficient redundancy, the impact on availability is obvious if the overall MTBF is too low.

4.4.2 Redundant configurations

A well established redundant configuration comprises four totally independent channels of sensors and computers in a parallel arrangement to give the required failure survival capability – such a configuration is referred to as a quadruplex system. The four independent channels are then configured to drive a failure survival actuation system with sufficient redundancy such that the overall FBW system of interconnected sensors, computers and actuators can survive any two failures from whatever cause.

The integrity of the electrical power supplies and the hydraulic power supplies is absolutely vital and adequate redundancy must be provided so that the system can survive failures in both the electrical and hydraulic power supplies. In fact the starting point in the design of any FBW system is the safety and integrity and redundancy levels required of the electrical and hydraulic power supply systems as their integrity dominates the overall FBW system integrity.

With four totally independent channels, the assumption is made that the probability of three or four channels failing at the same instant in time is negligible. This assumption will be discussed in Section 4.4.5. Thus if one system fails 'hard over' (ie demanding maximum control surface movement), the other three parallel 'good' channels can over-ride the failed channel. However to survive a second failure it is necessary to disconnect the first failed channel, otherwise it would be stalemate – two good channels versus two failed hard over channels so that the control surface does not move – a 'fail passive' situation.

Failures are therefore detected by cross comparison of the parallel channels and majority voting on the 'odd man out' principle. The quadruplex system is thus able to survive two failures by majority voting and disconnecting the failed channels, the system degrading to triplex redundancy after the first failure and duplex redundancy after the second failure. A third failure results in a fail passive situation, the 'good' channel counteracting the failed channel.

The incorporation of a monitoring system to check the correct functioning of a channel to a very high confidence level can also enable a failed channel to be identified and disconnected and this leads to an alternative failure survival configuration known as 'monitored triplex'. A monitored triplex configuration comprises three totally independent parallel channels with each channel monitored by a dissimilar system to detect a failure. Provided this monitoring is to a sufficiently high degree of integrity and confidence level, such a system can survive two failures.

The respective merits of the two systems are fairly evenly divided; the monitored triplex system has rather less hardware and so can be of lower cost. This is offset by the better 'visibility' in terms of failure survival confidence of the quadruplex system, particularly when it incorporates self monitoring to a high confidence level.

Figure 4.23 illustrates the quadruplex and monitored triplex redundancy configurations schematically.

4.4.3 Voting and consolidation

As already mentioned, failures are detected by cross comparison and majority voting. However, it is necessary to allow for normal variations in the outputs of the parallel sensors measuring a particular quantity due to inherent errors in the sensors and their manufacturing tolerances. Normal errors in sensors include such parameters as:

(a) Scale factor errors
(b) Linearity errors
(c) Null or zero offset errors
(d) Hysteresis errors
(e) Variation of the above parameters with temperature etc.

The absolute accuracy of the sensor is rather less important (within reason) than the variation, or 'tracking error' spread between individual sensors particularly under dynamic conditions. For instance, a situation where all the sensors were in error by nearly the same amount and the variations between the individual sensors were small would be preferable to a higher absolute accuracy but a larger variation. Variations in the dynamic response of the parallel sensors must be small. For

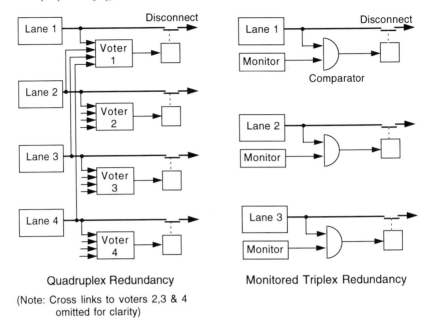

Quadruplex Redundancy   Monitored Triplex Redundancy

(Note: Cross links to voters 2,3 & 4
         omitted for clarity)

Fig. 4.23 Redundancy configurations.

instance, consider two sensors measuring a sinusoidally varying quantity but with a slightly different phase lag between the two sensor outputs. The difference between these two sensor outputs is a sinusoidally varying quantity 90° out of phase with the input quantity and with an amplitude directly proportional to this phase difference (see Fig. 4.24). Thus, a phase difference of 2° (1/30 rad approximately) results in a tracking spread of over 3% of the *peak* value of the input quantity.

This result is arrived at as follows: let θ_1 and θ_2 be the two sensor outputs measuring an input quantity θ, the first sensor having a phase lag ϕ and the second

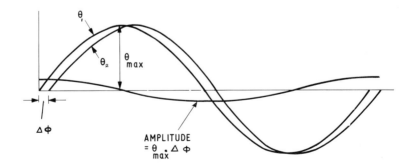

Fig. 4.24 Sensor tracking – effect of small phase difference.

sensor having a phase lag ($\phi + \Delta\phi$) and assume for simplicity that both sensors measure the peak amplitude θ_{max} without error.

Thus difference between the two sensor outputs

$$= \theta_{max} \sin(\omega t - \phi) - \theta_{max} \sin[\omega t - (\phi + \Delta\phi)]$$

$$\approx \Delta\phi \, \theta_{max} \cos(\omega t - \phi).$$

The sensors should thus have a high bandwidth so that the phase lags at aircraft control frequencies are small anyway and the phase variations between the sensors will thus be very small. The variations between the filters used to filter sensor output noise are also minimised by using accurate, stable components in the filters.

The failure detection algorithms enable the detection and isolation of a sensor whose output departs by more than a specified amount from the normal error spread. The value of this disconnect threshold in conjunction with the failure detection and isolation algorithm determine the magnitude of the transient which the aircraft may experience as a result of the failed sensor being disconnected.

There are thus two conflicting requirements:

a)   Low number of nuisance disconnections.
b)   Minimum transient on disconnecting a failed sensor.

If the disconnect threshold is set too low, a sensor can be deemed to have failed and be disconnected when it is merely at the edge of its specification tolerances, ie a nuisance disconnection. Conversely if set too high, when a real failure does occur the magnitude of the transient on disconnecting the failed sensor can be unacceptably large.

Failures can be divided into the following categories:

(i)   *'Hard over'* *failures* whereby the failed sensor output is hard over full scale and would demand full authority control surface movement with catastrophic results.
(ii) *Zero output failures* whereby the sensor output is zero and thus no control action would take place.
(iii) *'Slow over'* *failures* whereby the sensor output is slowly increasing or drifting with time, although the input is stationary, and would eventually cause a full scale hard over.
(iv) *Oscillatory failures* whereby the sensor output is oscillating, the amplitude and frequency being dependent on the type of failure. Loss of feedback, for instance, can result in a stop to stop oscillation.
(v)   *'Soft'* *failures* whereby the sensor is functioning but its output is outside the specification tolerances.
(vi) *Intermittent failures* whereby the sensor fails and then recovers intermittently.

There is a very large number of practical voting algorithms. A detailed discussion of voting algorithms is beyond the scope of this chapter and only one commonly used algorithm is therefore described. The basis of this algorithm is to select the middle value of the parallel sensor signals and compare the other signals with it. In

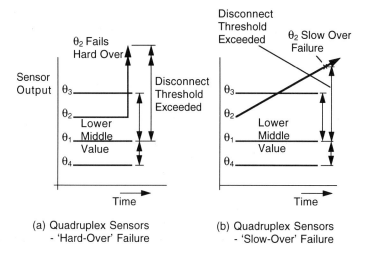

(a) Quadruplex Sensors
- 'Hard-Over' Failure

(b) Quadruplex Sensors
- 'Slow-Over' Failure

Fig. 4.25 Operation of failure detection algorithm – first failure.

the case of a quadruplex system, the lower of the two middle values is selected. The differences from the selected middle value of the highest and lowest sensor values are continually monitored. The four sensor outputs are valid providing these differences are less than the disconnect threshold value. Conversely if one sensor differs by more than the disconnect threshold it is deemed to have failed and its output is disconnected.

For example, suppose full scale sensor output is 100 and the normal sensor tolerance is ± 0.5%. The disconnect threshold is generally taken as (1.5 × maximum signal variation), and in this example would thus be equal to 1.5. Suppose for example sensor outputs for a particular quantity were as follows:

$$\theta_1 = 60, \ \theta_2 = 60.5, \ \theta_3 = 60.7, \ \theta_4 = 59.7$$

The lower middle value is 60 and the differences of the higher and lower value sensor outputs from this middle value are 0.5 and 0.3 respectively, so that all sensor outputs would thus be valid.

Consider now two failure examples: (a) a hard over failure and (b) a slow over failure, as shown in Fig. 4.25.

(a) The hard over failure results from, say, sensor θ_2 output going hard over to the full scale value of 100, i.e. $\theta_2 = 100$, with the input quantity stationary so that sensor outputs $\theta_1, \theta_3, \theta_4$ are unchanged. The difference between θ_2 and this lower middle value of 60 greatly exceeds the disconnect threshold and θ_2 would be detected as having failed and disconnected accordingly.

(b) The slow over failure results from say sensor θ_2 drifting with time from its original value of 60.5 and with the input quantity stationary so that $\theta_1, \theta_3, \theta_4$

are unchanged. When θ_2 reaches a value of just over 61.5 the difference from the new middle value of 60 exceeds the disconnect threshold value of 1.5 and θ_2 is disconnected.

After either of the above failures, the system would be at triplex level and the same two failure cases are considered for a second sensor failure, and are shown in Fig. 4.26.

Suppose sensor values are $\theta_1 = 60$, $\theta_3 = 60.7$, $\theta_4 = 59.7$

(a) Hard over failure results from, say, θ_3 failing hard over to full scale, i.e. $\theta_3 = 100$, with the input quantity stationary. The difference of θ_3 from the middle value of 60 greatly exceeds the disconnect threshold so that the θ_3 failure is detected and isolated.

(b) Slow over failure results from θ_3 slowly drifting off towards full scale from its original value of 60.7, with the input quantity stationary. θ_3 will be detected as having failed when its output exceeds a value of 61.5 and disconnected accordingly.

The adoption of digital computing technology enables more sophisticated failure isolation strategies to be adopted to reduce 'nuisance disconnections'. For instance, a sensor whose output differs from the other sensors during dynamic or transient conditions by an amount which just exceeds the disconnect threshold can be 'put on ice' and temporarily isolated. Its output can be continually compared with the other sensors and if it recovers within tolerance it can then be reconnected. The system can be designed to allow, say, up to 10 reconnections before completely isolating the sensor.

The use of integral terms in the control laws make it essential that the outputs of the individual parallel sensors are consolidated and only one signal is used for control purposes in each of the computing channels. This is because any differences

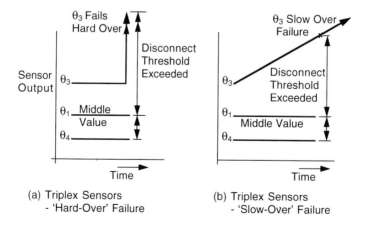

(a) Triplex Sensors
   - 'Hard-Over' Failure

(b) Triplex Sensors
   - 'Slow-Over' Failure

Fig. 4.26 Operation of failure detection algorithm – second failure.

between the sensor outputs would be integrated up with time so that the four computer outputs would diverge.

Thus if ε_1, ε_2, ε_3, ε_4 are the respective errors in the four sensor outputs, the four computer outputs would differ by $\int \varepsilon_1 dt$, $\int \varepsilon_2 dt$, $\int \varepsilon_3 dt$, $\int \varepsilon_4 dt$ and hence will diverge with time.

The lower middle value of the 4 sensor outputs (or middle of the 3 sensor outputs in the triplex case) is generally used as the consolidated output by the 4 computers.

4.4.4 Quadruplex system architecture

A typical quadruplex system architecture is shown in Fig. 4.27 with voters and output consolidation. The essential buffering and isolation of outputs (e.g. electro-optic isolators) is omitted for clarity, but is essential to prevent electrical faults in one channel propagating into another – common mode failures.

4.4.5 Common mode failures

As already mentioned the basis of the fault detection and isolation technique described relies on the probability of a single event causing all the parallel channels

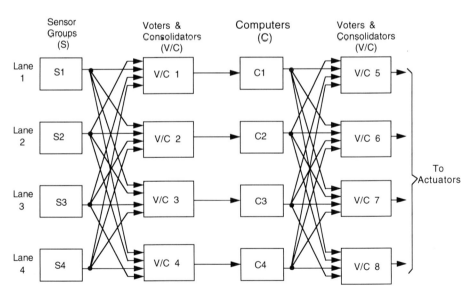

(Note: Cross lane data links are buffered)

Fig. 4.27 Quadruplex system architecture.

to fail simultaneously as being negligibly small. The type of failure which can affect all systems at the same time is termed a *Common Mode Failure*.

Examples of common mode failures are:

- Lightning strike
- Electro-magnetic interference
- Fire/explosion/battle damage
- Incorrect maintenance
- Common design errors – eg software

Every care is taken to minimise the probability of these failures occurring. For instance, very stringent electro-magnetic (EM) shielding practices are used including screened cables, segregation of cables and units, enclosure of all electronic components in EM shielded boxes with incoming wires terminated with EM filters etc. An attractive alternative is to transmit all the signals as coded light pulses suitably time division multiplexed along fibre optic cables, the optical transmission media being unaffected by any electro-magnetic interference. The fibre optic cable provides complete electrical isolation and eliminates the possibility of propagation of electrical faults between units. It also offers a much higher data rate transmission capability. The use of such technology is sometimes described as a 'fly by light' flight control system.

Hazards from fire, explosions or battle damage are minimised by the physical segregation and separation of the individual channels and is sometimes referred to as a 'brick wall' separation philosophy.

Very stringent control, inspection and maintenance disciplines are exerted to eliminate as far as is practical the possibility of common maintenance errors.

The common design error which could affect all the independent channels is a very major problem. One of the most difficult areas in the case of a digital FBW system is the possibility of an undetected error in the software which could affect all the channels. Very great care is taken in software generation to eliminate software errors as far as possible by the adoption of very stringent design procedures: these are briefly covered in Section 4.5.3.4.

The problem of eliminating the possibility of a common mode failure from whatever cause has thus led to the use of what is known as *dissimilar redundancy*.

4.4.6 Dissimilar redundancy

This requires a further control system which uses different hardware and technology and is able to take over control in the event of a common failure taking out all 3 or 4 parallel channels of the main system.

Civil FBW aircraft require such a dissimilar system to meet the safety and airworthiness requirements. Techniques which are used include:

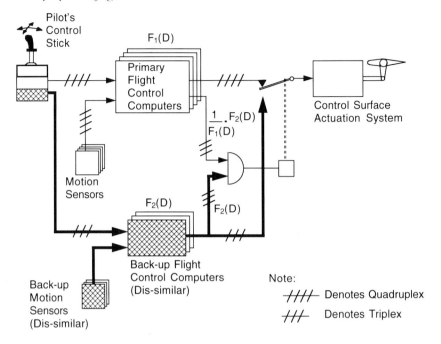

Fig. 4.28 Dissimilar redundancy.

(1) Use of two (or more) different types of microprocessors with dissimilar software written in different languages.
(2) Use of a backup analogue system in addition to the main digital system which is at quadruplex or triplex level of redundancy.
(3) Use of a backup system using different sensors, computing and control means e.g. separate control surfaces.
(4) Combinations of (1) to (3) above.

A possible means of implementing a dissimilar redundant system is shown in Fig. 4.28 to illustrate the concepts which can be employed. The parallel dissimilarly redundant 'back-up' system uses different technology components and computing to the primary channel (eg analogue computing) and may well use simplified 'get you home' type control laws with limited (or even fixed) air data gain scheduling.

The main FBW computer channels have the capacity and throughput to be able to accurately model the back-up system transfer function and hence compute what the output of the back-up system should be. The outputs of the primary and back-up systems can thus be continuously monitored by comparison.

For example, if the transfer function of the primary system relating the FBW computer output to stick and sensor inputs is $F_1(D)$ and the back-up system is

$F_2(D)$. Then the output of the back-up system should be equal to

$$F_2(D) = \frac{1}{F_1(D)} \times \text{(output of primary system computers)}$$

Now suppose there was an undetected common software error in the primary system computers and all four computers (assuming a quadruplex primary system) gave anomalous outputs under a rare combination of inputs coupled with an inherent systems design error or shortcoming. The comparators would indicate a major disparity between the primary and back-up systems and the decision would be made to switch control to the back-up system. This is because the redundant back-up system was operating correctly at the iteration period just prior to the event and the probability of all the redundant back-up system channels failing at the next iteration period was extremely low, the back-up system being inherently simpler and not subject to software errors.

It should be stressed again that this is just one possible way of achieving failure survival dissimilar redundancy.

4.5 DIGITAL IMPLEMENTATION

4.5.1 Advantages of digital implementation

Modern FBW flight control systems are implemented using digital technology, signals being transmitted as serial digital data using time division multiplexed data bus networks. The signal data are subsequently processed by digital microprocessors in the flight control computers which carry out the following tasks:

a) Voting, monitoring and consolidation
b) Control law implementation
c) Reconfiguration in the event of a failure
d) Built in test and monitoring.

Primary FBW flight control systems using analogue computing technology are still in service but nowadays analogue technology is mainly used to provide a dissimilarly redundant backup system for protection from common mode failures, as just explained.

The advantages of modern digital implementation compared with analogue are overwhelming with the technology now available. For instance:

Hardware economy One computer can control all three axes of control, whereas an analogue system requires dedicated hardware for each axis of control. The reduction in hardware weight and volume is of the order of 5:1 for a system of even modest complexity. The more complex systems could not be implemented economically using analogue technology.

Flexibility Control laws and gearings (or gains) can be changed by software changes as opposed to hardware modifications giving greater flexibility during the design and development phases. The introduction of modifications in service is also easier. The cost of software changes is not a trivial one, to say the least, but there is no doubt that it is still less than the cost of hardware modifications.

Reduced nuisance disconnects Digital computation allows sophisticated voting and consolidation algorithms to be used which minimise nuisance trip outs, or disconnects.

Smaller failure transients Sophisticated consolidation algorithms can be implemented which minimise the transient experienced on disconnecting a failed channel.

Built in test capability Very comprehensive self test capabilities can be incorporated into the system for pre-flight checkout and maintenance.

Digital data buses Very large reductions in the weight of cabling are achieved by the use of multiplexed data transmission and data bus networks. High integrity data transmission can be achieved with very comprehensive self checking and data validation capabilities. The use of three, or more, totally independent data bus

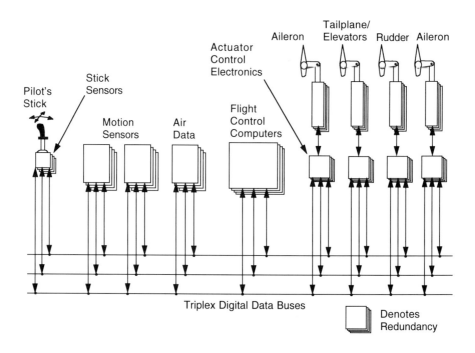

Fig. 4.29 Flight control system bus configuration.

networks enables the failure survival requirements to be met. Fig. 4.29 illustrates a flight control data bus configuration.

4.5.2 Digital data problems

The use of digital data introduces particular problems which need to be taken into account in a closed loop digital control system. These are essentially due to the need to sample data and the frequency of sampling.

4.5.2.1 Aliasing

Figure 4.30 illustrates the effect of sampling a time varying signal at too low a sampling frequency. The sampled output contains a much lower frequency signal which is not present in the real signal. The effect of sampling is to 'fold back' the high frequency content of the real signal above the sampling frequency so that it appears in the frequency range below the sampling frequency (see Fig. 4.31). This effect is known as 'aliasing'.

It is thus necessary to filter the signal to attenuate any noise with a frequency content above the sampling frequency before sampling – such filters are known as anti-aliasing filters (see Fig 4.32). Clearly, the higher the sampling frequency the better so that the lag introduced by the anti-aliasing filter is very small over the range of control frequencies.

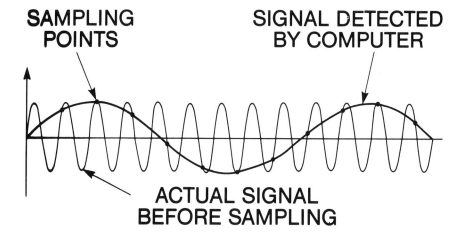

SAMPLING POINTS

SIGNAL DETECTED BY COMPUTER

ACTUAL SIGNAL BEFORE SAMPLING

Fig. 4.30 Signal aliasing due to sampling.

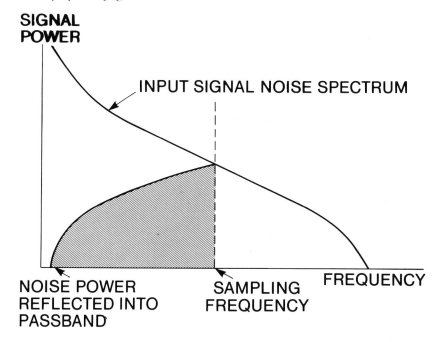

Fig. 4.31 Aliasing.

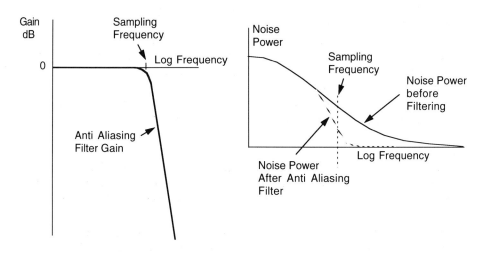

Fig. 4.32 Anti-aliasing filter.

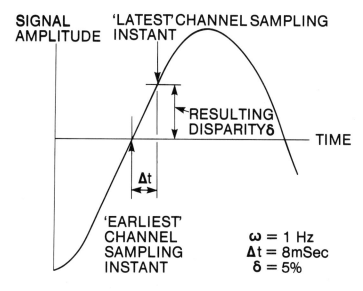

SIGNAL   'LATEST'CHANNEL SAMPLING
AMPLITUDE INSTANT

RESULTING
DISPARITYδ
TIME

Δt

'EARLIEST'
CHANNEL   ω = 1 Hz
SAMPLING   Δt = 8mSec
INSTANT   δ = 5%

Fig. 4.33 Differential data staleness.

4.5.2.2 Data Staleness

In comparing the corresponding outputs of say four independent parallel computing channels, or lanes, it is possible for the output of one channel to be delayed by one iteration period before its output can be compared with the others. Its information is thus 'stale' by one iteration period. Hence if the input quantity is changing during the iteration period the output of this 'stale' channel will differ from the others although no fault has occurred. The situation is illustrated in Fig. 4.33.

To overcome this problem it is necessary to have some form of time synchronisation of the computer iteration periods. This iteration period synchronisation is generally implemented through software.

The alternative is to allow longer iteration periods so that all computers are processing the same data. This however introduces a latency or time delay in the output which results in a phase lag and is destabilising.

4.5.2.3 Latency

Figure 4.34 illustrates the phase lag introduced by a time delay. This results in a phase lag in the output which increases as the frequency of the input increases. The amplitude, however, is not attenuated (as with a simple lag filter) and so the destabilising effect is increased.

If T is the time delay period, at a frequency of $1/2T$ there could be 180° lag. Thus if the time delay was 20 milliseconds, a frequency input of 25 Hz could result in a phase lag in the output of 180°. The corresponding phase lag for a 1 Hz input would be 7.2° and this can erode the phase margin of the overall loop.

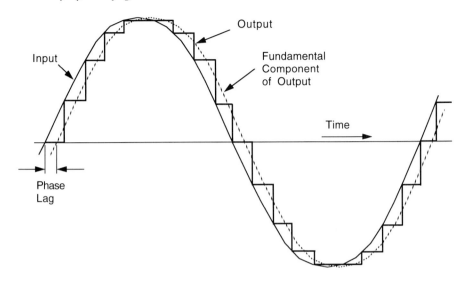

Fig. 4.34 Phase lag introduced by time delay – latency.

4.5.3 Software

4.5.3.1 Introduction
Software generation is one of the most challenging tasks in the design of a high integrity digital FBW flight control system and can account for between 60% to 70% of the total engineering development costs of the complete FBW system. This is because of the size of the software which is required to carry out all the flight control functions and the problems of establishing the safety of the software.

It is not possible in this book to do more than introduce the reader to the problems and the methods being used to overcome them. The subject of safety critical software is an evolving discipline and is generating a large and increasing bibliography of books and papers. A list of some appropriate references is set out at the end of the chapter.

4.5.3.2 The flight control software functions
The main functions that are carried out by the software for the flight control computers can be divided into three basic areas which have a degree of interaction and comprise:

(1) Redundancy management
(2) Control laws
(3) Built-in-test

Redundancy management can account for between 60% to 70% of the total software (and computer throughput) in the case of a modern civil FBW aircraft. The proportion is generally not quite so high in a military FBW aircraft – around 50% for a modern agile fighter.

Implementation of the flight control laws generally accounts for between 25% to 30% of the total software in a modern civil FBW aircraft. The proportion in a modern agile fighter, however, can amount to around 40%.

Built-in-test software for in-flight monitoring accounts for a smaller percentage of the total software in either case – around about 10%. However, during the preflight checking stage on the ground the proportion of the operating software taken up by Built-in-test is much higher and can be as high as 40%.

Some appreciation of the functions carried out by the redundancy management software can be gained by considering the tasks involved in failure detection and isolation and reconfiguration in the event of a failure in the case of a quadruplex system with four independent 'lanes' of sensors and computers.

This includes tasks such as:

Sensor data validation Checking the digital data from each individual sensor is correctly encoded and has not been corrupted within the data transmission system.

Sensor failure detection Detection of a failed sensor by cross comparison of sensor data and majority voting.

Sensor failure isolation and system reconfiguration Isolation of a failed sensor and system reconfiguration to survive the failure and also to minimise the resulting transient.

Sensor consolidation Consolidation of the data from a group of sensors to use a single representative value for subsequent computation of the control laws.

A first order list of sensors for a typical system is set out in Table 4.1 to show the numbers of sensors involved in these respective tasks. Most of these sensors would be at quadruplex or triplex level redundancy.

Cross lane data transfer Data transfer between the individual computing lanes to enable cross comparison of outputs.

Computer output voting and consolidation Cross comparison of the individual computer outputs to detect and isolate a computer failure. Consolidation of the 'good' computer outputs to provide the control surface demand signals to transmit to the control surface actuator servos.

Computer iteration period synchronisation the iteration periods of the individual computers need to be loosely synchronised to avoid data staleness problems.

Recording of fault data to enable the management of the maintenance function.

System status indication to the crew.

Control of overall system.

Considering the large number of sensors involved plus all the other tasks described enables some appreciation to be obtained of the redundancy management software task. The safety critical nature of this software can be appreciated.

Table 4.1 Pitch channel sensors

Pilot's controls	Control stick	Pitch command sensors
		Roll command sensors
	Rudder bar sensors	
	Nose wheel steering sensors	
	Undercarriage position (discrete signal)	
Primary control surface positions	Tailplane angular position sensors	
	Elevator angular position sensors	
	Aileron angular position sensors	
	Rudder angular position sensors	
Secondary control surface positions	Wing spoiler angular position sensors	
	Leading edge flaps / slats angular position sensors	
	Trailing edge flaps angular position sensors	
Aircraft motion inertial sensors	Pitch rate gyros	
	Roll rate gyros	
	Yaw rate gyros	
	Normal accelerometers	
	Lateral accelerometers	
Air data sensors	Total pressure sensors	
	Static pressure sensors	
	Longitudinal airstream incidence sensors	
	Lateral airstream incidence sensors	

The processes involved in carrying out just one of the sensor failure detection and output consolidation tasks is shown in the flow diagram in Fig. 4.35.

The control law software is, of course, of an equally critical nature together with the built in test software. The control laws now being implemented in the new generation of flight control systems are of much greater complexity than in existing conventional aircraft. The high speed microprocessors now available enable very sophisticated control laws to be implemented and the aircraft to have a higher performance and safer handling characteristics as a result. The control law software is thus of increased complexity. However, the cost of generating this software, although significant, can be amortised over the production run of what is a more competitive aircraft and hence can be cost effective. This is because the hardware costs are substantially unchanged from earlier systems because of the continuing progress in processor power and microcircuit design and manufacturing technology, such as the extensive use of ASICs (Application Specific Integrated Circuits).

The control law software tasks can comprise:

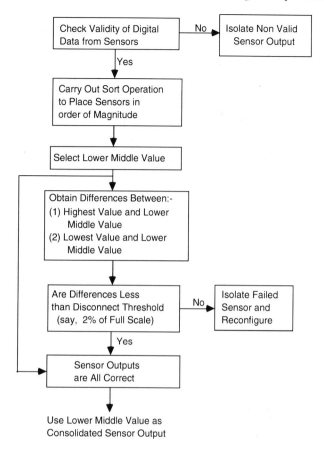

Fig. 4.35 Flow diagram of sensor voting and consolidation process.

- Complex gain scheduling of a large number of control terms as a function of height, airspeed, incidence and possibly attitude, aircraft configuration. This can be particularly important in a military aircraft which may be required to operate to the very edges of its flight envelope in combat when increased cross coupling effects etc. require changes in the control terms to be made. The control transfer functions involve proportional, derivative and integral terms which must be derived from the various motion sensors using suitable algorithms.
- Automatic limiting of the pilot's input commands to achieve 'carefree' manoeuvring to restrict the pilot from attempting to manoeuvre the aircraft into an unacceptable attitude or approach too near the stall or exceed the structural limits of the aircraft.
- Changing the control laws to maintain effective control in the event of sensor failures etc.

- Control of the secondary control surfaces – leading edge flaps/slats, trailing edge flaps, spoilers.
- Control of engine thrust and possibly thrust vectoring.

4.5.3.3 The software development process

The management of the software complexity is clearly one of, if not the major task and a phased development process as shown in Fig. 4.36 is followed. This is known as a 'top down' approach and starts with the requirements specification.

The correct specification of the flight control system requirements in a clear unambiguous form is the essential starting point and various methods and software tools are now being used in conjunction with work stations to assist in this process. They include the use of system description languages and formal methods. References to further reading on system description languages will be found at the end of the chapter, as it is beyond the scope of this book.

Formal methods will be discussed briefly in terms of their role in the development of safety critical software. It should also be pointed out that the use of formal methods in flight control system software design is still evolving.

Design techniques such as object orientated design (OOD) are used in both the top level design and detailed software design phases. OOD is an iterative design technique which is repeated until it results in both a description in English of the software and its task and a Booch diagram which is equivalent to a block diagram of the software packages and their relationship with the hardware. The technique is powerful in the early design stages to bridge the gap between the customer and the software engineer with a representation understandable (and workable) by both. The

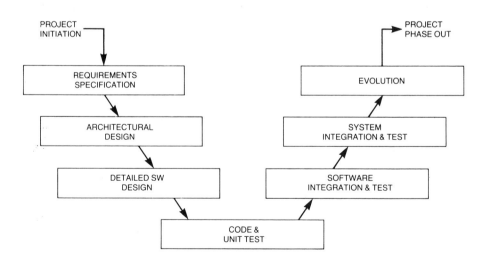

Fig. 4.36 Phased development software life cycle.

OOD method involves considering the design in terms of 'objects' and associated 'actions' on the objects and this can ease the translation into the high order language Ada because 'objects' within the design become Ada packages.

A number of computer assisted software engineering (CASE) tools are now available for running OOD on suitable workstations. Further reading on OOD is given at the end of the chapter.

The question of software languages is a crucial one in flight control systems. The software for digital flight control systems designed around the mid 1970s to the mid 1980s was generally written in Assembler language, usually for a specifically designed processor. (Available microprocessors at that time lacked the speed and throughput.) Assembler languages are essentially an assembly of specific instructions translated directly into machine code and have two major advantages – speed of execution and visibility of some of the computing processes within the computer from the point of view of failure modes and effects analysis. However, they lack flexibility, ease of use for handling complex problems, error checking capabilities and ease of transfer to run on other machines which a high order language (HOL) can provide.

The US DoD standard high order language is Ada and its use is mandatory in US military avionic systems. It is also now widely used in the UK and is gaining increasing acceptance for civil as well as military avionic systems. The widespread use and experience acquired with Ada brings benefits to all users, not least of which is the availability of validated compilers (which translate the HOL to machine executable instructions or code). However, 'nothing is for nothing', as the saying goes, and a HOL such as Ada can have penalties in the time it takes to execute certain instructions – significant increases in program run time compared with a simpler language can be incurred. The high speed microprocessors now available are generally sufficiently fast to accommodate this increase in program execution time.

4.5.3.4 Software validation and verification

The safety critical nature of flight control software makes the validation of its integrity essential. Clearly the software documentation and configuration control for all stages of the software life cycle must be fully compliant with the safety and airworthiness requirements of the regulatory authorities (eg CAA, FAA, MoD etc). Relevant standards are DO 178B, MIL STD 2167A and DEF STAN 00-55 (see Further Reading).

The stages of reaching this stage follow a carefully structured design approach and may include the use of formal methods. Formal methods of software development involve the use of mathematical set theory and logic to specify the functionality of software. Specifications in such mathematical notations have some major benefits over English or diagrammatic notations in that they are unambiguous, they can be checked for completeness and their properties can be examined by mathematical methods. This leads to specifications that are more detailed but contain fewer errors, being more consistent and less likely to be misinterpreted. The net effect is that certain aspects of the system are fully defined thereby reducing the development risks. The mathematics used in formal methods to describe information and systems is known as discrete mathematics, rather than the 'continuous' mathematics used to describe physical processes. Logic and set theory are used to represent system states and operations. One such method is a mathematical language known as 'Z' which is used to specify functional requirements of systems. This has been developed by the Oxford University Research Group of IBM. Z is based on Zermilo-Frankel set theory, from which it derives its name and is a flexible mathematical notation with constructs for creating abstract or concrete specifications in a well structured way. It should be pointed out, however, that the Z notation can take a little getting used to by less mathematically orientated engineers. The formal verification of the design process is shown in Fig. 4.37 and appropriate references to formal methods are given at the end of the chapter.

It should be stressed again that the whole subject of safety critical software is a very large and complex one and is still evolving and maturing. The objective of this

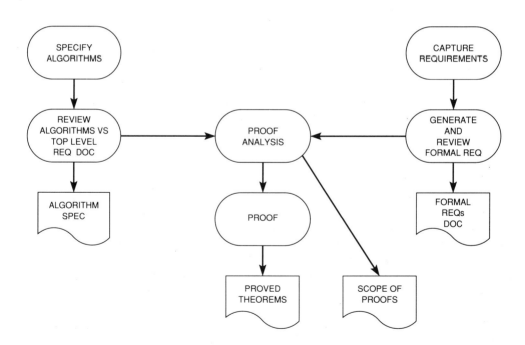

Fig. 4.37 Formal verification of design.

section is to try to introduce the reader to the subject and to indicate where appropriate references can be obtained.

4.5.3.5 *Dissimilar or multi-version software*

The extreme difficulty, if not impossibility, of proving the integrity of a system using common software in its parallel redundant channels (or lanes) to the safety levels required by the civil regulatory authorities has led to the need for dissimilar redundancy. Two (or more) totally independent failure survival flight control computing systems are installed. Each system uses a different type of microprocessor and the software is written in different software languages by totally independent software teams using the most stringent methods and procedures. The two (or more) systems may use different control laws and operate different sets of control surfaces and each system is capable of maintaining safe flight on its own. However, the degree of independence in writing dissimilar, or 'multi-version', software to meet a common system requirement is not 100%.

Some work carried out in the United States has shown that multi-version software written to meet a common stated requirement can contain errors which are not independent – ie common errors can occur.

Part of the problem resides in the understanding of the stated system requirements which may involve unforeseen problems and possible ambiguities in interpretation. The rigorousness and degree of control of the software development process is a further factor and the better this is, the fewer the errors anyway.

A conservative estimate is that multi-version software should achieve a 95% improvement in reducing the software errors present in a single software program – that is only 1 in 20 of the errors present would be shared. An improvement of 20:1 by using multi-version software is clearly well worth having. It is also considered that the use of formal methods to define the system requirements and ensure these are soundly stated should further improve the confidence level in reducing the number of shared errors.

4.5.4 Failure modes and effects analysis

An exhaustive 'failure modes and effects analysis' of the flight control system is required by the regulatory authorities before a Certificate of Airworthiness is granted. This embraces every part of the overall system including both hardware and software and power supplies. Any part of the system can fail from any cause and it must be shown that the overall system can survive this failure and maintain safe flight. The importance of dissimilar redundancy in this context is apparent.

The MTBF of each element in the system has to be established using well validated statistical data where possible and it must be shown that the system configuration will meet the overall safety and integrity requirements (as specified in Section 4.4.1).

4.6 FLY-BY-LIGHT FLIGHT CONTROL

4.6.1 Introduction

Mention has already been made of the common mode failures that can be caused by severe electro-magnetic interference (EMI) if the equipment is not effectively screened from EMI.

Electro-magnetic interference can arise from:

Lightning strikes Very large electro-magnetic pulses (EMPs) with electrical field strengths of hundreds of volts per metre can be produced and a very wide spectrum of electro-magnetic radiation frequencies generated.

Overflying high power radio/radar transmitters Several cases have occurred of aircraft experiencing severe transients in the automatic flight control system placing the aircraft temporarily out of control whilst overflying high power radar/radio transmitters. This has been due to the susceptibility of the flight control system analogue electronics to EMI because of inadequate electro-magnetic screening. Current digital flight control systems are designed to much higher electro-magnetic compatibility (EMC) specifications, (that is the ability to withstand high levels of EMI), and should be able to withstand such environments.

Failures in the electro-magnetic screening system so that the avionic equipment becomes susceptible to internal sources of electro-magnetic radiation such as radio and radar equipment, electrical generators etc, or, external sources of EMI. Failure of the electro-magnetic screening can result from breaks or high resistance in the earth connections to the aircraft structure and the electrical bonding of the structure due to corrosion, poor quality installation etc. Such failures can be of a dormant nature and might only become apparent when an abnormal EMI source is encountered.

Vicinity to a nuclear explosion and consequent emission of gamma radiation and an EMP of very high electric field strength. Military avionic equipment is specially designed to survive up to a certain (classified) level of these effects. This attribute is referred to as 'nuclear hardness' and is achieved by the design of the circuits, use of nuclear hard components, circumvention techniques and very efficient electro-magnetic screening. The subject is beyond the scope of this book and much of the information on the subject is of a highly classified nature. However, a military aircraft with a nuclear hard flight control system must have a very low susceptibility to EMI in order to meet the nuclear hardness specifications.

Electronic units and their enclosures, or 'boxes', can be designed to have very efficient electro-magnetic screening. However, the 'Achilles heel' of the system can be the interconnecting cables which link together all the elements of the system. These interconnecting cables must have very efficient electro-magnetic screening, which incurs a significant weight penalty. The incoming wires from the cables to the units (or boxes) must also be terminated with EM filters which present a low impedance to EMI induced transient voltages in the cables. The transient voltages are thus attenuated before they can affect the circuits inside the box. (It should be noted that the wires from the cable connectors and the EM filters are housed within a screened enclosure around the connector known as an 'EMC vault' so that there is no aperture for EMI radiation to enter the box.)

The use of digital data transmission enables error check encoding to be incorporated in the digital data word format so that any corruption of the digital data (eg by EMI) can be detected and the data ignored. However, the data are lost and the system 'frozen' for the period of the interference so that only a short transient loss could be tolerated. The integrity of the screening of the interconnecting cables must thus be very high indeed to meet the flight safety requirements.

The use of optical data transmission whereby digital (or analogue) data are transmitted as a modulated light intensity signal along a fibre optic cable overcomes these problems. The optical transmission media is unaffected by EMI and very much higher data rates can be achieved, eg GHz bandwidth (10^9 Hz). Current avionic data bus systems using fibre optic data transmission operate in the 20 Mbits/s to 100 Mbits/s bracket. (The principles of fibre optic data transmission and optical data buses are covered in Chapter 9.)

The fibre optic cable is also much lighter than the equivalent screened cables it replaces. The weight saving can be significant and as mentioned earlier, weight saved on equipment installation is geared up by a factor of about 10 on the overall aircraft weight. A further driver to the adoption of optical data transmission is the extensive use of composite materials in the aircraft structure and the consequent loss of the screening provided by an aluminium skin which forms a 'Faraday cage' as far as external electric fields are concerned.

The use of fibre optic links to interconnect the basic elements of the flight control system – pilot's stick, sensors, flight control computers, actuator control electronics, etc. has become known as a 'Fly-by-light' flight control system and is covered in more detail in Section 4.6.2.

It should be noted, however, that while a number of prototype aircraft have flown with flight control systems using fibre optic links, there are no production aircraft in service with a Fly-by-light flight control system at the time of writing this book. The advantages, however, are such that it is only a question of time before the technology is widely exploited.

4.6.2 Fly-by-light flight control systems

A Fly-by-light flight control system configuration is shown schematically in Fig. 4.38. The accompanying redundancy is omitted for clarity. The fibre optic links

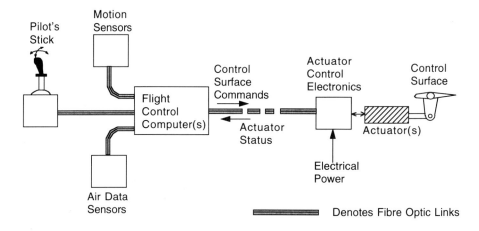

Fig. 4.38 Fly-by-light flight control system (redundancy omitted for clarity).

interconnecting the units of the flight control system eliminate the possibility of propagating electrical faults between units, as the optical fibre is an insulator.

Fibre optic links can be bi-directional and can also be used to convey the system status to the pilot's control/display panel. For instance, 'wrap round' tests can be carried out to check that the data have reached their destination and the status or 'health' of the control surface actuation system can be checked by monitoring the servo error signals and actuator hydraulic pressures.

The actuator control electronics and actuator are located in fairly close proximity to each other and the only connections required are the fibre optic cables and electric power supplies, the large wiring harness required in earlier analogue systems being eliminated. Such systems are sometimes referred to as 'smart actuators'.

A further advantage of fibre optic data transmission is the ability to use 'wave length division' multiplexing whereby a single fibre can be used to transmit several channels of information as coded light pulses of different wavelengths (or colours) simultaneously. The individual data channels are then recovered from the optically mixed data by passing the light signal through wavelength selective *passive* optical filters which are tuned to the respective wavelengths. WDM has a very high integrity as the multiplexed channels are effectively optically isolated.

It is interesting to note that one of the first applications of Fly-by-light technology has been on the earliest flying vehicle, namely the airship. Non-rigid airships may be acquiring a new lease of life as long duration platforms for airborne radar surveillance systems. The airship is able to carry a large high power radar antenna within its envelope and is able to remain airborne for several days (with occasional in-flight refuelling while hovering over a supply ship). It can also carry the relatively large crew required for 24 hour operation in considerable comfort.

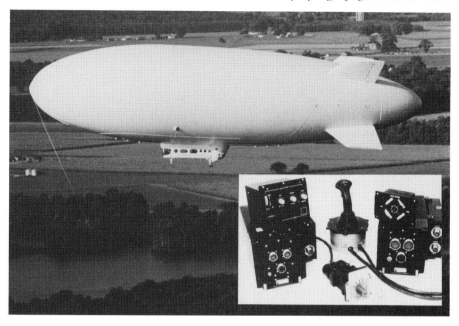

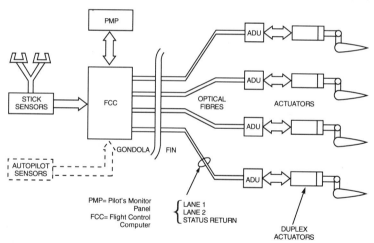

Fig. 4.39 'Skyship 600' airship with optically signalled flight control system (by courtesy of GEC-Marconi Avionics Ltd.).

Fully powered controls are required for the airship's large control surfaces because of the need for autopilot operation on long duration missions and autostabilisation to counter an inherent pitch instability and provide yaw damping. The long control cables required from the gondola to the control surfaces (100–200 metres on the projected large airships) and the probability of experiencing lightning strikes make an optically signalled flight control system an attractive solution. Fig. 4.39 illustrates the 'proof of concept' Skyship 600 airship which first flew with an optically signalled flight control system in October 1988.

4.6.3 Optical sensors

Figures 4.40 (a) and (b) show the large number of sensors involved in the flight control system for a representative civil and military aircraft respectively – many of these sensors being at quadruplex level. The approximate location of the sensors is indicated and it can be seen that many are located near the extremities of the aircraft and possibly less well screened than fuselage mounted equipment. The extensive use of composites can further degrade electro-magnetic screening in the vicinity of such sensors.

The use of what are referred to as passive optical sensors offers particular advantages in such applications as these sensors are not affected by EMI. Passive optical sensors are defined as sensors which do not require electrical supplies or any

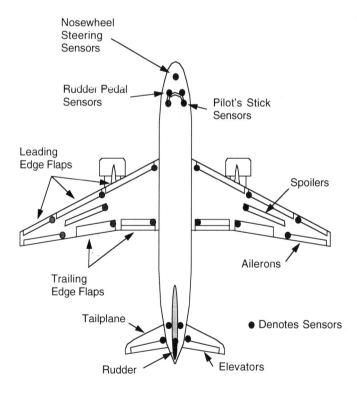

Fig. 4.40(a) Sensor locations – civil aircraft.

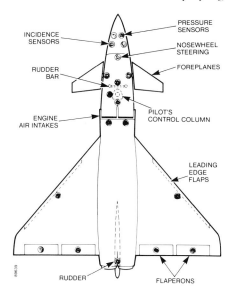

Fig. 4.40(b) Sensor locations – agile fighter.

electronic processing at the sensor, the output of the sensor being an optical signal which is modulated by the quantity being measured. The processing of this optical signal is carried out by a separate electronic unit which is fully screened from EMI, the only connection between the sensor and the electronic unit being an optical fibre cable. The basic concept of such sensors is shown in Fig. 4.41.

The exploitation of optical interferometric principles is also being actively pursued in the development of new passive sensors for measuring pressure, temperature, acceleration, etc. These passive sensors are also able to operate at high temperatures (up to 600°C) and have application in engine control systems as well as flight control.

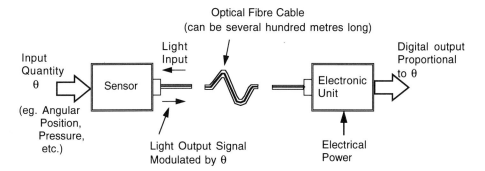

Fig. 4.41 Passive optical sensor concept.

Further reading

Barnes, J.G.P., *Programming in Ada*: Addison Wesley

Brière, D., Favre, C., Traverse, P., *A Family of Fault-Tolerant Systems: Electrical Flight Controls, from Airbus A320/330/340 to Future Military Transport Aircraft*: ERA Avionics Conference 1993

Coad, P., and Yourden, E., *Object Oriented Analysis*: Yourdon Press/Prentice Hall 1991

Corney, J.M., *The Evolution of the EAP Flight Control System*: Presented to the International Symposium on Aeronautical Science and Technology of Indonesia 24-26 June 1986

Davis, A.M., *Software Requirements - Analysis and Specification*: Prentice Hall 1990

D`Azzo, J.J., and Houpus, C.H., *Linear Control System Analysis and Design*: McGraw-Hill

DeMarco, T., *Structured Analysis and Design*: Yourdon Press Computing Press 1978

Harschburger, H.E., Glaser, B., and Hammel, J.R., *Back up modes for the F/A-18 Digital Flight Control System*: AIAA 84-2622 1984

Hatley, D.J., and Pirbhai, I.A., *Strategies for Real-Time System Specification*: Dorset House 1988

Ince, D.C., *An Introduction to Discrete Mathematics and Formal System Specification*: Oxford University Press 1988

Interim Defence Standard DEF STAN OO-55, *The Procurement of Safety Critical Software in Defence Equipment*: 5th April 1991

Marshall, R.E.W., Snelling, K.S., and Corney, J.M., *The Jaguar Fly by Wire Demonstrator Integrated Flight Control System*: Proceedings of the United States Air Force Academy Advanced Flight Controls Symposium 1981

McLean, D., *Automatic Flight Control Systems*: Prentice Hall.

Military Standard DOD-STD-2167A, *Defense System Software Development*: 29 February 1988

Mirza, N.A., *Primary Flight Computers for the Boeing 777*: ERA Avionics Conference 1992 ERA Report 92-0809

Ramage, J.K., *AFTI/F16 Automated Maneuvering Attack System Configuration and Integration*: Proceedings of the IEEE 1986 National Aerospace and Electronics Conference, Dayton, Ohio

RTCA-EUROCAE DO178B/ED-12B, *Software Considerations in Airborne Systems and Equipment Certification*: December 1992

Rushby, J., *Formal methods and the certification of critical systems*: Technical report CSL-93-7 December 1993 Computer Science Laboratory SRI International

Smith, R., *Flight Clearance of the Jaguar FBW Aircraft*: Royal Aeronautical Society Symposium April 1982

Smith, R.B., Tamagnone, B., *The AMX Fly by Wire System*: 14th Congress of the International Council of the Aeronautical Sciences, Toulouse, France, 9-14 September 1984 - Paper No. ICAS 84-4.4.3

5

Inertial sensors and systems

5.1 INTRODUCTION

Gyroscopes (hereafter abbreviated to gyro) and accelerometers are known as inertial sensors. This is because they exploit the property of inertia, namely the resistance to a change in momentum, to sense angular motion in the case of the gyro and changes in linear motion in the case of the accelerometer. They are fundamental to the control and guidance of an aircraft. For example, in a FBW aircraft the rate gyros and accelerometers provide the aircraft motion feedback which enables a manoeuvre command control to be achieved and an aerodynamically unstable aircraft to be stabilised by the flight control system (as explained in Chapter 4).

Gyros and accelerometers are also the essential elements of the spatial reference system or attitude/heading reference system (AHRS) and the inertial navigation system (INS). They largely determine the performance and accuracy of these systems and account for a major part of the system cost because of the precision required from them. The AHRS provides the vertical and directional reference information for the pilot's head up display (HUD), electronic flight instrument system (EFIS), the autopilot system and the navigation system.

The INS provides a completely self contained source of navigation information together with very accurate attitude and heading information. In addition a strap-down INS and also a strap-down AHRS can provide roll rate, pitch rate, yaw rate and normal and lateral acceleration (and forward acceleration if required) for the flight control system (FCS).

The AHRS and INS share common technology and operating principles. This chapter covers gyros and accelerometers (Section 5.2) and attitude/heading reference systems (Section 5.3) and provides the basic background to inertial navigation systems which are covered in Chapter 6 as part of navigation systems.

The particular objectives of this chapter are:-

(1) To explain the basic principles and characteristics of the gyros and accelerometers used in modern avionic systems (FCS, AHRS and INS).
(2) To show how the gyros are used in both a 'stable platform' and a 'strap-down' system mechanisation to derive the attitude of the aircraft as this is fundamental to both an AHRS and an INS.
(3) To explain the methods used for alignment and monitoring of an AHRS to constrain the vertical and heading errors.

5.2 GYROS AND ACCELEROMETERS

5.2.1 Introduction

The accuracy requirements for gyros and accelerometers can differ by several orders of magnitude depending on the application. Table 5.1 shows the accuracy requirements for a typical FBW flight control system and a 1 NM/hour strap down INS as representing the lower and upper ends of the performance spectrum respectively.

Table 5.1 Accuracy requirements

		Flight control system	Strapdown INS
Gyro	Scale factor	0.5%	0.001% (10ppm)
	Zero offset/rate uncertainty	1°/min	0.01°/hour
Acctr	Scale factor	0.5%	0.01% (100ppm)
	Zero offset/bias stability	$5 \times 10^{-3}g$	$5 \times 10^{-5}g$ (50µg)

Not surprisingly the costs of these devices can also differ by an order of magnitude or more.

The quest and attainment of the accuracies required for inertial navigation has involved many billions of dollars expenditure world-wide and the continual exploitation of state of the art technology. For example the world's first laser was demonstrated in 1960 and the first experimental ring laser gyro (RLG) was demonstrated in 1963 with an accuracy of a few degrees per hour. The first production RLG based inertial navigation systems (which requires 0.01°/hour accuracy) went into large scale civil airline service in 1981. RLG based IN systems are now dominating the INS market. A remarkable technical achievement by any standards.

Over 80% of the gyros used in current control and guidance systems are what is termed 'spinning rotor gyros' as they exploit the angular momentum of a spinning rotor to sense angular motion. However their mechanical complexity and inherent failure modes constrain their cost and reliability and hence cost of ownership. Power consumption and run-up time are further limiting parameters of this technology.

Gyros operating on 'solid state' principles are thus being actively developed because of their intrinsically higher reliability and lower cost of ownership. Optical gyros such as the ring laser gyro and the fibre optic gyro will progressively displace the spinning rotor types of gyro.

Already about 90% of new INS procurement is for RLG based systems because of the tenfold improvement in cost of ownership over the stable platform type of INS. The fibre optic gyro (FOG) is currently at the AHRS level of accuracy and the first large production order for FOG based attitude/heading reference systems was placed in 1991.

5.2.2 Angular momentum gyros

5.2.2.1 Introduction and basic principles

The spinning rotor or angular momentum type of gyroscope has enabled enormous advances to be made in control and guidance ranging from FBW flight control to precision self contained navigation. Although it is being gradually superseded by the 'solid state' type of gyro such as the ring laser gyro and the fibre optic gyro, spinning rotor type gyros are still being procured for new build aircraft systems. There are also very large numbers of spinning rotor gyros in service in existing systems and it will be many years before they are completely replaced. It is therefore necessary to have an appreciation of their characteristics.

The principle of the spinning rotor gyroscope was first demonstrated in 1852 by the French physicist Jean Foucault who used the property of the spin axis of a gimbal suspended spinning flywheel to maintain a fixed direction in space to measure the rotation of the Earth. He gave the device the name 'gyroscope' from the Greek words *gyros* meaning rotation and *skopein* meaning to view.

A literal translation is thus 'to view rotation' which is still a good description of the function of the instrument and is equally applicable to a ring laser gyro.

The spinning rotor type of gyroscope uses the fundamental characteristic of the angular momentum of the rotor to resist changing its direction to either provide a spatial reference or to measure the rate of angular rotation.

In the first case the spinning rotor is supported in such a way that the spin axis is free to point in any arbitrary direction and provided there are zero torques acting about its axes of freedom it will stay pointing in this fixed direction in space.

In the second case the rotor spin axis is constrained to follow the rotation the gyro experiences about its input axis or axes. The torque required to constrain the rotor is directly proportional to the input rate which can then be determined by measuring the torque. The first type is known as a free gyro and the second type is referred to as a rate gyro.

The torque required to change the direction of the rotor spin axis can be derived as follows.

Referring to Fig. 5.1, suppose the gyro rotor is rotated through a small angle $\Delta\theta$ about the OX axis in time Δt. The vector change in the angular momentum is equal to $H\Delta\theta$ and is in a direction parallel to OY (see inset vector diagram).

Where H = angular momentum of the rotor = $J\omega_R$

J = moment of inertia of rotor about spin axis

ω_R = angular velocity of rotor about spin axis.

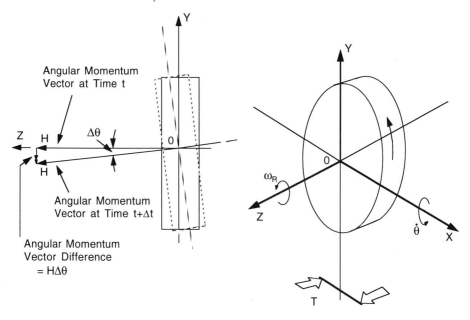

Fig. 5.1 Gyroscopic precession.

The rate of change of angular momentum is thus equal to $H\dfrac{d\theta}{dt}$ and this must be equal to the applied torque, T by Newton's second law.

Hence $T = H\dot{\theta}$ = gyroscopic reaction torque   (5.1)

or $\dot{\theta} = T/H$ (5.2)

This behaviour, where the application of a torque about an axis orthogonal to the spin axis causes the rotor to rotate about a third axis which is mutually orthogonal to both the applied torque axis and the spin axis, is called *Precession*.

It is instructive to show by means of a simple example how the exploitation of angular momentum to sense angular motion gives the gyro such sensitivity and accuracy.

Consider the use of a freely suspended inertial element to sense angular motion, for example, a sphere (or rotor) supported by some means of frictionless suspension so that it has complete angular freedom – for instance using an electro-static suspension system. The angular position of the rotor relative to the case is measured, say by optical pick-offs from a datum pattern on the rotor. Suppose the rotor has a moment of inertia $I = 50$ cgs units (gm cm^2) and it is subjected to a very small disturbing torque $T = 0.1$ cgs units (dyne cm; 1 dyne $= 10^{-5}$ Newtons) due to a change in the residual torques from the suspension system. Two cases (a) and (b) are considered.

Case (a) Rotor stationary
Angle θ rotor turns through in time t due to the action of the disturbing torque T is given by:

$$\theta = \int \int \frac{T}{I} \, dt \, dt = \frac{1}{2} \frac{T}{I} t^2$$

Suppose $t = 20$ seconds

$$\theta = \frac{1}{2} \times \frac{0.1}{50} \times 20^2 \times 57.3 = 23°$$

This is clearly unusable; the error being proportional to t^2.

Case (b) Rotor spinning at 2,500 rad/s (24,000 rpm)
Suppose now the rotor is run up to a speed of about 24,000 rpm and then allowed to coast.
Angular momentum of rotor H
$H = 50 \times 2,500$ cgs units
$H = 1.25 \times 10^5$ cgs units (gm cm^2/s)

$$\theta = \int \frac{T}{H} \, dt = \frac{T}{H} t$$

For $t = 1$ hour $= 3600$ s

$$\theta = \frac{0.1}{1.25 \times 10^5} \times 3,600 \times 57.3 = 0.17°$$

The error is thus proportional to t (as opposed to t^2 in case (a)).

The enormous improvement due to angular momentum can be seen – the effect is sometimes referred to as 'inertial multiplication'.

A 1 NM/hour accuracy INS requires that the gyro drift is less than 0.01°/hour. Typical gyros for airborne IN systems have angular momentums in the region of 0.5×10^6 cgs units. To meet the drift performance thus requires that sustained disturbing torques acting on the gyro from any cause (eg mass centre shifts) are less than 0.02 cgs units – an incredibly small figure. The task and cost of developing and producing a gyro with this performance and stability can thus be appreciated.

Space constraints and maintaining the balance of the chapter, particularly with the advent of solid state optical gyroscopes, have restricted the coverage of angular momentum gyros to the dynamically tuned gyro (DTG) only. This is a two axis gyro which can be operated as a free gyro, albeit of limited angular freedom, in a stable platform system, or, as a two axis rate gyro in a strap-down system. It is a widely used modern gyroscope which was initially developed for second generation IN systems (the first generation IN systems used floated gyros).

The DTG has been subsequently developed as a two axis rate gyro for strap-down applications covering a wide performance spectrum from 0.01°/hour to 30°/hour rate uncertainty. It is a compact gyro; typical size is about 40mm dia × 40mm. Applications range from strap-down IN systems, strap-down AHRS particularly for helicopters in view of the compact size and low weight, short term IN systems for missile mid-course guidance and FBW flight control systems. In FBW applications, the two axis rate measurement capability enables economical redundancy configurations to be achieved in terms of the number of gyros needed to meet the failure survival requirements.

5.2.2.2 The dynamically tuned gyro

The basic construction of the dynamically tuned gyro is shown in Fig. 5.2. The gyro

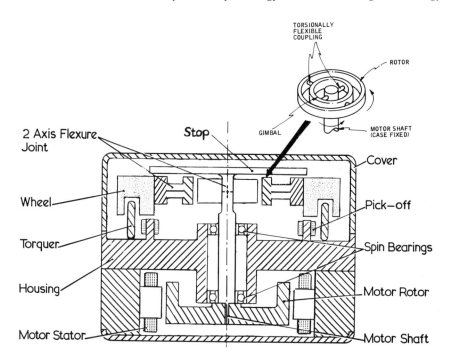

Fig. 5.2 Dynamically tuned two-axis gyro.

angular momentum wheel, or rotor, is coupled to the spin motor drive shaft by a flexible type of Hooke's joint, or universal coupling which provides a two axis gimbal system as shown in the inset sketch in Fig. 5.2. The flexural pivots used for the gimbal bearings are frictionless but have a torsional spring rate characteristic so that they will exert a torque on a rotor when there is an angular displacement between the normal to the plane of the rotor and the spin motor axis. This negative (i.e. restoring) spring rate can, however, be cancelled at a particular rotor speed by a dynamically created positive spring rate which arises from the torques exerted on the gimbal inertias when there is an angular displacement of the rotor plane. The condition for dynamic tuning is given by

$$K_G = \left(I_G - \frac{1}{2} J_G \right) \omega_R^2$$

where K_G = torsional spring rate of the flexural pivots,
I_G = moment of inertia of the gimbal about the gimbal axes,
J_G = moment of inertia of the gimbal about the polar axis,
ω_R = angular velocity of the rotor.

The gimbal inertias are adjusted (or tuned) so that the negative spring rate of the flexural pivots is exactly cancelled by the dynamically created positive spring rate at the design rotor speed (typically 320 rev/s). Hence there are zero torques exerted about the gimbal axes and the gyro behaves as an ideal free gyro over a limited angular movement. The plane of the rotor will thus maintain a fixed orientation in space for small angular movements of the gyro case about the input axes (see sketches in Fig. 5.3).

Inductive pick-offs mounted on the gyro case measure the angular displacement between the rotor and the case about the gyro input axes, these angles being referred to as the gyro output angles.

Electro-magnetic torque motors enable torques to be exerted on the rotor about the gyro input axes. These torque motors are basically of the D'Arsonval

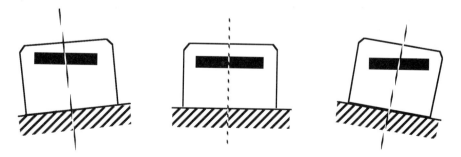

Fig. 5.3 Plane of rotor stays fixed in space.

galvanometer type with samarium cobalt permanent magnets incorporated into the rotor but with the torque coils fixed to the case. The magnetic field created by the rotor permanent magnets interacts with the current flowing through the coils and so exerts a torque on the rotor. The torque exerted is directly proportional to the torque motor current to a very high accuracy, very stable permanent magnets being used. Variations in the torque motor scale factor with temperature are also compensated within the system computer from measurement of the gyro temperature.

Only limited angular freedom is required for the rotor as the gyro output angles are maintained near null in both stable platform and strapped down operation. Stops which are integral with the spin motor drive shaft limit the angular freedom of the rotor to ± 0.5°, or less.

The gyro is operated as a two axis free gyro in a stable platform system. Two gyros are mounted on the gimbal suspended stable platform so as to sense angular movement about the three orthogonal platform axes. (The redundant axis available from one of the gyros is used as a system monitor.) The gyro output angle pick-offs control the gimbal servos so as to maintain the gyro output angles at or near null, irrespective of the airframe angular motion. Hence the platform maintains a fixed orientation with respect to space axes. The platform is rotated about its axes at the appropriate rates of rotation so that it stays aligned with Earth's axes, eg North–South, East–West and local vertical axes by exerting the required precession torques on the gyro rotor through the gyro torque motors. The stable platform concept is shown in Fig. 5.4.

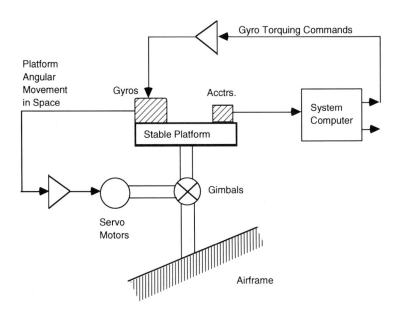

Fig. 5.4 Stable platform concept.

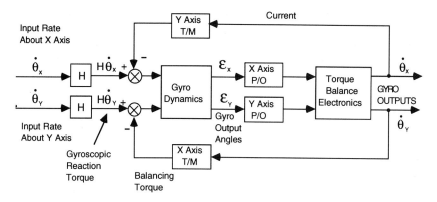

Fig. 5.5 Torque balance rate gyro block diagram.

The rate gyro version of the DTG is basically similar to the stable platform version. The rotor angular momentum, however, is generally lower because of the torque motor power requirements at high input rates which can be as high as 400°/s. The torque motors must also be capable of exerting higher maximum torques.

The two axis rate gyro operation is achieved by 'torque balance' closed loop control. The torque balance loops control the gyro torque motors from the gyro output angle pick-off signals so as to maintain the output angles at, or near null. The torques exerted by the torque motors then exactly balance the gyroscopic reaction torques acting on the rotor when the gyro is rotated about its input axes. As explained earlier, the gyroscopic reaction torques act about the axis which is mutually perpendicular to the rotor spin axis and the input axis about which the plane of the rotor is being rotated and are directly proportional to these input angular rates of rotation. Measurement of the appropriate torque motor currents hence enables the gyro input rates to be accurately determined. The basic principles are shown in the block diagram in Fig. 5.5.

The electronic configuration of the torque balance loops is shown in Fig. 5.6. These loops are frequently called 'caging loops' as they effectively 'cage' the rotor to the gyro case. The two loops must cage the rotor in two axes simultaneously and this couples the two caging loops together due to the gyroscopic cross axis reaction of the rotor to applied torques. There is also no inherent viscous damping present (unlike fluid filled or floated gyros) as the DTG is a 'dry gyro'. It is thus necessary to cross feed components of the gyro pick-off signals to each control loop and suitable dynamic compensation is required to achieve a wide bandwidth with adequate stability margins. The dynamic compensation typically comprises proportional plus integral and phase advance control terms. Typical bandwidths range from 50 Hz to 75 Hz. The torque balance loops can be analogue loops with the outputs subsequently digitised in an A to D converter or the loop can be operated in a digital pulse torque mode.

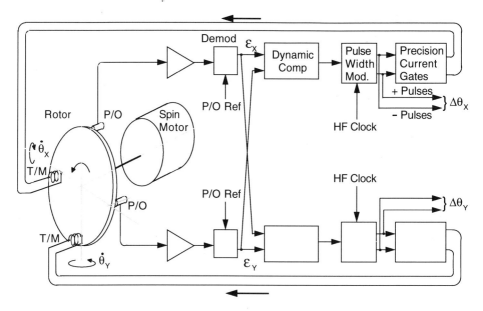

Fig. 5.6 DTG rate gyro torque balance loops.

Pulse torque operation of the caging loops is frequently used as this gives a simple digital interface to the system computer and avoids torque motor linearity errors as the torque motor is operated by constant amplitude positive or negative current pulses of constant time duration. The torque exerted is proportional to the net pulse frequency, f, that is the average number of net positive (or negative) pulses per unit time.

Mean torque exerted by torque motor $= Kf$

where K is a constant dependent on the torque constant of the torque motor (Nm/amp), the amplitude of the current pulse (amp) and the pulse duration (s).

In the steady state torque balance condition

$$Kf = H\frac{d\theta}{dt} \tag{5.3}$$

Where H is the rotor angular momentum (kg m^2/s) and $\dfrac{d\theta}{dt}$ is the rate of change of the input angular rotation (radian/s).

Hence

$$\int_0^T f\,dt = \frac{H}{K}\int_0^\theta d\theta \tag{5.4}$$

Hence the number of torque pulses in a given time period, T, is directly proportional to the incremental angular rotation about the gyro input axis in that time interval. Each pulse thus represents a fixed incremental angular rotation about the input axes – typical values range from 1 to 10 seconds of arc. A re-scaling of the amplitude of the current pulse amplitude may be required at high input rates – eg the output pulse quantisation is 1 second of arc at low input rates and, say, 5 seconds of arc at high input rates.

5.2.3 Optical gyroscopes

5.2.3.1 Introduction

Optical gyroscopes such as the ring laser gyro and the fibre optic gyro measure angular rate of rotation by sensing the resulting difference in the transit times for laser light waves travelling around a closed path in opposite directions. This time difference is proportional to the input rotation rate and the effect is known as the 'Sagnac effect' after the French physicist G. Sagnac. Sagnac, in fact, demonstrated that rotation rate could be sensed optically with the Sagnac interferometer as long ago as 1913. Figure 5.7(a), (b) and (c) illustrate schematically the basic configurations of the Sagnac interferometer, ring laser gyro and the fibre optic gyro.

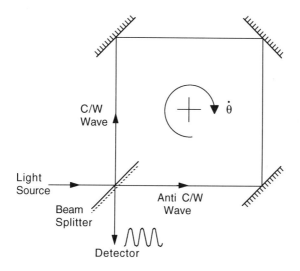

The clockwise and anti-clockwise waves interfere to produce a fringe pattern which shifts when Interferometer is subjected to Input Rate, $\dot{\theta}$.

Fig. 5.7(a) Sagnac interferometer.

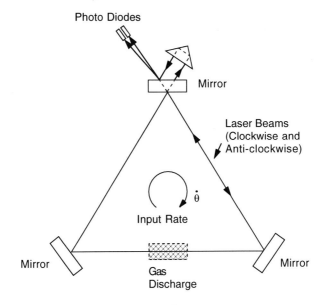

Fig. 5.7(b) Laser gyro schematic.

The time difference, ΔT, between the clockwise (cw) and anti-clockwise (acw) paths is given by

$$\Delta T = \frac{4A}{c^2}\dot\theta \qquad (5.5)$$

Where A is the area enclosed by the closed path, c the velocity of light and $\dot\theta$ the angular rate of rotation about an axis normal to the plane of the closed path.

The difference in optical path length ΔL, where L is the perimeter of the path, is given by

$$\Delta L = c\Delta T$$

$$\Delta L = \frac{4A}{c}\dot\theta \qquad (5.6)$$

A rigorous derivation of the above formulae requires the use of the general theory of relativity. A simpler kinematic explanation, however, is given below for the case of a circular path in vacuo.

Referring to Fig. 5.8, consider a photon of light starting from P and travelling round the perimeter in a cw direction and a photon starting from P travelling in the acw direction.

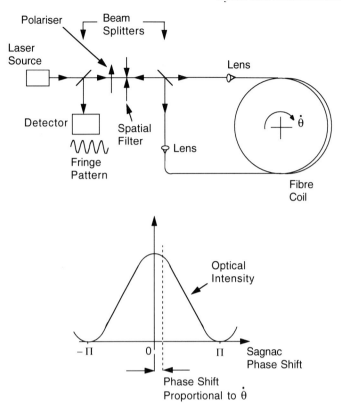

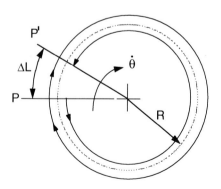

Fig. 5.7(c) Basic fibre optic gyro.

Fig. 5.8 Sagnac effect.

In the absence of an input rate, the transit times of the two photons will be identical and equal to $2\pi R/c = T$. Now consider the path rotating at a rate $\dot{\theta}$. In time T, P has moved to P^1 and the path length for the cw photon is equal to $(2\pi R + R\dot{\theta} T)$ and the path length for the acw photon is equal to $(2\pi R - R\dot{\theta} T)$

The difference in transit time, $\Delta T = \dfrac{(2\pi R + R\dot{\theta} T)}{c} - \dfrac{(2\pi R - R\dot{\theta} T)}{c}$

substituting $T = 2\pi R/c$ yields

$$\Delta T = \frac{4\pi R^2}{c^2}\dot{\theta} \tag{5.7}$$

and

$$\Delta L = \frac{4\pi R^2}{c}\dot{\theta} \tag{5.8}$$

$A = \pi R^2$ hence formulae (5.7) and (5.8) are identical with (5.5) and (5.6).

It can be shown that these formulae are unchanged when the optical path comprises a medium of refractive index n such as when an optical fibre provides the closed path.

The Sagnac effect is very small for low rates of rotation. Michelson and Gale in 1925 used a Sagnac interferometer with a rectangular cavity 600m × 330m to measure the earth's rotation rate and measured a path difference of only 1/4 fringe.

The invention of the laser in 1960 radically changed the situation and enabled optical gyroscopes based on the Sagnac effect to be developed by exploiting the laser's ability to provide a source of highly coherent light with very stable wave length characteristics.

The first experimental ring laser gyro was demonstrated in the US by Macek and Davis in 1963 and has since been developed by a number of companies and establishments world-wide to large scale production status. Many tens of thousands of RLG's are operating in strap-down INS systems and have established high accuracy with better than 0.01°/hour bias uncertainty and MTBFs in excess of 60,000 hours. The RLG is described in the next section and is basically an active resonant system with the laser cavity forming the closed optical path. Input rotation rates are measured by the difference in the resonant frequencies of the clockwise and anti-clockwise paths resulting from the difference in the path lengths produced by the rotation.

The development of low loss single mode optical fibre in the early 1970s for the telecommunications industry enabled Sagnac effect fibre optic gyros to be developed. These use an external laser diode source together with suitable beam splitting optics to launch the laser light so that it travels in cw and acw directions through a cylindrical coil comprising many turns of optical fibre. The effective area of the closed optical path is thus multiplied by the number of turns in the coil. Path lengths of hundreds of metres are achievable. The first FOG was demonstrated in the US by Vali and Shorthill in 1976. Development of both the passive interferometer type of FOG, or IFOG, and the passive ring resonator type of FOG, or RFOG, is proceeding in many companies and establishments world-wide.

The IFOG operates basically as a Sagnac interferometer and measures the phase shift in the fringe pattern produced by the input rotation rate. The accurate measurement of very small Sagnac phase shifts has required the development of special techniques which will be explained. The IFOG has now been developed to production status to meet an AHRS level of accuracy and as mentioned earlier the first large production orders for IFOG based AHRS were placed in 1991. Development to inertial accuracy is proceeding in many organisations and IFOG based IN systems are anticipated in the not too distant future.

The RFOG operates in a similar manner to the RLG but with the resonant ring of optical fibre driven by an external laser diode. Two acousto-optical modulators are used to shift the laser frequency injected into the cw and acw paths respectively with the modulation frequencies being controlled by servo loops to maintain the resonance peak condition. The difference in the resonant frequencies of the two paths is then directly proportional to the input rotation rate as with the RLG.

The RFOG is still under development and time will tell which mechanisation will provide the best solution from a cost and performance standpoint.

Attention has thus been concentrated on the RLG and the IFOG as these are devices which are currently available.

5.2.3.2 *The ring laser gyro*

The basic elements of the RLG are shown schematically in Fig. 5.7(b). The two counter propagating laser beams are generated from the lasing action of a helium–neon gas discharge within the optical cavity, the triangular closed path being formed by reflecting mirrors at each corner of the triangle. This closed path forms the resonant cavity and the longitudinal mode frequency, f, is determined by the cavity optical path length, L, being given by $f = nc/L$ where n is an integer and c the velocity of light. At zero input rotation rate, the cw and acw path lengths are equal and there is zero difference between the frequencies of the cw and acw waves. When the RLG is rotated about an axis normal to the plane of the closed path there is a difference in the path length of the cw and acw travelling waves, as shown earlier, which causes a frequency difference between the two waves. This frequency difference is measured by allowing a small percentage of the two laser beams to be transmitted through one of the mirrors. A corner prism is generally used to reflect one of the beams so that it can be combined with the other to generate a fringe pattern at the read-out detector photo-diodes. An input rotation rate causes the fringe pattern to move relative to the read-out photo-diodes at a rate and in a direction proportional to the frequency difference (positive or negative). A sinusoidal output signal is generated by each fringe as it passes by the photo-diodes. These are spaced so that there is a 90° phase difference between their outputs so that the direction of rotation can be determined from which photo-diode output is leading. The photo-diode outputs are then converted into positive (or negative) pulses by suitable pulse triggering and direction logic circuits.

The frequency difference, Δf, resulting from a difference in optical path length, ΔL, is given by

$$\frac{\Delta L}{L} = \frac{\Delta f}{f}$$

From equation (5.6)
$$\Delta L = \frac{4A}{c}\dot{\theta}$$

whence
$$\Delta f = \frac{4Af}{cL}\dot{\theta} \qquad (5.9)$$

The wavelength of the laser transition $\lambda = c/f$

hence
$$\Delta f = \frac{4A}{\lambda L}\dot{\theta} \qquad (5.10)$$

ie
$$\Delta f = K_0\dot{\theta} \qquad (5.11)$$

where K_0 = gyro scale factor = $4A/\lambda L$
The gyro thus behaves as an integrating rate gyro

$$\int_0^T \Delta f \, dt = K_0 \int_0^\theta d\theta \qquad (5.12)$$

As with the pulse torque balance DTG, the angle turned through about the gyro input axis in the time period, T, is equal to the net number of positive (or negative) pulses counted in that period. The RLG thus provides a direct digital output of the input angular rotation with a process of inherent integration carried out in the optical domain.

The high sensitivity of the RLG can be seen from the following example. A typical RLG with an equilateral triangular cavity of perimeter, $L = 0.2$m (20cm) has an area, $A = 1.9245 \times 10^{-3}$ m^2 (19.245cm^2) and laser transition wavelength, $\lambda = 0.633$ microns (0.633×10^{-6}m).

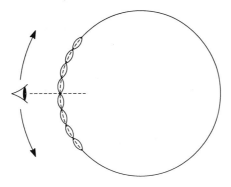

Fig. 5.9 Laser gyro principle.

RLG scale factor, $= \dfrac{4 \times 1.9245 \times 10^{-3}}{0.633 \times 10^{-6} \times 0.2} = 60825$ Hz/rad per s and

hence 1 pulse count $= \dfrac{1}{60825}$ rad $= 3.39$ arc seconds

This resolution can be increased by a factor of 4 by triggering on the positive and negative going zero crossings of both detectors.

An alternative and equally valid explanation of the RLG which some readers may prefer is as follows. The two counter propagating waves set up by the lasing action will beat together and set up a standing wave pattern in the cavity. This standing wave pattern remains fixed in space irrespective of the angular rotation of the cavity. An observer rotating with the cavity, in this case the read out photo-diodes, will thus see a succession of light and dark fringes as the read-out moves past the spatially fixed standing wave. Figure 5.9 illustrates the principle. The RLG behaves in an analogous way to an incremental optical encoder with the encoder shaft angle fixed in space and so measures the angle through which the gyro has been rotated. The fact that the standing wave pattern remains fixed in space as the cavity rotates can be interpreted as a manifestation of the inertia which electro-magnetic energy possesses in view of its mass equivalence.

Figure 5.10 illustrates the construction of a typical RLG. The equilateral triangular cavity is machined from a solid block of a vitro-ceramic material known as 'CERVIT' which has a very low coefficient of thermal expansion. The three mirrors are mounted on the block by optical contact for stability and ruggedness. The mirrors have multi-dielectric coatings and have a reflectivity of more than 99.9%. The mirror system, being a critical area of RLG technology, absorption, and above all scattering, must be minimised.

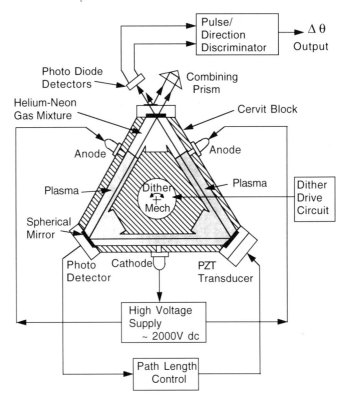

Fig. 5.10 Ring laser gyro.

The optical path length of the cavity is adjusted by means of a piezo-electric transducer attached to one of the cavity mirrors. The position of the mirror being controlled by a servo loop so that the laser oscillates at its peak average power.

High voltages of the order of 2,000 volts dc are applied between the two separate anodes and the common cathode to ionise the He–Ne gas mixture and provide the required lasing action. The system uses a common cathode and two separate anodes so that Langmuir flow effects in each plasma arm are cancelled out by control currents which balance the currents in the two discharges. These flow effects can cause bias shifts due to differential changes in the refractive index of the light travelling along the cw and acw paths.

A major problem which has had to be overcome with the RLG is the phenomenon known as 'lock-in'. This arises because of imperfections in the lasing cavity, mainly in the mirrors, which produce back scattering of one beam into the other. The resulting coupling action tends to pull the frequencies of the two beams together at low rotation rates producing a scale factor error. For input rates below a threshold known as the 'lock-in rate', the two beams lock together at the same frequency so that there is zero output and a dead zone results. Figure 5.11(a) illustrates the effect of lock-in under steady input rate conditions. This lock-in dead zone is of the order of 0.01 to 0.1°/s compared with 0.01°/hour accuracy required for an INS. A very effective method of overcoming this problem is to mechanically dither the laser block about the input axis at a typical frequency of about 100 Hz with a peak velocity of about 100°/s (corresponding to an amplitude of 1.5 arc seconds approximately). The amplitude of the dither rate and acceleration are chosen so that the dwell time in the lock-in zone is so short that lock-in cannot occur. Figure 5.11(b) illustrates the effect of dither on removing the dead zone. The dither signal can be removed from the output by mounting the read-out reflector

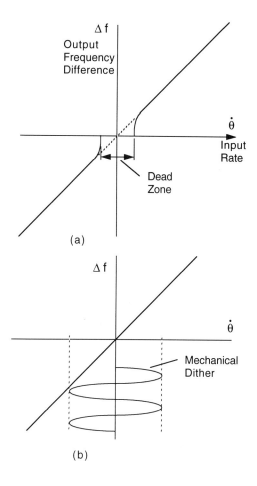

Fig. 5.11 Lock-in and effect of mechanical dither.

prism on the gyro case and the read-out photo-diodes on the block so as to produce an optical cancellation of the dither signal. Alternatively the read-out prism and read-out photo-diodes can both be mounted on the block and the unwanted dither signal can be removed by a digital filter. Some reduction in gyro bandwidth is incurred, however, because of the digital filter response.

Alternative 'solid state' techniques are also employed to overcome lock-in. The mechanical dither technique, however, is by far the most widely used technique at the current time. Although a criticism could be made in terms of 'technical elegance' of a solid state system which depends on mechanical dither, the facts are that zero failures in the dither system in over 100 million hours have been reported.

RLG performance characteristics The RLG performance characteristics are summarised below.

- High accuracy – The RLG is the only strap-down gyro currently available which meets the dynamic range requirements for a pure IN system of being able to measure angular rates from .01°/hour to 400°/s to the required accuracy – a dynamic range of 10^8:1.
- Insensitivity to acceleration – The RLG has no acceleration sensitive bias errors (unlike angular momentum gyros) as it is based on optical effects rather than inertial effects.
- Very high rate range – This is limited only by the noise/bandwidth characteristics of the read out electronics: ± 1,000°/s is no problem.
- Very high scale factor accuracy – Errors are in the 5 to 10 ppm bracket.
- Negligible warm up time – Full gyro operation is attained at the instant of turn-on.
- Excellent turn-on to turn-on performance – Performance capabilities can be maintained over several years without calibration.
- Random noise uncertainty – This is measured in 'degrees per √hour', and is one of the RLG's most important error characteristics. The error is significantly higher than experienced with angular momentum gyros. It affects the system heading determination in the gyro compassing phase during the initial alignment process. This is because it extends the time required to filter the Earth's rate signal from the gyro noise in order to determine the initial heading. (Gyro compassing is explained later.)
- Very high reliability – MTBF in excess of 60,000 hours are being demonstrated in large scale service. 100,000 hours MTBF is being aimed at.
- Life – The laser beams ultimately destroy the mirrors but this is over a very long period. A life equivalent to over 100 years usage is being claimed.
- Volume – The only real deficiency of current RLGs is their size in comparison with strap-down angular momentum gyros. The volume occupied by three orthogonally mounted RLG's is a factor of two or more greater than the volume

occupied by two DTG's. The RLG performance in terms of accuracy, reliability and cost of ownership, however, more than compensates for this shortcoming.

Figure 5.12 is a photograph of a typical modern RLG (by courtesy of GEC-Marconi Avionics Ltd).

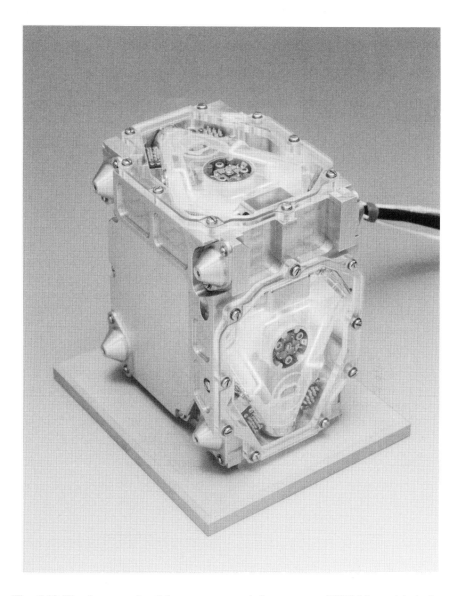

Fig. 5.12 Ring laser gyro inertial measurement unit (by courtesy of GEC-Marconi Avionics Ltd.).

5.2.3.3 The interferometric fibre optic gyro

The implementation of the interferometric type of FOG is best explained in a series of stages starting with the simple basic system shown in Fig. 5.7(c). Light from the laser diode source is passed through a first beam splitter and a single optical mode is selected. The light passes through a second beam splitter and propagates in both directions around the fibre coil. In the absence of rotation, as already explained, the transit times are identical so that when the light arrives back at the second beam splitter, perfect constructive interference occurs with accompanying fringe pattern. The gyro output signal is obtained by directing the returning light via the first beam splitter to a photo-detector. As explained, an input rotation rate about an axis normal to the plane of the coil results in a difference in the transit times between the cw and acw beams as given by equation (5.5)

viz
$$\Delta T = \frac{4A}{c^2}\dot{\theta}$$

If the fibre coil has N turns, $A = \pi R^2 N$ where R = mean radius of the coil, N can be expressed in terms of the length of the coil, L and the mean radius R

ie
$$N = \frac{L}{2\pi R}$$

hence
$$\Delta T = \frac{LD}{c^2}\dot{\theta} \qquad (5.13)$$

when D = coil diameter = $2R$
This transit time difference results in the Sagnac phase shift, given by

$$\Phi_s = \omega\Delta T$$

where ω = angular frequency of the laser source = $2\pi c/\lambda$
      λ = laser wavelength

hence
$$\Phi_s = \frac{2\pi LD}{\lambda c}\dot{\theta} \qquad (5.14)$$

This phase shift between the cw and acw travelling light waves results in a reduction in the intensity of the light at the detector. This change in intensity is very small for useful rotation rates and has required the development of special techniques.

For example a typical FOG with an optical path length L = 200m, coil diameter D = 70mm and source laser wavelength = 1.3 microns when subjected to an input rate of 10°/sec will have a Sagnac phase shift

$$\Phi_s = \frac{2\pi \times 200 \times 70 \times 10^{-2}}{1.3 \times 10^{-6} \times 3 \times 10^{-8}} \times 10 = 2.26°$$

Resolution of Earth's rate (15°/hour) requires the measurement of a phase shift of 3.38 arc seconds.

Resolution of 0.01°/hour seems barely credible at first sight but is achievable as will be shown.

The next problem to be overcome with the simple interferometer shown in Fig. 5.7(c) is that the relationship of the intensity of the fringe pattern to the Sagnac phase shift (which is proportional to the input rate) is a 'one plus cosine' law (see Fig. 5.7(c)). This results in a minimum sensitivity at low input rates and it is thus necessary to introduce a phase shift of 90° so that the sensor is operated in the region of maximum sensitivity.

The technique generally adopted is an alternating phase bias modulation method whereby a 'phase modulator' is placed at the end of the fibre loop (Fig. 5.13). The two counter propagating waves receive the same phase modulation but at different times (cw wave at the end of the transit round the coil, acw wave at the beginning of the transit). The phase modulator is driven at a frequency such that half a cycle corresponds to the time delay for the light to travel round the coil. (For a 200m FOG this corresponds to a modulation frequency of about 500 kHz.)

A non-reciprocal modulation is thus produced – ie a modulation which affects the cw direction of propagation differently from the acw direction of propagation. Application of an input rotation results in a modulated output signal whose amplitude is proportional to the input rate (see Fig. 5.13). This improved system as it stands is however an open loop system with attendant disadvantages such as:

(a) Output stability and scale factor is dependent on the stability of the gains of the various amplifiers preceding the demodulator, the demodulator stability and the stability of the intensity of the laser light source.

(b) The non-linear output characteristic with the phase shift resulting from large input rates (output characteristic is sinusoidal).

The answer is closed loop control, shown in Fig. 5.14. The output of the demodulator is passed through a servo amplifier which then drives a non-reciprocal phase transducer placed within the fibre interferometer. In this way the sensor is always operated at null by generating a suitable non-reciprocal phase shift that is equal and opposite to that generated by the input rate. The output of the system is then the output of the non-reciprocal phase transducer.

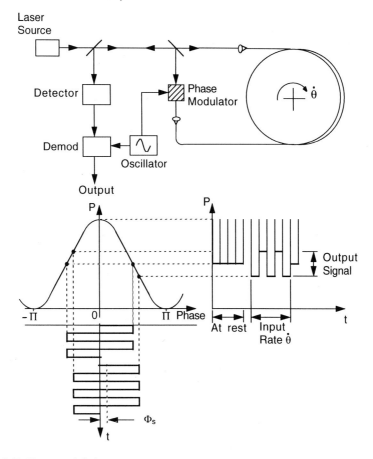

Fig. 5.13 Phase modulation.

The advantages of the closed loop system are:

(a) Output is independent of light source intensity variations as the system is always operated at null.
(b) Output is independent of the gains of individual components in the measurement system as long as a very high open loop gain is maintained.
(c) Output linearity and stability depend only on the phase transducer.

Typically an acousto-optic phase shifter is used in the phase transducer to generate a frequency difference to null the phase difference produced by the input rate.

Frequency difference and phase shift are directly related. The phase shift, Φ, that light experiences in propagating along a single mode optical fibre of length, L, and refractive index, n, is given by

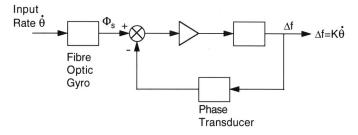

CONCEPT DIAGRAM

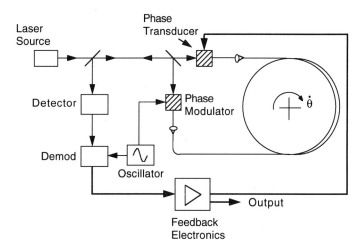

Fig. 5.14 Closed loop interferometric FOG (polariser and spatial filter omitted for clarity).

$$\Phi = 2\pi f \frac{L}{c/n}$$

(f = frequency of the light, velocity of propagation in the fibre = c/n).

hence

$$d\Phi = \frac{2\pi nL}{c} df$$

Sagnac phase shift,

$$\Phi_s = \frac{2\pi LD}{\lambda c} \dot{\theta}$$

hence at null
$$\frac{2\pi n L}{c}\Delta f = \frac{2\pi LD}{\lambda c}\dot{\theta}$$

from which
$$\Delta f = \frac{D}{n\lambda}\dot{\theta} \tag{5.15}$$

ie
$$\Delta f = K_1\dot{\theta}$$

where
$$K_1 = D/n\lambda$$

The output characteristic is thus the same as the RLG and the IFOG behaves as an integrating rate gyro.

The scale factor stability is dependent on the stability of D, n and λ.

For the example FOG already used $D = 70$mm, $\lambda = 1.3$mm, $n = 1.5$. Substituting these values, one output pulse corresponds to an angular increment of 5.75 arc seconds.

Equation (5.15) can be shown to be the same as the RLG output equation (5.10) by multiplying the numerator and denominator of (5.15) by $\pi D/4$

viz
$$\Delta f = \frac{D}{n\lambda}\frac{\pi D/4}{\pi D/4}\dot{\theta} = \frac{4A}{\lambda(n\pi D)}\dot{\theta}$$

The equivalent optical perimeter ($n\pi D$) corresponds to the RLG path length, L.

The inherent simplicity of the FOG and the use of *Integrated Optic Chips* where the various electro-optical elements are integrated on to a substrate leads to low manufacturing costs and the IFOG promises to be a highly competitive gyro in terms of cost, reliability, ruggedness and performance. Integrated three axis versions are being produced with the three FOGs sharing the same laser diode source.

Figure 5.15 is a photograph of the inertial measuring unit (IMU) of the Smiths Industries' fibre optic gyro based Attitude Heading Reference System. The IMU incorporates three interferometric fibre optic gyros and three accelerometers; the three FOGs share the same laser source. The FOG triad system is configured to share many of the optical and electrical components between the three FOG axes by time sequencing the processing of the optical signals for each axis using optical switches. By time sharing components, the cost, weight and volume are significantly reduced while reliability is increased by reducing the number of parts. The optical switches also replace the conventional beam splitters and increase the signal to noise ratio in the detected signal.

The fibre coils of the gyro comprise 400 metres of non-polarisation preserving single mode optical fibre wound on a 3.5cm diameter spool. The quoted gyro performance is shown in Table 5.2.

Fig. 5.15 Fibre optic gyro inertial measurement unit (by courtesy of Smiths Industries).

Table 5.2 Gyro performance

Bias error	0.8°/hour, 1σ
Scale factor	± 60 ppm, 1σ
Random walk	0.04°/√hour

The IMU size, weight and power are shown in Table 5.3.

Table 5.3 IMU characteristics

Weight	420 g
Volume	245 cm³
Power	7.5 Watts

The major advance in AHRS technology is apparent.

It is worth commenting briefly that the effect of environmental noise sources must be minimised in a FOG design in order to achieve inertial quality performance. This is because the non-reciprocal effects introduced would otherwise swamp the small Sagnac phase shifts being measured. Environmental noise sources include:

- Temperature variations – Introducing time dependent temperature gradients along the optical fibre.
- Acoustic noise – Optical fibre is acoustically sensitive (sensitive to sound pressure) and in fact fibre optic sensors are widely used in sonar arrays because of this sensitivity.
- Mechanical vibration – Introducing mechanical stresses in the fibre with accompanying non-reciprocal effects.
- Magnetic fields – A magnetic field acting parallel to the optical fibre results in a Faraday rotation of the state of polarisation in the fibre. This is a non-reciprocal effect which results in an unwanted phase shift.

5.2.4 Accelerometers

5.2.4.1 Introduction – specific force measurement

The acceleration of a vehicle can be determined by measuring the force required to constrain a suspended mass so that it has the same acceleration as the vehicle on which it is suspended, using Newton's law : force = mass × acceleration.

The measurement is complicated by the fundamental fact that it is impossible to distinguish between the force acting on the suspended mass due to the Earth's gravitational attraction and the force required to overcome the inertia and accelerate the mass so that it has the same acceleration as the vehicle. The vehicle acceleration, $\underline{a}$, being produced by the vector sum of the external forces acting on the vehicle, namely, the propulsive thrust, $\underline{T}$, lift, $\underline{L}$, drag, $\underline{D}$, and the gravitational force, $m\underline{g}$, acting on the aircraft mass, m.

(The bar under the symbols denotes the quantities are vector quantities).

$$\underline{T} + \underline{L} + \underline{D} + m\underline{g} = m\underline{a}$$

$$\underline{a} = \frac{(\underline{T} + \underline{L} + \underline{D})}{m} + \underline{g} \tag{5.17}$$

The vector sum of the external forces excluding the gravitational force divided by the aircraft mass, that is $(\underline{T} + \underline{L} + \underline{D})/m$, is known as the 'specific force'. The force, $\underline{F_a}$, required to constrain the suspended mass, m_a, is thus given by

$$\underline{F_a} + m_a \underline{g} = m_a \underline{a}$$

$$\frac{\underline{F_a}}{m_a} + \underline{g} = \underline{a} = \frac{(\underline{T} + \underline{L} + \underline{D})}{m} + \underline{g}$$

hence
$$F_a = m_a \frac{(\underline{T} + \underline{L} + \underline{D})}{m}$$
(5.18)

ie the accelerometer will thus measure the specific force component along its input axis and not the vehicle acceleration component.

It is thus essential to know the magnitude and orientation of the gravitational vector with respect to the accelerometer input axes in order to compute the vehicle acceleration components from the accelerometer outputs.

Only if the accelerometer input axis is exactly orthogonal to the gravity vector (ie horizontal) so that there is zero gravitational force component will the accelerometer measure the vehicle acceleration component along its input axis.

5.2.4.2 Simple spring restrained pendulous accelerometer

A simple spring restrained pendulous accelerometer is shown schematically in Fig. 5.16. This comprises an unbalanced pendulous mass which is restrained by the spring hinge so that it can only move in one direction, that is along the input axis. The spring hinge exerts a restoring torque which is proportional to the angular deflection from the null position. When the case is accelerated the pendulum deflects from the null position until the spring torque is equal to the moment required to accelerate the centre of mass of the pendulum at the same acceleration as the vehicle. This simple type of accelerometer is typically oil filled to provide viscous damping so that the transient response is adequately damped. An electrical position pick-off measures the deflection of the pendulum from the null position and provides the output signal.

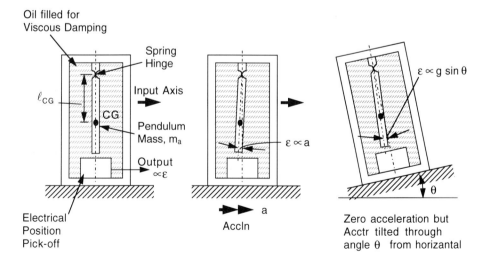

Fig. 5.16 Simple spring restrained pendulous accelerometer.

The transfer function of the simple accelerometer described is a simple quadratic lag filter of the type

$$\frac{\text{Output}}{\text{Input Accln.}} = \frac{K_0}{D^2 + 2\zeta\omega_0 D + \omega_0^2} \qquad (5.19)$$

K_0 = accelerometer scale factor, and the undamped natural frequency $\omega_0 = \sqrt{\frac{K}{I}}$, where K = torsional spring stiffness of hinge and I = moment of inertia of pendulum about hinge axis.

The simple spring restrained accelerometer is used in applications where the accuracy requirements are not demanding, e.g. 1% to 2% bracket.

The maximum angular deflection of the pendulum from the null position is limited to about $\pm 2°$ to minimise cross coupling errors. This results in a maximum cross axis coupling error of sin 2°, that is approximately 3.3% of any acceleration acting along an axis normal to the input axis. The acceleration input range thus determines the spring stiffness, or rate, and hence the bandwidth of the sensor. For example, consider an accelerometer with a maximum input range of say $\pm 10g$ and an undamped natural frequency of 25 Hz. The same basic accelerometer designed for an input range of $\pm 3g$ would have a bandwidth of only 14 Hz approximately, being reduced by a factor $\sqrt{3/10}$.

5.2.4.3 Torque balance pendulous accelerometer

The above shortcomings in terms of accuracy and bandwidth can be overcome by closed loop torque balance operation of the accelerometer. The deflection of the pendulum from its null position under an input acceleration is sensed by the position pick off and the pick off signal after amplification and suitable dynamic compensation is used to control a precision torque motor to maintain the pendulum at or very near to its null position. Measurement of the torque exerted by the torque motor to balance the inertia torque resulting from accelerating the pendulous mass so that it has the same acceleration as the vehicle then enables the acceleration to be determined. (Strictly speaking, the specific force is measured.) The operating principle is basically similar to the torque balance operation of the rate gyro version of the DTG described earlier. Pulse torque operation of the caging servo loop is frequently used, each pulse representing a velocity increment.

The construction of a typical torque balance pendulous accelerometer is shown schematically in Fig. 5.17. The pendulum consists basically of a beam which is suspended within the case by a very low stiffness flexural hinge. A very sensitive position pick-off detects the deflection of the pendulum from its null position. An inductive pick-off is shown in the diagram, however, capacitive pick-offs are also widely used. Torques are applied by means of a moving coil/permanent magnet

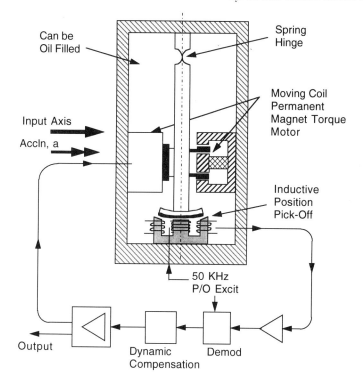

Fig. 5.17 Torque balance pendulous accelerometer schematic.

force motor (basically similar to a loud-speaker driver) with the coils fixed to the beam and the permanent magnets to the case. Thin flexible conducting ligaments enable electrical connections to be made to the coils.

Inertial quality pendulous accelerometers are frequently filled with a viscous oil which fills the cavity between the pendulum and the case. The oil filling has two advantages:

(a) Partial flotation of the pendulum increases the ruggedness of the instrument to shock and high vibration environments and enables a thinner flexural hinge of lower stiffness to be used for a given pendulosity. This reduces hysteresis effects in the hinge and bias effects due to residual spring torques caused by small shifts in the null position of the pick-off.

(b) Provides damping in the torque balance loop.

The disadvantages are in terms of additional cost of manufacture in areas such as seals, expansion bellows, avoidance of bubbles during the oil filling process, etc.

'Dry' torque balance pendulous accelerometers, generally of slightly lower accuracy, are also widely used exploiting techniques such as fabricating the pendulum and hinge from fused quartz by an etching process. Capacitive pick-offs are used to measure the pendulum deflection from null and electro-magnetic torque motors maintain the pick-off at null through the 'capture' or caging amplifier.

'Solid state' accelerometers fabricated from silicon using semiconductor manufacturing technology are now being widely used for lower accuracy applications. The new technology offers very small size and low manufacturing costs. Figure 5.18 illustrates the construction of a 'solid state' silicon accelerometer. The capacitive pick-off senses the deflection of the pendulum under acceleration and electrostatic forces are exerted to maintain the pendulum near the null position by applying dc voltages to the case mounted electrodes by means of a torque balance loop.

The torque balance loop of an electro-magnetic torque balance type of pendulous accelerometer is illustrated in block diagram form in Fig. 5.19.

The accelerometer loop gain must be sufficiently high to keep the pendulum excursions from the null position very small under the vibration environment. Significant rectification errors can otherwise be introduced as the accelerometer will sense a component of the vibration at right angles to its input axis which is proportional to the pendulum displacement from null. A simplified explanation of this is set out below.

Assume for simplicity (and worst case) that the vibration is at 45° to the

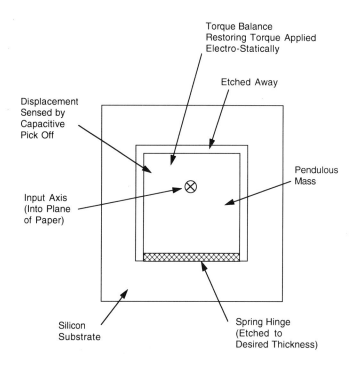

Fig. 5.18 'Solid state' accelerometer construction.

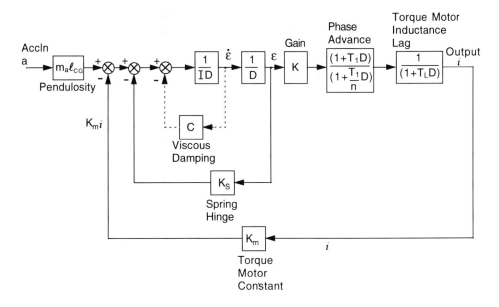

Fig. 5.19 Torque balance accelerometer loop.

accelerometer input axis so that the along and cross axis vibration components are equal and are given by:

$$a = a_{max} \sin \omega t$$

where a_{max} is the peak value of the vibration acceleration, a, and ω is the angular frequency.

The displacement of the pendulum from null is given by:

$$\varepsilon = \frac{a_{max}}{K} \sin(\omega t + \Phi)$$

where K is dependent on the loop gain at frequency ω and Φ is the phase shift between ε and a.

Sensed cross axis acceleration $= \varepsilon \cdot a_{max} \sin \omega t$

$$= \frac{a_{max}^2}{K} \sin \omega t \cdot \sin(\omega t + \phi)$$

The product term $\left[\sin \omega t \cdot \sin(\omega t + \phi) \right]$ has an average value of $\dfrac{\cos \Phi}{2}$ so that a proportion of the input vibration is rectified. This error source contributes to the

'g^2 error sensitivity' of the accelerometer as it is proportional to the square of the vibration acceleration.

5.3 ATTITUDE/HEADING REFERENCE SYSTEMS

5.3.1 Introduction

The measurement of the aircraft's attitude with respect to the horizontal plane in terms of the pitch and bank angles and its heading, that is the direction in which it is pointing in the horizontal plane with respect to North, is essential.

This information is vital for the pilot in order to fly the aircraft safely in all weather conditions, including those when the normal visibility of the horizon and landmarks is poor or not available, for example in haze or fog conditions, flying in cloud and night flying. Attitude and heading information is also essential for the key avionic systems which enable the crew to carry out the aircraft's mission. These systems include the autopilot system (eg 'Attitude and Heading Hold' modes, 'Auto-land' etc.), navigation system and the weapon aiming system. The information is also required for pointing radar beams and infra-red sensors. The accurate measurement of the attitude of the accelerometer input axes with respect to a set of inertial axes, that is an axis set which points in a fixed direction in space, is also fundamental to inertial navigation.

The spatial attitude of an aircraft is specified by the three Euler angles which are defined in Chapter 3, Section 3.4.2. Referring to Fig. 3.17, the aircraft is first rotated in the horizontal plane through the heading (or azimuth) angle, ψ, then rotated about the pitch axis through the pitch angle, θ, and then rotated about the roll axis through the bank angle, Φ. The rotations must be made in that order as they are non-commutative, that is a different orientation would be obtained if the rotations were made in a different order.

There are two basic inertial mechanisations which are used to derive the Euler angles to the required accuracy. The first method, which is known as a *stable platform* system, has the gyros and accelerometers mounted on a platform which is suspended in a set of gimbals. The gyros then control the gimbal servos so that the platform maintains a stable orientation in space irrespective of the aircraft manoeuvres (refer to Fig. 5.4). Angular position pick-offs on the gimbals then provide a direct read-out of the Euler angles.

The second method is known as a *strap-down* system as it has the gyros and accelerometers mounted on a rigid frame or block which is directly fixed, that is 'strapped-down', to the airframe. The gyros and accelerometers thus measure the angular and linear motion of the aircraft with respect to the aircraft's body axes. The Euler angles are then computed from the body rate information by the system computer.

It should be noted that a stable platform system and a strap-down system are mathematically equivalent systems. The function of the mechanical gimbals is carried out by the strap-down system computer which in effect contains a mathematical model of the gimbals.

Sections 5.3.2 and 5.3.3 cover the derivation of the aircraft's attitude by means of stable platform and strap-down gyro systems respectively and are equally applicable to an AHRS or an IN system.

The gyros measure the aircraft's angular motion with respect to inertial axes and it is therefore necessary to convert this information so that it is with respect to Earth referenced axes. The initial alignment and methods of monitoring a medium accuracy AHRS using gyros in the 0.3° to 10°/hour drift bracket are discussed in Section 5.3.4. More accurate AHRS using gyros in the 0.1°/hour or better performance bracket exploit Schuler tuning, however, and this is covered in the next chapter under Inertial Navigation.

5.3.2 Stable platform systems

A typical gimbal suspended stable platform is shown schematically in Fig. 5.20. The four gimbals are required to give full manoeuvring capability, in particular vertical or near vertical pitch attitudes and looping manoeuvres. The fourth gimbal prevents what is known as a 'gimbal lock' condition when the aircraft roll axis approaches coincidence with the platform azimuth axis. Without the fourth gimbal, any slight yawing motion would cause the platform to 'topple' as it would have no axis of freedom.

Typical gimbal freedoms are:-

Azimuth axis	Unlimited
Inner roll axis	± 20°
Pitch axis	Unlimited
Outer roll axis	Unlimited

Three single degree of freedom gyros, or 2 two degree of freedom gyros are used to stabilise the platform. Fig. 5.20 shows 2 two axis gyros such as the DTG described in Section 5.2.3.2.

The *vertical gyro* is mounted on the platform with its spin axis parallel to the platform azimuth axis which is normally aligned to the local vertical. The gyro input axes are thus normally in the horizontal plane so the gyro senses angular movements of the platform from the horizontal. The output angle signals, ε_X and ε_Y, from the vertical gyro are used to control the pitch and inner roll gimbal servos after first resolving ε_X and ε_Y through the azimuth angle, ψ, so as to maintain the correct feedback as the heading changes, *viz*:

Pitch gimbal servo error = $\varepsilon_X \cos \psi - \varepsilon_Y \sin \psi$

Inner roll gimbal servo error = $\varepsilon_X \sin \psi + \varepsilon_Y \cos \psi$

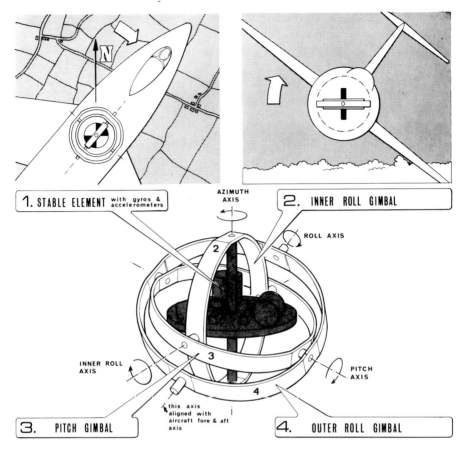

Fig. 5.20 Stable platform schematic.

These gimbal servos maintain the servo error signals near null so that the platform stays horizontal irrespective of the aircraft manoeuvres.

The *azimuth gyro* is mounted with its spin axis orthogonal to the vertical gyro spin axis and is thus normally in the horizontal plane. One input axis is aligned parallel to the azimuth axis so that it senses angular movements about the azimuth axis and the other input axis is aligned at 45° to the vertical gyro input axes. The azimuth output signal from the gyro is used to control the azimuth servo which maintains the signal near null so that the platform stays pointing in a fixed direction in the horizontal plane. The output signal from the '45°' input axis is used to cage the gyro about this axis by feeding it back after amplification to the gyro torque motor. The torque motor current of this 45° axis can be used to monitor the vertical gyro as the gyro measures a component of the angular motion about each input axis of the vertical gyro and so will detect any abnormal drift in the vertical gyro.

The outer roll gimbal servo motor is controlled from an angular position pick-off on the inner roll axis which measures the deviation from orthogonality between the inner roll and pitch gimbals. The outer roll servo loop operates to keep this pick-off signal at null so as to maintain the azimuth axes orthogonal to the pitch axis under all conditions and thus prevent 'gimbal lock' if the aircraft is vertical. eg during a loop manoeuvre. The operation of the fourth gimbal during a vertical manoeuvre is shown in Fig. 5.21.

Angular position pick-offs such as synchros or resolver synchros mounted coaxially with the gimbal axes measure the Euler angles directly.

High gain gimbal servo loops with suitable dynamic compensation. such as phase advance and integral of error terms, are used to maintain the gyro output angles to within a few seconds of arc irrespective of the aircraft manoeuvres. Typical servo bandwidths range from 50 Hz to 100 Hz depending on the platform size and inertia. It should be noted that the platform inertia assists the spatial stabilisation when the platform is subjected to high frequency disturbing torques as the platform behaves as a seismic mass and the spatial excursions become smaller and smaller as the frequency increases. The spatial stabilisation of the platform to within a few

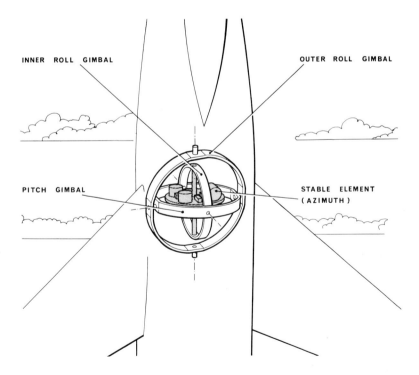

Fig. 5.21 Operation of fourth gimbal.

seconds of arc is necessary to minimise the effects of airframe conical motion which would otherwise result in significant gyro drift due to the rectification of this vibration.

5.3.3 Strap-down systems

Figure 5.22 illustrates the strap-down AHRS or INS concept. Such a mechanisation eliminates the mechanical complexity of the stable platform with its gimbals, gimbal bearings, slip-rings, gimbal servo motors, synchros etc. The reliability and cost of ownership are greatly improved, power consumption is lower and the weight and volume reduced. Hence the attractions of the strap-down configuration.

The Euler angles can be derived from the body rates of rotation measured by the strap-down gyros by using suitable algorithms.

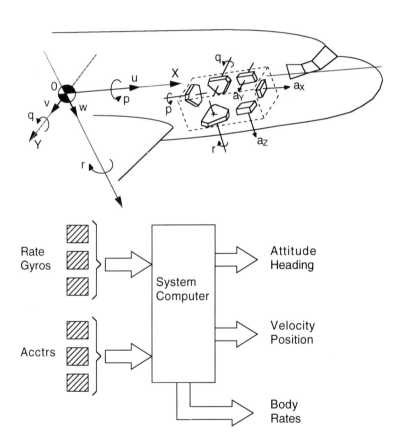

Fig. 5.22 Strap-down system schematic.

5.3.3.1 Attitude algorithms

The Euler angles, Φ, θ, ψ are defined in Chapter 3, Section 3.4.2 and the equations relating the body angular rates p, q, r to the Euler angle rates $\dot{\Phi}$, $\dot{\theta}$, $\dot{\psi}$ are derived and comprise equations (3.16), (3.17), (3.18).

Equations for $\dot{\Phi}$, $\dot{\theta}$, $\dot{\psi}$ can be derived from the above equations by suitable algebraic manipulation and are set out below:

$$\dot{\Phi} = p + q\sin\Phi\tan\theta + r\cos\Phi\tan\theta \qquad (5.20)$$

$$\dot{\theta} = q\cos\Phi - r\sin\Phi \qquad (5.21)$$

$$\dot{\psi} = q\sin\Phi\sec\theta + r\cos\Phi\sec\theta \qquad (5.22)$$

These equations can be expressed more compactly in matrix form.

$$\begin{bmatrix} \dot{\Phi} \\ \dot{\theta} \\ \dot{\psi} \end{bmatrix} = \begin{bmatrix} 1 & \sin\Phi\tan\theta & \cos\Phi\tan\theta \\ 0 & \cos\Phi & -\sin\Phi \\ 0 & \sin\Phi\sec\theta & \cos\Phi\sec\theta \end{bmatrix} \begin{bmatrix} p \\ q \\ r \end{bmatrix} \qquad (5.23)$$

The Euler angles can then be derived from the Euler angle rates by a process of integration using as initial conditions a known attitude at a given point in time. This process becomes meaningless, however, at $\theta = 90°$ when $\tan\theta$ and $\sec\theta$ become infinite. (This is the mathematical equivalent of 'gimbal lock'.) The use of the three-parameter Euler algorithms is therefore generally limited to pitch angles between $\pm 30°$ as the error equations are unbounded, and to avoid mathematical singularities.

A fully manoeuvrable system is therefore required where there are no restrictions on the pitch angles. The limitations of the three-parameter Euler system are overcome by the use of what is known as the *Euler four symmetrical parameters* to define the vehicle attitude.

It can be shown that an axis set may be moved to any required orientation by a single rotation about a suitably positioned axis. Let this axis make angles $\cos^{-1}\alpha$, $\cos^{-1}\beta$ and $\cos^{-1}\gamma$ with the inertial axes OX_0, OY_0 and OZ_0 respectively. Let a single rotation, μ, about this axis bring a moving axis set from OX_0, OY_0, OZ_0 into coincidence with OX, OY, OZ, the set whose orientation it is desired to specify.

The Euler four symmetrical parameters are given by

$$
\left.
\begin{aligned}
e_0 &= \cos \mu / 2 \\
e_1 &= \alpha \sin \mu / 2 \\
e_2 &= \beta \sin \mu / 2 \\
e_3 &= \gamma \sin \mu / 2
\end{aligned}
\right\}
\tag{5.24}
$$

Then e_0, e_1, e_2, e_3 can be used to specify the attitude of the vehicle with respect to OX_0, OY_0, OZ_0.

The following relationships with the Euler angles can be derived

$$
\sin \theta = 2(e_0 e_2 - e_3 e_1)
\tag{5.25}
$$

$$
\tan \psi = \frac{2(e_0 e_3 + e_1 e_2)}{e_0^2 + e_1^2 - e_2^2 - e_3^2}
\tag{5.26}
$$

$$
\tan \Phi = \frac{2(e_0 e_1 + e_2 e_3)}{e_0^2 - e_1^2 - e_2^2 + e_3^2}
\tag{5.27}
$$

It can be shown that

$$
\begin{bmatrix} \dot{e}_0 \\ \dot{e}_1 \\ \dot{e}_2 \\ \dot{e}_3 \end{bmatrix} = \frac{1}{2}
\begin{bmatrix}
-e_1 & -e_2 & -e_3 \\
e_0 & -e_3 & e_2 \\
e_3 & e_0 & -e_1 \\
-e_2 & e_1 & e_0
\end{bmatrix}
\begin{bmatrix} p \\ q \\ r \end{bmatrix}
\tag{5.28}
$$

Because four parameters are being used to describe the orientation when only three are necessary, a constraint equation exists of the form

$$
e_0^2 + e_1^2 + e_2^2 + e_3^2 = 1
\tag{5.29}
$$

These two equations have great advantages over the equivalent Euler angle equations.

(1) They apply to all attitudes.
(2) The error equations are bounded by the constraint equation.
(3) The numerical value of each parameter always lies in the range −1 to +1, so easing the scaling problems in the computing mechanisation.

Equation (5.28) can be re-arranged into the form

$$
\begin{bmatrix} \dot{e}_0 \\ \dot{e}_1 \\ \dot{e}_2 \\ \dot{e}_3 \end{bmatrix} = \frac{1}{2} \begin{bmatrix} 0 & -p & -q & -r \\ p & 0 & r & -q \\ q & -r & 0 & p \\ r & q & -p & 0 \end{bmatrix} \begin{bmatrix} e_0 \\ e_1 \\ e_2 \\ e_3 \end{bmatrix}
\tag{5.30}
$$

This can be written more compactly as

$$
\dot{\mathbf{X}} = \mathbf{AX}
$$

Where $\mathbf{X} = \begin{bmatrix} e_0 \\ e_1 \\ e_2 \\ e_3 \end{bmatrix}$ and $\mathbf{A} = \frac{1}{2} \begin{bmatrix} 0 & -p & -q & -r \\ p & 0 & r & -q \\ q & -r & 0 & p \\ r & q & -p & 0 \end{bmatrix}$

(The bold capital letters denote that $\mathbf{X}$ and $\mathbf{A}$ are matrices).

This equation can be solved by approximate integration techniques by assuming p, q, r are constant over a short period of time Δt from time t_n to time t_{n+1}.

The predicted value of $\mathbf{X}$ at time t_{n+1}, $\mathbf{X}_{n+1}$ is given by

$$
\mathbf{X}_{n+1} = \mathbf{X}_n + \dot{\mathbf{X}}_n \Delta t = \mathbf{X}_n + \mathbf{AX}_n \Delta t
$$

ie

$$
\mathbf{X}_{n+1} = (1 + \mathbf{A}\Delta t)\mathbf{X}_n
$$

(**1** is the unity matrix)

$(1 + \mathbf{A}\Delta t)$ is in fact an approximation to the *transition matrix* which relates the value of the state vector at time t_{n+1} to the value at time t_n.

The incremental angular rotations measured about the roll, pitch and yaw axes are denoted by ΔP, ΔQ, ΔR respectively. (The pulse torqued DTG, the RLG and the IFOG all function as integrating rate gyros.)

$$
\Delta P = \int_{t_n}^{t_{n+1}} p\,dt = p\Delta t
$$

$$
\Delta Q = \int_{t_n}^{t_{n+1}} q\,dt = q\Delta t
$$

$$
\Delta R = \int_{t_n}^{t_{n+1}} r\,dt = r\Delta t
$$

Hence the approximate transition matrix is

$$
\begin{bmatrix}
1 & 0 & 0 & 0 \\
0 & 1 & 0 & 0 \\
0 & 0 & 1 & 0 \\
0 & 0 & 0 & 1
\end{bmatrix}
+ \tfrac{1}{2}
\begin{bmatrix}
0 & -\Delta P & -\Delta Q & -\Delta R \\
\Delta P & 0 & \Delta R & -\Delta Q \\
\Delta Q & -\Delta R & 0 & \Delta P \\
\Delta R & \Delta Q & -\Delta P & 0
\end{bmatrix}
$$

Hence

$$
\begin{bmatrix}
e_0 \\
e_1 \\
e_2 \\
e_3
\end{bmatrix}_{t_{n+1}}
=
\begin{bmatrix}
1 & -\Delta P/2 & -\Delta Q/2 & -\Delta R/2 \\
\Delta P/2 & 1 & \Delta R/2 & -\Delta Q/2 \\
\Delta Q/2 & -\Delta R/2 & 1 & \Delta P/2 \\
\Delta R/2 & \Delta Q/2 & -\Delta P/2 & 1
\end{bmatrix}
\begin{bmatrix}
e_0 \\
e_1 \\
e_2 \\
e_3
\end{bmatrix}_{t_n}
\tag{5.31}
$$

The performance of the numerical integration algorithm can be further improved by using second order Runge-Kutta algorithms or fourth order Runge-Kutta algorithms where very high accuracy is required, and by decreasing the integration time increment, Δt with a more powerful computer.

The constraint equation $e_0^2 + e_1^2 + e_2^2 + e_3^2 = 1$ is used to correct the transition matrix for the accumulated computational errors in the integration process to maintain orthogonality of the computed axes.

The 4 symmetrical Euler parameters are also mathematical *Quaternions* as it can be shown that they are made up of the sum of a scalar quantity and a vector with orthogonal components.

The values of the Euler angles ψ, θ and Φ are then calculated from equations (5.26), (5.25) and (5.27).

The process is shown in block diagram form in Fig. 5.23.

Worked example on attitude algorithm. A simple worked example is set out below to show how the attitude is derived from the Euler four parameter transition matrix to give a practical appreciation of the maths involved in evaluating the matrix for those readers less familiar with matrices.

Assume at time t_n the values of the four Euler parameters are:-

$$
\begin{bmatrix}
e_0 \\
e_1 \\
e_2 \\
e_3
\end{bmatrix}
=
\begin{bmatrix}
0.9 \\
0.3 \\
0.1 \\
0.3
\end{bmatrix}
$$

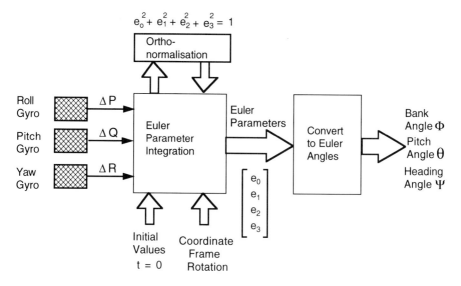

Fig. 5.23 Attitude derivation.

(Check $e_0^2 + e_1^2 + e_2^2 + e_3^2 = 0.81 + 0.09 + 0.01 + 0.09 = 1$)

The values of the body angular rates are:-

$p = 1$ rad/s	(57.3°/s)
$q = 0.04$ rad/s	(2.3°/s approx)
$r = 0.05$ rad/s	(2.9°/s approx)
iteration period	$\Delta t = 0.01$ s
Whence	$\Delta P = 0.01$ rad
	$\Delta Q = 0.0004$ rad
	$\Delta R = 0.0005$ rad

The Euler angles at time t_n are calculated using equations (5.25), (5.26) and (5.27) and substituting the appropriate values of e_0, e_1, e_2, e_3

from which
$$\begin{cases} \psi = 36.869898° \\ \theta = 0° \\ \Phi = 36.869898° \end{cases}$$

The transition matrix =

$$
\begin{bmatrix}
1 & -\Delta P/2 & -\Delta Q/2 & -\Delta R/2 \\
\Delta P/2 & 1 & \Delta R/2 & -\Delta Q/2 \\
\Delta Q/2 & -\Delta R/2 & 1 & \Delta P/2 \\
\Delta R/2 & \Delta Q/2 & -\Delta P/2 & 1
\end{bmatrix}
=
\begin{bmatrix}
1 & -0.005 & -0.0002 & -0.00025 \\
0.005 & 1 & 0.00025 & -0.0002 \\
0.0002 & -0.00025 & 1 & 0.005 \\
0.00025 & 0.0002 & -0.005 & 1
\end{bmatrix}
$$

hence at time t_{n+1}

$$e_0 = 1\,(0.9) - 0.005\,(0.3) - 0.0002\,(0.1) - 0.00025\,(0.3)$$

ie $e_0 = 0.898405$

Evaluating the updating matrix yields

$$
\begin{bmatrix} e_0 \\ e_1 \\ e_2 \\ e_3 \end{bmatrix}_{t_{n+1}}
=
\begin{bmatrix} 0.898405 \\ 0.304465 \\ 0.101605 \\ 0.299785 \end{bmatrix}
$$

Check $e_0^2 + e_1^2 + e_2^2 + e_3^2 = 1.000025$

The small error in the constraint equation is due to using a first order approximate solution.

The Euler angles at time t_{n+1} are:-

$$\psi = 36.918525°$$
$$\theta = 0.000963°$$
$$\Phi = 37.442838°$$

The process is repeated at each iteration period.

5.3.3.2 Digital processing of attitude algorithms

The strap-down system accuracy is critically dependent on the accuracy of the digital processing involved in the numerical solution of the attitude algorithms. Computer performance was in fact a major limitation in the implementation of strap-down systems until the advent of high speed microprocessors.

The major error sources in the digital processing are briefly discussed below, together with the methods used to minimise these errors.

(1) Commutation. Errors arise in digital processing because of the non-commutativity of angular motion. A different attitude is attained if a series of rotations about the body axes are made in a different order. Commutation errors are introduced in the numerical processing because for each solution update of the transformation matrix, the processor operates sequentially on the input angular rotation increments in a fixed order. This may not correspond to the actual time

ordered sequence in which the angles were accumulated in the vehicle. Commutation errors can thus arise in deriving the vehicle attitude particularly when subjected to cyclical or repetitive inputs such as conical motion, which can result in a rectification drift of the transformation matrix. These commutation errors can be minimised by decreasing the angle increment size involved in each computation. Coning motion is briefly covered in the next section.

(2) Integration. Errors arise in the integration because of using discrete approximate solutions of a continuous process and equations which have no analytic solution. These errors are minimised by using sophisticated integration algorithms (eg Runge Kutta fourth order) and by increasing the update rate.

(3) Round off. These errors are due to the finite resolution of data and rounding off the computations to the value of the least significant bit. This causes a random walk error build up in multiple computations. Double precision working (32 bit) for critical operations is used to minimise these errors.

(4) Quantisation. These errors result from the process of converting the analog output of the sensors into discrete increments which can be input as a series of pulses into the digital computer. This quantisation process causes the input to the computer to always lag the outputs of the sensors. The average value of this quantisation (or sensor storage error) is one half of the quantisation angle. This information is not lost but received in the next update cycle and results in a quantisation noise in the output that needs to be modelled for the best performance. These errors are minimised by decreasing the quantisation angle and mathematical modelling.

5.3.3.3 Coning motion

A brief explanation of coning motion is set out below in view of its importance as a possible source of error. Coning motion results from the vehicle being subjected to angular oscillations of the same frequency about two orthogonal axes with a 90° phase difference between the two motions. The motion of the third orthogonal axis of the vehicle describes the surface of a cone – hence the term conical motion (see Fig. 5.24).

Suppose the motion measured by the gyros is as follows:

$$p = 0$$
$$q = A\omega \cos \omega t$$
$$r = A\omega \sin \omega t$$

referring to equation (5.21)

$$\dot{\theta} = q\cos\Phi - r\sin\Phi$$

assuming the steady state Euler angles are all zero and θ and Φ are small angles

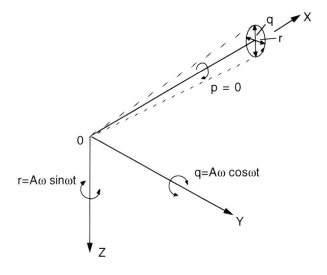

Fig. 5.24 Conical motion.

$$\dot{\theta} = q = A\omega\cos\omega t$$

integrating yields

$$\theta = A \sin \omega t$$

referring to equation (5.20)

$$\dot{\Phi} = p + q \sin\Phi\tan\theta + r\cos\Phi\tan\theta$$

For small angles sin $\Phi = \Phi$, tan $\theta = \theta$, cos $\Phi = 1$ approximately and neglecting second order terms

$$\dot{\Phi} = r\theta = A\omega\sin\omega t \cdot A\sin\omega t$$

ie
$$\dot{\Phi} = \frac{A^2\omega}{2}(1 - \cos 2\omega t) \tag{5.32}$$

hence, average value of rectified rotation rate, $\dot{\Phi} = A^2\omega/2$ \qquad (5.33)

The effect of conical motion about two orthogonal axes of an angular momentum type rate gyro produces a rectification drift due to the gimbal excursions from the null position because of the bandwidth limitations in the torque balance loops. In such cases it may be necessary to compute the coning motion and apply a correction to the sensor outputs.

In the case of a stable platform system, it is possible to isolate the gyros from airframe motion by suitably designed high performance gimbal servo loops coupled with the inherent stabilisation conferred by the inertia of the stable element. The degree of isolation achieved is designed to reduce the coning rectification drift to negligible proportions.

It should be stressed that the errors introduced by coning motion in strap-down systems arise essentially from errors in computing the motion profile due to the non-commutativity of angular rotations.

For example, a circular coning motion resulting from angular motions of 1° amplitude and 1 Hz frequency measured by gyros with a scale factor error of 0.01% results in a rectified drift rate equal to

$$\frac{0.01}{100} \times \frac{1}{2} \times \left(\frac{1}{57.3}\right)^2 \times 2\pi \times 1 \times 57.3 \times 60 \times 60$$

degrees/hour, that is 0.02°/hour, which is significant in an INS.

5.3.4 Alignment and monitoring

5.3.4.1 Introduction

The aircraft *Pitch* and *Bank* angles are required with respect to the local level plane, that is a plane which is normal to the local vertical, defined as a line through the aircraft to the Earth's centre.

The aircraft *Heading* angle is generally required with respect to true North, that is the direction of the local meridian pointing towards the North pole. (The local meridian being a circle round the Earth which passes through the North and South Poles and the aircraft's present position.)

It is therefore necessary to convert the gyro data, which is with respect to inertial axes, to an Earth referenced axis frame.

There are three basic directional references which are used to align an AHRS or INS. These comprise:-

(1) Earth's gravitational acceleration vector – This is sensed by the accelerometers and enables the local vertical to be determined.
(2) Earth's angular velocity vector – This can be measured by the gyros, provided they are of sufficient accuracy, and enables the aircraft heading to be determined by a process known as 'gyro compassing'. It will be shown that a heading accuracy of 0.1°, however, requires the gyro bias uncertainty to be less

than 0.01°/hour. The technique is therefore not suitable for AHRS using lower accuracy gyros.
(3) Earth's magnetic field – This enables a heading reference to generally within 0.7° accuracy to be established at latitudes below 60° North or South of the equator.

Figure 5.25 illustrates the basic Earth reference axis frame, generally refered to as a 'local level, North slaved axis frame' or 'North, East, Down axes – NED'.

The angular rate corrections for the Earth's rotation in order to maintain a local vertical reference and heading reference are explained in Section 5.3.4.2.

It is also necessary to correct for the angular rotation of the local vertical as the aircraft flies over the earth's surface because the Earth is spherical. These corrections which are known as the vehicle rate corrections are explained in Section 5.3.4.3.

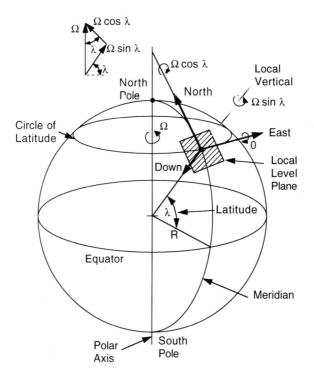

Fig. 5.25 Earth rotation components – local level, North slaved axes.

5.3.4.2 Angular rate corrections for the earth's rotation

Figure 5.25 and accompanying inset diagram shows the Earth's angular velocity vector, $\underline{\Omega}$, can be resolved into two components.

(a) Component along the North pointing axis = $\Omega \cos \lambda$
     Where λ = latitude angle at the aircraft's position.
     This correction must be applied to the *Vertical Gyro* to precess it about the North pointing axis in the case of a stable platform system. In the case of a strap-down system the correction is applied within the computer. As an example, at a latitude of 51°30′N (London area), the Earth's rate component about the North axis = 15 cos 51°30′ degrees per hour = 9.3377°/hour (Ω = 15°/hour).
(b) Component about the local vertical axis = $\Omega \sin \lambda$
     This correction must be applied to the *Azimuth Gyro* in the case of a stable platform system, if it is desired to maintain the platform pointing North.

In the case of a strap-down system the correction is applied within the computer.

It should be noted that there is zero component of the Earth's angular velocity about the East pointing axis and this forms the basis of the gyro compassing alignment technique which is explained later.

5.3.4.3 Vehicle rate corrections

Figure 5.26 illustrates how the local vertical rotates in space as the aircraft flies over the surface of the Earth because the Earth is spherical. The angular rates of rotation of the local vertical with respect to inertial axes are equal to:

V_N/R about the East pointing axis, where V_N = northerly component of the aircraft's velocity and R = radius of the Earth.

V_E/R about the North pointing axis, where V_E = easterly component of the aircraft's velocity.

These angular rates are referred to as the vehicle rates. It is therefore necessary to precess the vertical gyro about its East pointing and North pointing axes at angular rates of V_N/R and V_E/R respectively in order to maintain the platform locally level.

In the case of a strap-down system the vehicle rate corrections are made within the computer.

The magnitude of these vehicle rate correction terms is more easily appreciated by expressing the aircraft's velocity in knots (nautical miles/hour) bearing in mind 1 nautical mile is equal to 1 minute of arc at the Earth's surface.

For example, an aircraft with a ground speed of 600 knots and a track angle of 030° has a northerly velocity component V_N = 600 cos 30°, that is 519.615 knots and an easterly velocity component V_E = 600 sin 30°, that is 300 knots.

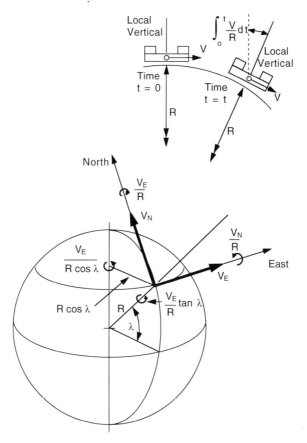

Fig. 5.26 Vehicle rate components – local level, North slaved axes.

The vertical gyro must therefore be precessed at a rate equal to $\dfrac{519.615}{60}$ that is 8.66°/hour, about the East axis and $\dfrac{300}{60}$ that is 5°/hour about the North axis in order to maintain the platform locally level.

An external source of velocity such as a Doppler radar, or less accurately air data, can be used to derive the northing and easting velocity components for the vehicle rate correction terms. The Doppler or air data derived velocity components would generally be combined with the inertially derived velocity components from the accelerometers as explained later.

The use of very high accuracy gyros and accelerometers enables the velocity components to be derived directly from the accelerometers and fed back to correct the vertical reference so as to form a Schuler tuned system, and is explained in the next chapter.

There is also a vehicle rate correction, $(V_E/R) \tan \lambda$, about the local vertical (down) axis which must be applied to the azimuth gyro in order to maintain the north reference. Figure 5.26 also shows the three vehicle rate correction terms V_N/R, V_E/R, $(V_E/R) \tan \lambda$. The $(V_E/R) \tan \lambda$ term is derived by resolving the vehicle rate component about the earth's polar axis $V_E/(R \cos \lambda)$, through the latitude angle.

viz
$$\frac{V_E}{R \cos \lambda} \cdot \sin \lambda = \frac{V_E}{R} \tan \lambda$$

It can be seen that the $(V_E/R) \tan \lambda$ term increases rapidly at high latitudes and becomes infinite at the poles when $\lambda = 90°$. A 'free azimuth' system (ie no corrections made) is used to overcome this limitation at high latitudes ($>75°$). This is explained in the next chapter.

5.3.4.4 Vertical monitoring

As mentioned in the introduction, the basic vertical reference is the Earth's gravitational vector. The platform can thus be initially aligned locally level using the platform mounted accelerometers to measure the tilt errors from the local vertical. The accelerometers act in the sense of a plumb bob. This assumes the process is carried out when the aircraft is stationary on the ground, or in straight and level flight at constant speed and the aircraft is not accelerating. However, any aircraft accelerations would result in the accelerometers defining a deflected or 'dynamic' vertical because the accelerometers are unable to distinguish between the gravity component and an acceleration component.

An ideal vertical reference system using gyros with negligible drift or bias uncertainties once initially aligned with the local vertical would continue to measure the aircraft's attitude with respect to the local vertical using the gyros alone, assuming the Earth's rate and vehicle rate corrections were applied without error. In practice, however, gyro drift and errors in applying the necessary correction terms would cause a progressive build up in the vertical error unless some means for monitoring the vertical error and applying correcting feedback were incorporated. In fact a compromise between accelerometer measurement and gyro measurement of the vertical is always used.

Figure 5.27 illustrates a simple gravity monitoring system whereby the accelerometer outputs are used to precess the vertical gyro so as to reduce the tilt errors to near zero. The system is also referred to as a 'gravity erection system' as it erects the platform to an upright position. The system is a simple first order system, one axis only being considered for simplicity. Assuming the aircraft is stationary on the ground so that there are zero acceleration components, the platform will erect exponentially to the local vertical from an initial tilt with a time constant, $T = 1/Kg$ seconds. Where $K =$ gyro precession rate (rad/sec) per m/s^2 acceleration.

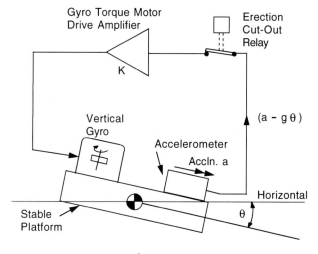

Platform Precession Rate = $\dot{\theta}$ = K(a - g θ)

$$(1 + TD)\,\theta = a/g \qquad \text{where } T = 1/Kg$$

If a = 0 and $\theta = \theta_0$ at time t = 0 , $\theta = \theta_0\, e^{-t/T}$

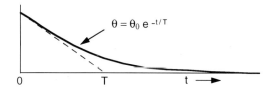

Fig. 5.27 Simple gravity erection system (one axis only shown for clarity).

The effect of gyro drift, W, is to produce a steady state tilt error equal to WT to back off this drift. The steady state accuracy is thus determined by the gyro drift and the erection time constant. For example, a requirement for a steady state tilt error of less than 0.2° and a vertical gyro drift of 4°/hour (1/900°/s) requires an erection time constant

$$T = \frac{0.2}{1/900} = 180 \text{ s}$$

A constant aircraft acceleration, a, initiated when the platform was initially level would cause the platform to erect exponentially to the dynamic vertical which is at an angle $\tan^{-1} a/g$ to the local vertical. Hence, for the above example, a steady acceleration of $0.2g$ ($12°$ dynamic vertical error) would result in a vertical error after 60 seconds of $12(1 - e^{-60/180})$ degrees, that is $3.4°$.

An erection cut-out system is used to minimise the acceleration induced errors by cutting out the gravity monitoring system when the aircraft acceleration exceeds a given threshold, say $0.1g$.

The monitoring is thus cut out during manoeuvres exceeding $0.1g$ and the vertical gyro allowed to drift. A low gyro drift and a long erection time constant are thus required. This type of simple gravity monitoring system is used on older types of stable platform AHRS and vertical gyro units of the two axis free gyro type which are still used as standby, or reversionary, vertical reference systems. Low cost liquid level sensors are generally used in these systems in place of conventional accelerometers. The electrolytic level sensors function like a spirit level, and provide a voltage output which is proportional in magnitude and sign to the tilt angle; they are in fact simple accelerometers.

The errors in the vertical reference resulting from the effects of accelerations during manoeuvres can be virtually eliminated by the use of Schuler tuning. This, however, requires high accuracy gyros with very low drift rates and accelerometers with very small bias errors. Schuler tuning is explained in the next chapter.

An alternative method of monitoring the vertical reference which enables lower performance and hence lower cost gyros and accelerometers to be used in an AHRS is based on the use of an independent velocity source. The technique is known as *Doppler/inertial* or *air data/inertial* mixing depending on the source of the aircraft's velocity. The errors in the vertical reference resulting from the effects of accelerations during manoeuvres can be kept small with such a monitoring system, given an accurate velocity source. Small transient errors can be induced in the vertical reference due to changes in gyro drift rates or accelerometer bias errors. The gyros and accelerometers should thus be of reasonable accuracy although they can be up to two orders of magnitude below inertial standards.

The method uses the same techniques as used in IN systems to derive the aircraft velocity components from the AHRS accelerometers by integrating the suitably corrected accelerometer outputs with respect to time from known initial conditions.

The major source of error in the inertially derived velocity components arises from the vertical errors (or tilt errors) in the AHRS producing gravitational acceleration errors which are integrated with time. The inertially derived velocity components are, therefore, compared with the velocity components measured by the reference velocity system. The velocity differences are then fed back to correct the vertical errors in the attitude reference and the inertial velocity errors. Figures 5.28 and 5.29 illustrate the basic concepts.

The accelerometer outputs must be corrected for the gravitational acceleration components acting along their input axes. It is also necessary to correct the accelerometer outputs for the acceleration terms introduced by the rotation of the axis frame which is being used for comparing the respective velocities. In the case of a stable platform system the axis frame would typically be a North slaved, locally

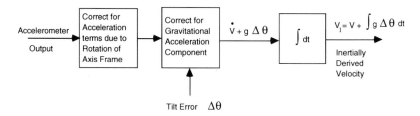

Fig. 5.28 Derivation of velocity from accelerometer outputs.

level set of axes. A strap-down AHRS using air data velocity monitoring would compare the velocity components along the aircraft's body axes. The rotation of the earth referenced axis frame introduces accelerations known as *Coriolis accelerations.*

The Coriolis accelerations arise because in order to travel in a straight line with respect to the earth's surface, which is itself rotating in space, the actual trajectory of the aircraft in space is curved so that an acceleration must be produced. The Coriolis acceleration is equal to $2V\omega$ where V is the vehicle velocity with respect to the rotating axes and ω is the rate of rotation of the reference axes in space and it is mutually at right angles to $\underline{V}$ and $\underline{\omega}$. For example, the value of the Coriolis acceleration component for an aircraft velocity component of 500 knots (250m/s approx.) and an earth's rotation component about the local vertical of 10°/hour (4.85×10^{-5} rad/s) is equal to 0.02425m/s^2. If uncorrected this would cause a tilt error of $0.14°$.

In order to derive the rate of change of the aircraft velocity components along the body axes from the strap-down accelerometer outputs (and hence the velocity

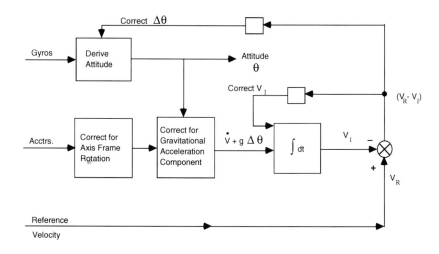

Fig. 5.29 Vertical monitoring using Doppler or air data/inertial velocity mixing.

components) it is necessary to correct for the following acceleration terms:

(i) Centrifugal acceleration terms due to the rotation of the body axes
(ii) Gravitational acceleration components
(iii) Coriolis acceleration terms due to the earth's rotation.

The computing mechanisation of the correction terms for an air data velocity monitored strap-down AHRS is covered later in the section.

Doppler velocity sensor systems provide very accurate velocity information over land with errors of around 0.1% in the ground speed measurement. The drift angle, that is the angle between the ground speed vector and the aircraft's fore and aft axis, is also measured to an accuracy of between 0.1° to 0.2°. Accuracy over water is degraded as it is influenced by complex sea state errors and typically involves a 4 knot uncertainty. A Doppler velocity monitored AHRS using gyros in the 0.3°/hour drift uncertainty bracket can, however, achieve a vertical accuracy of the order of 1 to 2 minutes of arc.

The use of the air data system to derive the aircraft's velocity components results in a somewhat lower vertical accuracy compared with a Doppler velocity monitored system as the air data velocity errors are of the order of 1%. An air data velocity monitored AHRS with gyros in the few degrees/hour uncertainty bracket would have a typical vertical accuracy of around 0.2° under favourable conditions.

Inertial velocity mixing exploits the technique of *complementary filtering* for combining the data from two independent sources of information. Other examples include:

- Combining a magnetic heading reference with a gyro heading reference
- Combining barometrically measured altitude or radar altimeter measured altitude, with inertially derived altitude.

Sophisticated mixing systems employing Kalman filters to control the gains of the mixing loops are frequently used, particularly in view of the computing power now available to implement such systems.

It is, therefore, appropriate at this point to discuss the basic principles of complementary filtering. A simple first order mixing system is used as an example to keep the maths relatively simple. The build up to the more powerful second order, third order, fourth order and above systems with variable mixing gains can then be appreciated.

In this simple example, the outputs from the two independent systems measuring a quantity θ are designated θ_1 and θ_2 respectively.

Their characteristics are:

θ_1 has a low noise content and a fast response to changes in θ. It is also subject to drift with time.

θ_2 has a high noise content but has good long term accuracy.

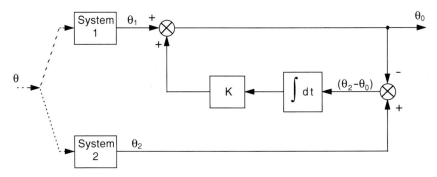

Fig. 5.30 Simple first order mixing.

A method of combining the two sources with a simple first order mixing system is shown in the block diagram in Fig. 5.30. It can be seen from Fig. 5.30 that:

$$\theta_0 = \theta_1 + K\int(\theta_2 - \theta_0)dt \tag{5.34}$$

from which

$$\theta_0 = \frac{TD}{1+TD}\theta_1 + \frac{1}{1+TD}\theta_2 \tag{5.35}$$

where $T = 1/K$

That is

$$\theta_0 = F_1(D)\theta_1 + F_2(D)\theta_2 \tag{5.36}$$

where $F_1(D) = \dfrac{TD}{1+TD}$ which is the transfer function of a first order high pass filter

(or 'wash-out' filter) and $F_2(D) = \dfrac{1}{1+TD}$ which is the transfer function of a first order low pass filter.

The output, θ_0, is thus the sum of two components:

(i) the quantity θ_1 is coupled through a simple high pass filter with a transfer function $F_1(D) = \dfrac{TD}{1+TD}$ which passes dynamic changes in θ without lag but has zero output at dc. A constant drift is, therefore dc blocked and has no effect on θ_0. A change in drift will, however, cause a transient change in θ_0.

(ii) the quantity θ_2 is coupled through a simple low pass filter with a transfer function $F_2(D) = \dfrac{1}{1+TD}$ which filters the noise present in θ_2 but preserves the dc accuracy.

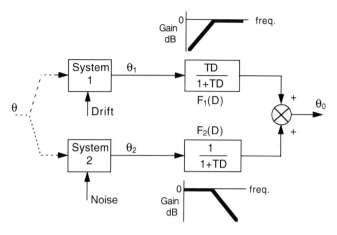

Fig. 5.31(a) Complementary filtering.

Figure 5.31(a) and (b) shows the basic concept and the response of the various elements of the system to a step input in θ.

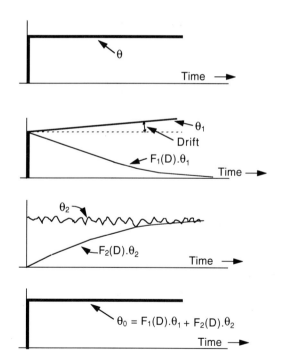

Fig. 5.31(b) Complementary filtering.

The output θ_0 thus reproduces the input θ from the sum of the two complementary filtered outputs of θ_1 and θ_2 without lag and with a low noise content.

A constant drift in θ_1 is dc blocked but will produce a steady state offset in θ_0 to back-off this drift. To eliminate this offset requires a parallel integral term leading to a second order system. Other errors such as bias errors may require a third integral term to be added to the system leading to a third order mixing system, and so on.

It can be seen with this simple example that the complementary filtered output is superior to either source on its own.

The derivation of the transfer functions for a second order mixing loop for a gyro/magnetic heading reference monitoring system is shown in Section 5.3.4.5, to show how a more complex mixing system can be analysed.

The mechanisation of an air data/inertial velocity mixing system to monitor the vertical reference of a strap-down AHRS is described below to show:

(a) The use of air data derived velocity as opposed to Doppler.
(b) The use of aircraft body axes as a reference frame of axes.
(c) The application to a lower accuracy AHRS with lower cost gyros in the few degrees/hour bias uncertainty performance bracket.

The basic stages involved in the vertical monitoring system are set out below.

Derivation of velocity components from the air data system The air data system provides outputs of true airspeed, V_T and the angle of attack/incidence, α, and the sideslip incidence angle, β.

Referring to Fig. 5.32, it can be seen that:-

$$U = V_T (1 + \tan^2 \alpha + \tan^2 \beta)^{-\frac{1}{2}} \tag{5.37}$$

$$V = V_T (1 + \tan^2 \alpha + \tan^2 \beta)^{-\frac{1}{2}} \tan \beta \tag{5.38}$$

$$W = V_T (1 + \tan^2 \alpha + \tan^2 \beta)^{-\frac{1}{2}} \tan \alpha \tag{5.39}$$

The air data derived forward velocity, U_A, sideslip velocity, V_A, and vertical velocity, W_A, are thus computed using equations (5.37), (5.38), (5.39) from the air data system outputs V_T α and β, the suffix A being used henceforth to denote the source of U, V, W (ie $U = U_A$, $V = V_A$, $W = W_A$)

Inertial derivation of forward velocity, and side-slip velocity The aircraft acceleration components along the body axes are derived from first principles in Chapter 3, Section 3.4.1. Referring to equations (3.13), (3.14), (3.15):

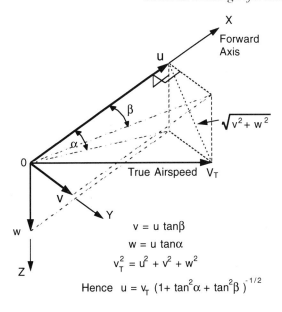

Fig. 5.32 Resolution of true airspeed on to OX, OY, OZ axes.

Acceleration along forward axis, $\mathrm{OX} = \dot{U} - Vr + Wq$
Acceleration along sideslip axis, $\mathrm{OY} = \dot{V} + Ur - Wp$
Acceleration along vertical axis, $\mathrm{OZ} = \dot{W} - Uq + Vp$

Referring to Fig. 5.33 it can be seen that the gravitational acceleration components g_x, g_y, g_z along OX, OY, OZ respectively are :

$$g_x = -g \sin \theta \tag{5.40}$$

$$g_y = g \cos \theta \sin \Phi \tag{5.41}$$

$$g_z = g \cos \theta \cos \Phi \tag{5.42}$$

The outputs of the forward, sideslip and vertical accelerometers are denoted by a_X, a_Y, and a_Z respectively

$$a_X = \dot{U} - Vr + Wq + g_x \tag{5.43}$$

$$a_Y = \dot{V} + Ur - Wp + g_y \tag{5.44}$$

$$a_Z = \dot{W} - Uq + Vp + g_z \tag{5.45}$$

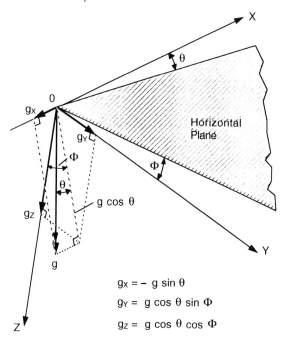

The equations shown in the figure:

$$g_X = - g \sin \theta$$
$$g_Y = g \cos \theta \sin \Phi$$
$$g_Z = g \cos \theta \cos \Phi$$

Fig. 5.33 Resolution of gravity vector.

The accelerometer bias errors are ignored at this stage for simplicity.

From equations (5.43) and (5.44) it can be seen that $\dot{U}$ and $\dot{V}$ can be obtained by computing the centrifugal acceleration component terms $(- Vr + Wq)$ and $(Ur - Wp)$ and the gravitational acceleration component terms g_x and g_y and subtracting these computed components from the accelerometer outputs a_X and a_Y respectively. The inertially derived velocity components U_I and V_I are then obtained by integrating $\dot{U}$ and $\dot{V}$ with respect to time.

The air data/inertial velocity mixing loops The overall air data/inertial velocity mixing loops are shown in Fig. 5.34. The computed centrifugal acceleration component terms $(-Vr + Wq)$ and $(Ur - Wp)$ are derived from the air data velocity components U_A, V_A, W_A and so are computed with respect to wind axes. The fact that the air mass is used as a reference frame and the wind axes are moving does not affect the values of these computed centrifugal acceleration terms provided the wind velocity is constant. Any acceleration of the reference frame due to changes in the wind velocity, however, can introduce errors. The assumption is made that the air mass accelerations generally appear as gusts of comparatively short duration which can be effectively smoothed by the air data/inertial mixing.

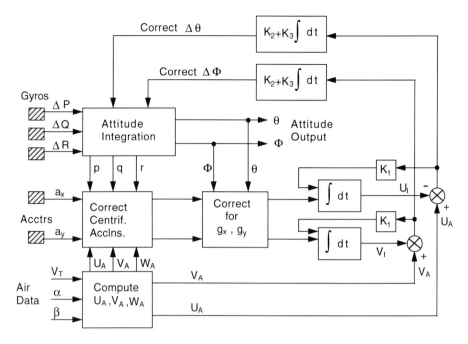

Fig. 5.34 Air data vertical monitoring.

The gravitational acceleration components g_x and g_y are computed from the values of the pitch angle, θ and bank angle Φ, derived from the attitude computations using the incremental angular data from the strap-down gyros as explained earlier.

Integrating $\dot{U}$ and $\dot{V}$ with respect to time yields U_I and V_I and these values are subtracted from the air data derived values of U_A and V_A to yield $(U_A - U_I)$ and $(V_A - V_I)$.

The Coriolis acceleration components have been ignored for simplicity. (The air mass, of course, rotates with the Earth.) The major errors in deriving U_I and V_I are due to the tilt errors $\Delta\theta$ and $\Delta\Phi$ in deriving the pitch angle, θ, and bank angle, Φ, from the gyro outputs, assuming the errors in the computed centrifugal accelerations are small. The accelerometer bias errors are also assumed to be small.

These tilt errors, $\Delta\theta$ and $\Delta\Phi$, result in acceleration errors $g\Delta\theta$ and $g\Delta\Phi$ (assuming $\Delta\theta$ and $\Delta\Phi$ are small angles) in the computation of g_x and g_y which are integrated with time causing a divergence between U_I and U_A and V_I and V_A unless corrected. The air data/inertial velocity errors $(U_A - U_I)$ and $(V_A - V_I)$ are thus fed back as shown in Fig. 5.34.

(a) To the U_I and V_I integrators to slave U_I to U_A and V_I to V_A.
(b) To the attitude computer to correct the pitch and bank angle tilt errors and reduce these to zero.

Proportional plus integral control,

$$viz, \qquad K_2 \text{ (Velocity error)} + K_3 \int \text{(Velocity error)} \, dt$$

is used in the feedback to the attitude computer to eliminate steady state tilt errors due to gyro bias.

The values of the feedback gains, K_1, K_2 and K_3 are selected to ensure good damping of the closed loops and good overall accuracy. These loops are also used to align the AHRS when the aircraft is stationary on the ground as U_A and V_A are both zero. Any velocity errors result from initial tilt errors and so are fed back to correct the vertical reference. It should be noted that the aircraft is usually fairly level when on the ground so that the pitch and bank angles are generally small. The coarse values of pitch and bank angles used to initialise the attitude computation from the gyro incremental outputs can thus be set at zero and subsequently corrected by the alignment process using the accelerometer outputs.

The gyro derived attitude information is coupled in through a high pass filter with pass frequencies of the order of 0.003 Hz, or lower so that the short term measurement of the attitude changes of the aircraft is provided mainly by the gyros. The attitude reference derived from the accelerometers is filtered by a low pass filter which smooths the noise present and attenuates short period fluctuations but retains the long term accuracy of the gravitational reference.

It should be noted that there are considerable variations in the implementation and sophistication of the mixing systems used by strap-down AHRS manufacturers. The above description is a simplified outline only of the basic principles used; the corrections for earth's rate and vehicle rate have also been omitted for clarity.

5.3.4.5 Alignment and monitoring of the heading reference

Given inertial quality gyros with drift/bias uncertainties of less than 0.01°/hour, it is possible to determine the aircraft heading with respect to true North to within 0.1° (or better depending on the latitude) by the technique known as gyro compassing. (This will be explained in the next chapter.) Gyro compassing, however, is not suitable for an AHRS with lower accuracy gyros – alignment errors of 1° would result from using gyros with 0.1°/hour bias uncertainty. Typical AHRS gyros are in the 0.3°/hour to 5°/hour bias uncertainty bracket so that it can be seen that other methods must be used for initial alignment and subsequent monitoring of the aircraft heading. (The heading can be unmonitored for several hours if the gyro drift is less than 0.01°/hour.)

Magnetic monitoring using the Earth's magnetic field as the directional reference is generally used for the alignment and monitoring of the heading output of an AHRS using gyros which are not of inertial quality. This is a self-contained and simple system of relatively low cost and high reliability. Using modern magnetic sensors with computer compensation of the sensor errors together with an accurate vertical reference from the AHRS enables the heading errors to be constrained generally to less than 0.7° for latitudes up to about 60°. A magnetic heading reference is also used for coarse initial alignment of an INS with the fine alignment carried out by gyro compassing.

It should be noted that the Earth's magnetic poles are offset from the Earth's polar axis so that the direction of *magnetic North* differs from the direction of *true North*, the difference between the two is known as the *magnetic variation*. This is not constant and varies over the Earth's surface. Concentrations of iron ore in the Earth's crust, for instance, can cause significant changes (or 'magnetic anomalies') in the local magnetic variation. There are also long term changes in the magnetic variation over several years. Magnetic variation has been accurately measured over large areas of the Earth's surface so that the magnetic heading can be corrected to give true heading by adding (or subtracting) the local magnetic variation, which can be stored within the system computer.

The presence of magnetic material on the aircraft causes local distortions in the Earth's magnetic field measured within the aircraft introducing a further error called 'magnetic deviation'. This is defined as the difference between the direction of the magnetic North measured in the aircraft and that at the same location in the absence of the aircraft. Extensive measurements are made with the aircraft on the ground to establish the magnitude of the errors due to the aircraft's internal magnetic environment.

The Earth's magnetic field is inclined at an angle to the horizontal in general. This angle is known as the *angle of dip* and is dependent on the latitude, being roughly 0° near the equator and 90° at the North and South magnetic poles. The magnetic heading is determined by measuring the direction of the horizontal component of the Earth's magnetic field. Errors in measuring the orientation of the magnetic sensors with respect to the vertical will result in a component of the vertical magnetic field being measured and will hence introduce an error in the magnetic heading. The vertical magnetic field component is equal to the horizontal magnetic field component multiplied by the tangent of the angle of dip and at high latitudes the vertical component can be several times greater than the horizontal component.

The magnetic heading error caused by a vertical error is thus 'geared up' by the tangent of the angle of dip. For example an angle of dip of 68° (North America) results in a magnetic heading error which is *2.5 × vertical error*. The importance of an accurate vertical reference can thus be seen. It should be noted that the magnetic heading reference becomes unusable at very high latitudes as the angle of dip approaches 90° and the horizontal component of the magnetic field approaches zero. The unmonitored gyro heading output must be relied upon for flight sectors over the polar regions.

The magnetic heading is measured by an instrument known as a *three axis fluxgate detector* in conjunction with an AHRS which provides the vertical reference for the fluxgate detectors. The system comprises three magnetic field sensors, or *'fluxgates'*, which are mounted orthogonally in a block directly fixed to the airframe (ie 'strapped-down') with the sensor axes parallel to the principal axes of the aircraft. The location of the fluxgate triad is chosen to be as far away as practical from magnetic field sources (AC and DC) within the aircraft. Possible locations may be in one of the wings or in the fin, the small size and electrical output of the fluxgates giving a high degree of flexibility in the installation. The typical fluxgate detector consists of primary and secondary coils wound on a pair of soft iron cores. The primary is excited by an audio frequency sinusoidal current of sufficient amplitude to drive the cores into saturation at the current peaks. The secondary which is wound differentially about the two cores picks up the second harmonic content caused by the change in inductance at saturation. This output is approximately proportional to the magnetic field strength for applied static fields less than the saturation field of the cores. To improve the linearity of the output, the fluxgates are operated in a feedback mode. The output of the secondary is demodulated and fed back to the primary so that the net field in the cores is zero and the DC current in the primary is then a measure of the impressed magnetic field. The fluxgates, however, also measure any low frequency magnetic fields and not just the component of the Earth's magnetic field along the measuring axis of the fluxgate. The fluxgate outputs are thus in error from what is known as 'hard iron' effects due to unwanted AC magnetic fields and also from 'soft iron' effects that distort the Earth's magnetic field through the magnetic materials in the aircraft.

The components of the Earth's magnetic field with respect to the aircraft's principal axes are denoted H_X, H_Y and H_Z along the forward, sideslip and vertical axes respectively (refer to Fig. 5.35).

These true components H_X, H_Y, H_Z of the Earth's magnetic field along each axis can be expressed in terms of a set of equations which relate the total measurements x_m, y_m, z_m of the three fluxgates. This is to account for axis misalignment and the 'hard iron' and 'soft iron' effects which are characteristic of the fluxgates and their environment in the aircraft.

This set of equations is of the form

$$H_X = a_1 x_m + b_1 y_m + c_1 z_m + d_1 \tag{5.46}$$

$$H_Y = a_2 x_m + b_2 y_m + c_2 z_m + d_2 \tag{5.47}$$

$$H_Z = a_3 x_m + b_3 y_m + c_3 z_m + d_3 \tag{5.48}$$

The coefficients a_1, a_2, a_3 and b_1, b_2, b_3 and c_1, c_2, c_3 and d_1, d_2, d_3 are measured and stored in the AHRS computer so that the fluxgate measurements can be corrected and the true components H_X, H_Y and H_Z of the Earth's magnetic field determined.

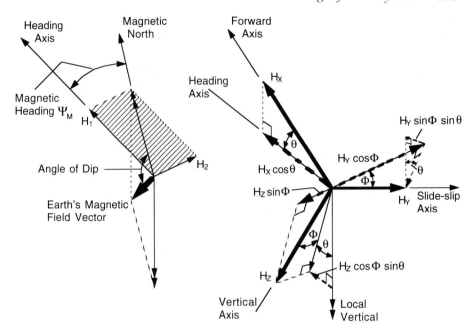

Fig. 5.35 Components of Earth)s magnetic field.

Referring to Fig. 5.35, the resolved components of the Earth's magnetic field in the horizontal plane along the heading axis, H_1, and at right angles to the heading axis, H_2, are given by:

$$H_1 = H_X \cos\theta + H_Y \sin\Phi\sin\theta + H_Z \cos\Phi\sin\theta \qquad (5.49)$$

$$H_2 = H_Y \cos\Phi - H_Z \sin\Phi \qquad (5.50)$$

The horizontal component of the Earth's magnetic field, H_H, is therefore equal to

$$\sqrt{H_1^2 + H_2^2}$$

The aircraft's magnetic heading, Ψ_M, is thus given by:

$$\Psi_M = \cos^{-1}\frac{H_1}{H_H} \quad \text{or} \quad \sin^{-1}\frac{H_2}{H_H} \qquad (5.51)$$

The computational processes in deriving the magnetic heading, Ψ_M, are shown in Fig. 5.36 and comprise:

(1) Computation of the true components H_X, H_Y, H_Z of the Earth's magnetic field with respect to aircraft axes from the measured outputs of the three fluxgates x_m, y_m, z_m and the stored coefficients a_1, a_2, a_3, b_1, b_2, b_3, c_1, c_2, c_3, d_1, d_2, d_3 using equations (5.46), (5.47) and (5.48).

(2) Computation of the horizontal components of the Earth's magnetic field along the heading axis, H_1, and at right angles to the heading axis, H_2, using equations (5.49) and (5.50).

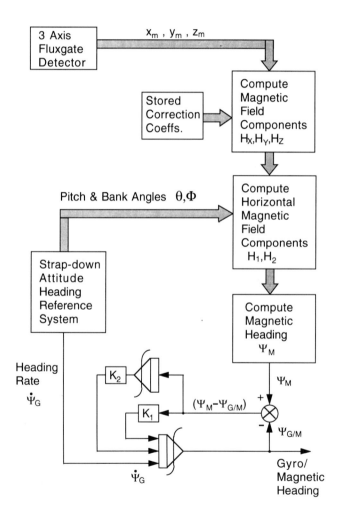

Fig. 5.36 Magnetic monitoring of gyro heading.

(3) Computation of the magnetic heading angle, Ψ_M, from

$$\Psi_M = \cos^{-1}\frac{H_1}{\sqrt{H_1^2+H_2^2}} \quad \text{or} \quad \sin^{-1}\frac{H_2}{\sqrt{H_1^2+H_2^2}}$$

The gyro/magnetic heading monitoring system is illustrated in Fig. 5.36, a basic second order mixing system being shown for simplicity. (More complex mixing systems using a Kalman filter can be employed.) Referring to Fig. 5.36:

$\dot{\Psi}_G$ = Gyro derived heading rate from the strap-down AHRS

$\dot{\Psi}_G = \dot{\Psi} + W$

where Ψ = true heading angle and W = gyro bias uncertainty

Ψ_{GM} = gyro/magnetic heading

K_1 = mixing gain for $(\Psi_M - \Psi_{GM})$ feedback

K_2 = mixing gain for $\int(\Psi_M - \Psi_{G/M})dt$ feedback

From the block diagram it can be seen that

$$\Psi_{G/M} = \int\left[\dot{\Psi}_G + K_1\left(\Psi_M - \Psi_{G/M}\right) + K_2\int\left(\Psi_M - \Psi_{G/M}\right)dt\right]dt \qquad (5.52)$$

Differentiating equation (5.52) twice and substituting $\dot{\Psi} + W$ for $\dot{\Psi}_G$ gives

$$\left(D^2 + K_1 D + K_2\right)\Psi_{GM} = \left(K_2 + K_1 D\right)\Psi_M + D^2\Psi + DW \qquad (5.53)$$

The dynamic response of the system is that of a simple second order system with an undamped natural frequency, $\omega_0 = \sqrt{K_2}$ and damping ratio, $\zeta = K_1/2\sqrt{K_2}$. The integral feedback gain, K_2, determines the undamped natural frequency and the proportional feedback gain, K_1 determines the damping. K_1 and K_2 are determined by the value of the gyro bias uncertainty, W, and the need to have a well damped closed loop response with ζ equal to or near to 1.

From equation (5.53):

$$\Psi_{G/M} = \frac{\omega_0^2\left(1+\dfrac{2\zeta}{\omega_0}D\right)}{\left(D^2 + 2\zeta\omega_0 D + \omega_0^2\right)}\cdot\Psi_M + \frac{D^2}{\left(D^2 + 2\zeta\omega_0 D + \omega_0^2\right)}\cdot\Psi$$
$$+ \frac{D}{\left(D^2 + 2\zeta\omega_0 D + \omega_0^2\right)}\cdot W \qquad (5.54)$$

Ignoring the gyro bias uncertainty term W, as it has zero contribution to $\Psi_{G/M}$ if W is constant.

$$\Psi_{G/M} = F_1(D)\Psi_M + F_2(D)\Psi \tag{5.55}$$

where $F_1(D) = \dfrac{\omega_0^2\left(1 + \dfrac{2\zeta}{\omega_0}D\right)}{\left(D^2 + 2\zeta\omega_0 D + \omega_0^2\right)}$ and $F_2(D) = \dfrac{D^2}{\left(D^2 + 2\zeta\omega_0 D + \omega_0^2\right)}$

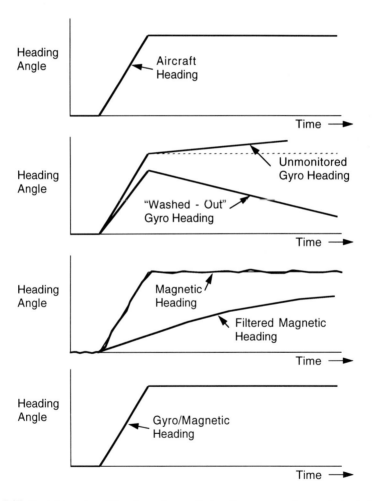

Fig. 5.37 Complementary filtered gyro/magnetic heading response to constant rate turn.

The magnetic heading, Ψ_M, component of Ψ_{GM} is coupled through a low pass filter of transfer function $F_1(D)$. This greatly attenuates the magnetic noise content of the Ψ_M output and short term transient errors such as would be caused by flexure of the structure at the fluxgate sensor location under manoeuvre loads.

The gyro measured component of Ψ_{GM} is coupled through a high pass, or, 'wash-out' filter of transfer function $F_2(D)$.

This enables the changes in heading to be measured without lag but ensures that the steady gyro bias, W, is 'DC blocked' (or 'washed-out').

The complementary filtering thus enables the 'best of both worlds' to be achieved. The basic accuracy and repeatability of the magnetic heading sensor is retained but any noise and transient errors during manoeuvres are heavily smoothed and filtered. The excellent dynamic response of the gyro system is retained and the gyro bias is DC blocked. There are also no steady state magnetic heading errors due to gyro bias.

Fig. 5.37 illustrates the system response to a constant rate turn (assuming a near critically damped system).

Further reading

AGARD Lecture Series No.95, *Strap-Down Inertial Systems*.

Aronowitz, F., 'Laser Gyros', *Laser Applications Vol.1*: edited by M.Ross, Academic Press, New York 1971

Bergh, R.A., Lefevre, H.C., and Shaw, H.J., *An Overview of Fiber Optic Gyroscopes*: Journal of Light Wave Technology, Vol LT-2 No.2 April 1984

Ezekiel, S., and Arditty, H.J., *Fiber Optic Rotation Sensors and Related Technologies*: Springer Series in Optical Sciences Vol.32 Springer-Verlag New York 1982

Matthews, A., *Fiber Optic Based Inertial Measuring Unit*: SPIE Vol.1173 Fiber Optic Systems for Mobile Platforms III, 1989

Page, J.L., *Multiplexed Approach to the Fiber Optic Gyro Inertial Measuring Unit*: OE/FIBERS 1990 SPIE Conference, San Jose, California September 1990

Roland, J.J., Agrawal, G.P., *Optical Gyroscopes*: Optics and Laser Technology, October 1981 pp 239-244

Siouris, G.M., *Aerospace Avionic Systems*: Academic Press 1993

Wrigley, W., Hollister, W.M., and Denhard, W.G., *Gyroscopic Theory, Design and Instrumentation*: MIT Press 1969

6

Navigation systems

6.1 INTRODUCTION AND BASIC PRINCIPLES

6.1.1 Introduction

The objective of this chapter is to explain the basic principles of aircraft navigation systems and the technologies which are used to meet the very exacting requirements.

The need for reliable, accurate navigation can be briefly summarised as follows:

(1) Reliable and accurate navigation is essential for the effective operation of any aircraft – to get lost could be catastrophic. Aircraft navigation must be automatic because of the speed of the aircraft. For instance, an aircraft travelling at a speed of 500 knots covers over two nautical miles in 15 seconds. The need to determine the aircraft's position in as short a time as possible can be clearly seen.

(2) The density of air traffic on major air routes requires the aircraft to fly in a specified corridor or 'tube in the sky', these air routes being defined by the Air Traffic Control authorities. As explained earlier, not only must the aircraft follow the defined three dimensional flight path with high accuracy, but there is also a fourth dimension, namely that of time, as the aircraft's arrival time must correspond to a specified time slot. High accuracy navigation systems are thus essential and form a key part of the flight management system. This is covered in Chapter 8.

(3) For military operations, very accurate navigation systems are essential to enable the aircraft to fly low and take advantage of terrain screening from enemy radar, to avoid known defences and in particular to enable the target to be acquired in time. The aircraft flies fast and very low so that the pilot cannot see the target until the aircraft is very near to it. There may be then only about six to ten seconds in which to acquire the target, aim and launch the weapons. It is thus essential to know the aircraft's position near the target area to within 0.5 km or better. Otherwise the pilot could be looking at the wrong area and fail to acquire the target in time for a successful attack. Having alerted the enemy, a second pass attack is at very high risk.

The use of stand-off weapons which are released several kilometres away from the target also requires an accurate knowledge of the aircraft's position in order to initialise the mid-course inertial guidance system of the missile. (The terminal homing phase is achieved with a suitable infra-red or microwave radar seeker system.)

It is appropriate at this stage to distinguish between the two basic methods of navigation which are used, namely, dead reckoning (or DR) navigation systems and position fixing navigation systems. The art of navigation is basically to combine the two methods in an optimum manner which is where Kalman filters are now used to great effect.

As briefly explained in Chapter 1, DR navigation is the process of continually estimating a vehicle's position as the journey proceeds from a knowledge of the starting point position and the vehicle's speed and direction of motion and the time elapsed. It is essentially an incremental process of continually estimating the changes that have occurred and updating the present position estimate accordingly.

The basic quantities required by an automatic DR navigation system are the northerly and easterly velocity components of the aircraft in the local horizontal plane. This enables the rates of change of latitude and longitude to be derived, as will be shown. The present position latitude and longitude co-ordinates are then computed by integrating the rates of change of latitude and longitude with respect to time, knowing the latitude and longitude co-ordinates of the starting point. The main types of airborne DR navigation systems are categorised below on the basis of the means used to derive the velocity components of the aircraft.

In order of increasing accuracy these are:

(1) *Air data based DR navigation* The basic information used comprises the true airspeed (from the air data computer) with wind speed and direction (forecast or estimated) and the aircraft heading (from the AHRS).

(2) *Doppler/heading reference systems* These use a Doppler radar velocity sensor system to measure the aircraft's ground speed and drift angle. The aircraft heading is provided by the AHRS.

(3) *Inertial navigation systems* These derive the aircraft's velocity components by integrating the outputs of the accelerometers, as will be explained.

(4) *Doppler/inertial navigation systems* These combine the Doppler and INS outputs, generally by means of a Kalman filter, to achieve increased DR navigation accuracy.

A fundamental characteristic of DR navigation systems is that their accuracy is time dependent. For example, a good quality INS has an accuracy of 1 NM/hour so that the aircraft position uncertainty after five hours would be 5 NM.

A position fixing system is thus required to constrain the error growth of a DR navigation system and correct the DR position errors.

A position 'fix' can be obtained by recognising a prominent landmark or set of terrain features either visually, possibly using an infra-red imaging sensor, or from a radar-generated map display. Alternatively, a suitable position fixing navigation system can be used. In general, several dissimilar systems may be used to provide the necessary confidence in the overall navigation accuracy.

Position fixing navigation systems depend on external references to derive the aircraft's position. For example, radio/radar transmitters on the ground, or satellites whose orbital positions are precisely known, or the terrain elevation in the vicinity of the aircraft. Unlike DR navigation systems, their errors are not time dependent. The errors are also independent of the aircraft's position in most position fixing systems.

The main types of position fixing navigation systems used in modern aircraft are briefly summarised below.

(1) Range and bearing ('R/θ') radio navigation aids
These comprise VOR/DME and TACAN.

VOR (VHF omni-directional range) is an internationally designated short-distance radio navigation aid and is an integral part of Air Traffic Control procedures.

DME (distance measuring equipment) is co-located with VOR and provides the distance from an aircraft to the DME transmitter.

TACAN is the primary tactical air navigation system for the military services of the USA and NATO countries. It is often co-located with civil VOR stations, such combined facilities being known as VORTAC stations. The systems are line of sight systems and provide the slant range of the aircraft from the ground station using a transmitter/transponder technique. A rotating antenna system at the TACAN ground station (or beacon) enables the aircraft to measure the bearing of the TACAN beacon to an accuracy of about 2°, using a principle analogous to a lighthouse.

Navigation position accuracy is of the order of one to two miles.

(2) Hyperbolic radio navigation systems
These comprise LORAN C and OMEGA.

LORAN C provides a positional accuracy of around 150m in regions covered by the system. Eight LORAN C chains comprising 34 ground station transmitters provide coverage over approximately 10% of the earth's surface. The system is a pulse transmission system operating in the frequency range 90 kHz to 110 kHz

OMEGA is also a pulse system and operates in the VLF range at 10 kHz using eight ground station transmitters to provide essentially global coverage – 88% by day and 90% by night. Positional accuracy is around 2 NM world wide.

LORAN C and OMEGA determine the aircraft's position from the measurements of the *phase differences* between the radio signals received from the ground station transmitters. The locus of the points on the earth's surface where there is a constant phase difference between the signals received from two ground station transmitters is a hyperbola. The position of the aircraft can thus be determined from the intersection of the appropriate hyperbolae corresponding to the measured phase differences, hence the name 'hyperbolic radio navigation systems'.

(3) Satellite navigation systems – GPS (global position system)

GPS is the most important and accurate position fixing navigation system developed to date. It can be used by every type of vehicle – aircraft, ships and land vehicles. The equipment required by the GPS user is entirely passive and requires a GPS receiver only. Electronic miniaturisation has enabled very compact and lightweight GPS receivers to be produced. The 2-D position errors for civil users of GPS are less than 100m world wide. Military users of GPS, who have access to the encrypted codes in GPS satellite transmissions, can achieve nearly an order higher accuracy with a 3-D positional accuracy of 16m, and a velocity accuracy of 0.1m/s. Precise time to within a few billionths of a second is also available.

(4) Terrain reference navigation (TRN) systems

Terrain reference navigation systems derive the vehicle position by correlating the terrain measurements made by a sensor in the vehicle with the known terrain feature data in the vicinity of the DR estimated position. The terrain feature data is obtained from a stored digital map data base.

The principal navigation systems used in modern civil and military aircraft are summarised in Table 6.1.

6.1.2 Basic navigation definitions

A very brief review of the terms and quantities used in navigation is set out below for those readers who may be a little 'rusty' on some of the basic definitions. Position on the earth's surface is generally specified in terms of *latitude* and *longitude* co-ordinates which provide a circular grid over the surface of the earth. The earth is basically a sphere – the variation in the radius of the earth is only about 40 NM in a radius of 3,438 NM at the equator, being slightly flattened at the poles. (This variation is taken into account in the navigation computations.)

Referring to Fig. 6.1, latitude and longitude are defined with respect to the polar axis, the equator and the prime meridian. The *prime meridian* is the meridian passing through Greenwich which provides the datum for measuring the longitude. The *latitude* of a point on the earth's surface is the angle subtended at the earth's centre by the arc along the meridian passing through the point and measured from the equator to the point. The range of latitude angles is from 0° to 90° North and 0° to 90° South. The *longitude* of a point on the earth's surface is the angle subtended at the earth's centre by the arc along the equator measured East or West of the prime meridian to the meridian passing through the point. The range of longitude angles to cover all points on the earth's surface is thus 0° to 180° east of the prime meridian and 0° to 180° west of the prime meridian. Latitude and longitude are expressed in degrees, minutes of arc, and seconds of arc.

Great circles are circles on the surface of a sphere with their centre at the centre of the sphere, that is, the plane of a great circle passes through the earth's centre.

Table 6.1 Navigation systems – civil and military aircraft

Navigation system	Method used	Co-ordinates provided	Accuracy	Coverage	Passive user	Velocity data	Remarks
Inertial	Integration of accelerometer outputs	Latitude, longitude, height (with baro. data)	Time variable typically 1 NM hour	Global with position updates	Yes	3-D velocity (with baro.)	Accuracy dependent on gyros and accelerometrs Unjammable
Doppler heading reference	Doppler measurement of velocity and drift angle	Latitude, longitude	Time variable typically 1% of distance flown	Global with position updates	No	3-D velocity	Accuracy dependent on heading reference Doppler degraded over water
LORAN C	Hyperbolic ranging	Latitude, longitude	150m	Regional coverage - approx 10% of earth	Yes	Crude	8 LORAN C chains with 34 transmitters
OMEGA	Hyperbolic ranging	Latitude, longitude	1 to 2 NM	Essentially global 88% day, 98% night	Yes	No	8 transmitters in world-wide operation
VOR DME TACAN	Lighthouse signal + spherical ranging	Bearing, slant range	Variable - typically 2 NM	Regional	No	No	Line-of-sight coverage along present air routes
GPS	Spherical ranging	Latitude, longitude, altitude and precise time	Civil users 100m (2-D) Military users 16m (3-D)	Global 24 hours per day	Yes	3-D military 0.1m/s accuracy	24 GPS satellites in geo-synchronous orbits (3 orbiting spares) High resistance to jamming

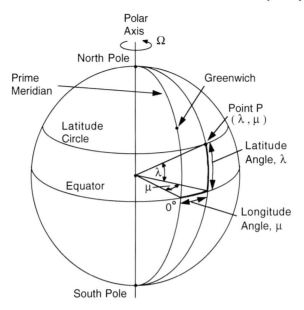

Fig. 6.1 Latitude/longitude co-ordinates.

Meridians and the equator are thus great circles. Parallels of latitude which are circles round the earth parallel to the equator, are, however, small circles.

The shortest distance between two points on the surface of a sphere is a great circle, hence navigation routes try to follow a great circle path. Navigation between points on the earth's surface thus involves the solution of spherical triangles as shown in Fig. 6.2. (A spherical triangle being defined as a triangle on a sphere whose sides are part of great circles. Their solution is achieved using the well established formulae of spherical trigonometry.)

Referring to Fig. 6.3, it can be seen that the distance along the arc of a meridian between the two circles of latitude going through point A (λ_1, μ_1) and point B (λ_2, μ_2) can be obtained by subtracting the latitude co-ordinates of A and B,

viz $(\lambda_1 - \lambda_2)$

eg   Point A lat/long = 52° 33′ 10″ N, 5° 25′ 15″ E
Point B lat/long = 61° 40′ 15″ N, 20° 37′ 30″ E
A′B = AB′ = 9° 7′ 5″ = 547.08 mins of arc = 547.08 NM

This simple relationship, however, does not apply to the longitude co-ordinates because of meridian convergency. The distance along a circle of latitude between two meridians decreases as the latitude increases and becomes zero at the poles.

Difference in longitude between A and B = $(\mu_2 - \mu_1)$
Distance along equator = $E_1 E_2 = (\mu_2 - \mu_1)R$
Distance along latitude circle λ_1 = AA′ = $(\mu_2 - \mu_1)R \cos \lambda_1$

Fig. 6.2 Spherical triangles.

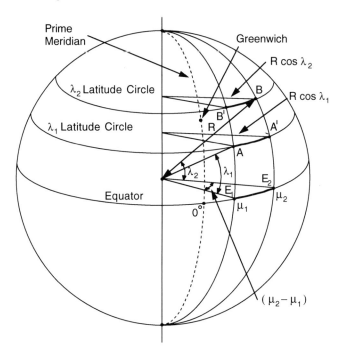

Fig. 6.3 Meridian convergency.

Distance along latitude circle $\lambda_2 = B'B = (\mu_2 - \mu_1)R \cos \lambda_2$

For the above example

$\mu_2 - \mu_1 = 15° \ 12' \ 15'' = 912.25'$
$E_1E_2 = 912.25$ NM
AA$' = 912.25 \cos 52° \ 33' \ 10'' = 554.67$ NM
BB$' = 912.25 \cos 61° \ 40' \ 15'' = 432.9$ NM

The importance of meridian convergency in mechanising an automatic navigation system will be shown in the next section.

6.1.3 Basic automatic DR navigation systems

The basic principles of deriving a DR navigation position estimate are explained below. The following quantities are required:

(1) Initial position – latitude/longitude
(2) The northerly and easterly velocity components of the aircraft, V_N and V_E

Referring to Fig. 6.4 it can be seen that

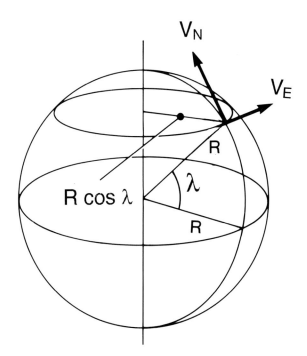

Fig. 6.4 Derivation of rates of change of latitude and longitude.

rate of change of latitude $\qquad\qquad \dot{\lambda} = \dfrac{V_N}{R}$ $\qquad\qquad$ (6.1)

rate of change of longitude $\qquad \dot{\mu} = \dfrac{V_E}{R\cos\lambda}$

ie $\qquad\qquad\qquad\qquad\qquad \dot{\mu} = \dfrac{V_E}{R}\sec\lambda$ $\qquad\qquad$ (6.2)

The change in latitude over time, t, is thus equal to $\dfrac{1}{R}\displaystyle\int_0^t V_N dt$ and hence the present latitude at time t can be computed given the initial latitude λ_0. Similarly, the change in longitude is equal to $\dfrac{1}{R}\displaystyle\int_0^t V_E \sec\lambda dt$ and hence the present longitude can be computed given the initial longitude, μ_0.

viz: $\qquad\qquad\qquad\qquad \lambda = \lambda_0 + \dfrac{1}{R}\displaystyle\int_0^t V_N dt$ $\qquad\qquad$ (6.6)

$$\mu = \mu_0 + \dfrac{1}{R}\displaystyle\int_0^t V_E \sec\lambda dt \qquad\qquad (6.7)$$

It can be seen that a mathematical singularity is approached as λ approaches $90°$ and $\sec\lambda$ approaches infinity. This method of computing the latitude and longitude of the DR position is hence limited to latitudes below $80°$. A different co-ordinate reference frame is used to deal with high latitudes as will be explained later.

The basic computational processes in a DR navigation system using a Doppler/heading reference system are shown in Fig. 6.5. In the case of a Doppler/heading reference system, the ground speed V_G and drift angle δ are measured directly by the Doppler radar velocity sensor system.

The AHRS system provides an accurate measurement of the heading angle, ψ, and hence the track angle, ψ_T can be obtained from

$$\psi_T = \psi + \delta$$

The northerly velocity component of the aircraft, V_N, and the easterly velocity component, V_E, are then derived by resolution of the ground speed vector, V_G (see inset diagram on Fig. 6.5).

Hence

$$V_N = V_G \cos\psi_T$$

$$V_E = V_G \sin\psi_T$$

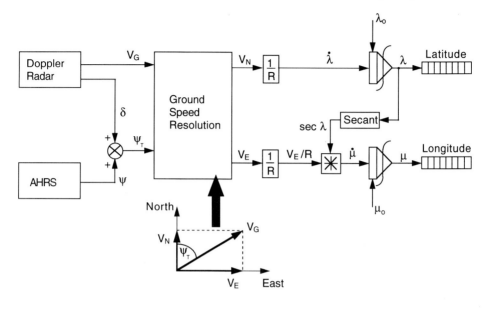

Fig. 6.5 Doppler/heading reference DR navigation system.

In the case of an air data based DR navigation system the northerly and easterly velocity components of the aircraft can be derived as follows:

1. The horizontal velocity component V_H of the true airspeed V_T is obtained by resolving V_T through the aircraft pitch angle θ

   viz $$V_H = V_T \cos \theta$$

2. The northerly and easterly velocity components of the airspeed are then derived by resolving V_H through the aircraft heading angle ψ

   viz   Northerly airspeed $= V_H \cos \psi$
                     Easterly airspeed $= V_H \sin \psi$

3. The forecast (or estimated) wind speed V_W and direction ψ_W is resolved into its northerly and easterly components

   viz   Northerly wind component $= V_W \cos \psi_W$
                     Easterly wind component $= V_W \sin \psi_W$

4. The northerly and easterly velocity components of the aircraft are then given by:

$$V_N = V_H \cos\psi + V_W \cos\psi_W \qquad (6.5)$$

$$V_E = V_H \sin\psi + V_W \sin\psi_W \qquad (6.6)$$

Such a system provides a reversionary DR navigation system in the absence of Doppler (or an INS) and would generally be used in conjunction with a radio navigation system.

6.1.4 Brief overview of Doppler radar velocity sensors

The importance of a Doppler radar velocity sensor as a self-contained and accurate source of the aircraft's ground speed and drift angle merits a brief overview of the operating principles.

The principle of the Doppler effect is named after the Austrian physicist C. J. Doppler. Doppler discovered in 1842 that the pitch or frequency of a sound changes when there is relative motion between the observer and the source. The frequency increases as the source moves towards the observer and decreases as the source moves away, the change in frequency being directly proportional to the relative velocity between the observer and the source. The effect applies to all types of transmitted wave energy from sound waves to electro-magnetic waves across the frequency spectrum.

As a very simple example, suppose a source emits waves at a frequency of 10 per second and the waves travel with a velocity of 1m/s. An observer then counts the number of waves passing per second under the following conditions:

(a) Observer stationary. In this case the observer would count 10 waves/s.
(b) Observer moves towards the source at 0.5m/s. In moving towards the source, in one second the observer would count 5 waves that have passed by plus 10 emitted, that is 15 waves/s.
(c) Observer moves away from the source at 0.5m/s. This time the observer would only count 5 waves/s as 5 would not have arrived.

Consider an aircraft flying over the ground with a transmitter on the aircraft emitting radar frequency waves from an antenna which is pointing at the ground as shown in Fig. 6.6 (a). Some of the radar energy will be reflected from the ground back to the aircraft. Because of the Doppler effect, the frequency of the received signal will increase relative to the transmitted frequency by an amount which is proportional to aircraft velocity.

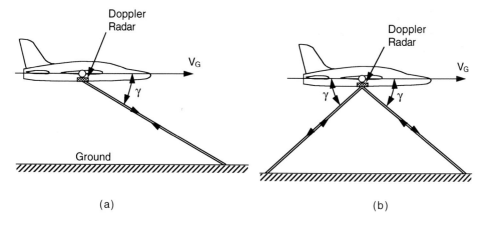

Fig. 6.6 Basic concepts of airborne Doppler radar velocity sensor system.

The relationship between the received signal frequency f_{R1} and the transmitted frequency, f_T is given by:

$$f_{R1} = f_T + kV_G \cos \gamma$$

where k = constant, V_G = aircraft ground speed and γ = depression angle of the radar beam, that is the angle between the radar beam and the aircraft's forward axis. (The aircraft pitch angle, θ, is assumed to be zero for simplicity, so that the angle the radar beam is depressed from the horizontal is equal to γ.)

Now consider the configuration shown in Fig 6.6 (b), where a second antenna transmits the same frequency radar energy rearwards and downwards.. The received signal frequency f_{R2}, from this rearward transmission will decrease by an amount proportional to the aircraft's ground speed and is given by:-

$$f_{R2} = f_T - kV_G \cos \gamma$$

It is assumed that the side-slip velocity, pitch angle and bank angle of the aircraft are all zero.

Subtracting the forward and rearward received signal frequencies results in a difference frequency, Δf which is directly proportional to the aircraft's ground speed and is equal to:

$$\Delta f = 2kV_G \cos \gamma$$

There is thus a doubling of the frequency shift due to the Doppler effect resulting in a doubling of the sensitivity.

A practical configuration adopted for many airborne Dopplers comprises a set of four antennas mounted rigidly on the aircraft. These transmit forward and rearwards to the left and right of the aircraft through a dielectric panel in the underside of the aircraft. Such an array is often called a 'Janus' array, after the Roman god of doorways, who was able to face both ways at once.

The antenna configuration and transmitting lobe patterns are shown in Fig. 6.7. The Doppler velocity components in the signals received from the four beams reflected from the ground contain a proportion of the aircraft's sideways velocity component V_S in the horizontal plane as well as its forward velocity V_F in the horizontal plane.

Comparing the received signals from the forward left and right beams with the rearward left and right beams then enables the aircraft's forward speed V_F to be derived, taking into account the antenna geometry and the aircraft's pitch and bank angles in resolving the appropriate velocity components. (The attitude information is provided from an AHRS or INS.)

The side-slip velocity V_S is similarly derived by comparing the received signals from the left side forward and rearward beams with the right side forward and

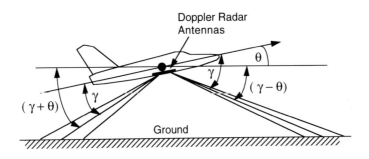

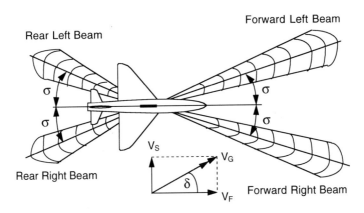

Fig. 6.7 Doppler radar antenna geometry.

rearward beams, taking into account the antenna geometry and the aircraft's pitch and bank angles.

The ground speed, V_G, is then computed from V_F and V_S

viz: $$V_G = \sqrt{V_F^2 + V_S^2}$$

The drift angle δ is computed from $\delta = \tan^{-1} \dfrac{V_S}{V_F}$

6.2 INERTIAL NAVIGATION

6.2.1 Introduction

It is instructive briefly to review the reasons for the development of inertial navigation and its importance as an aircraft state sensor.

The attributes of an ideal navigation and guidance system for military applications can be summarised as follows:-

- High accuracy
- Self contained
- Autonomous – does not depend on other systems
- Passive – does not radiate
- Unjammable
- Does not require reference to the ground or outside world

In the late 1940s these attributes constituted a 'wish list' and indicated the development of inertial navigation as the only system which could be capable of meeting all these requirements. It was thus initially developed in the early 1950s for the navigation and guidance of ballistic missiles, strategic bombers, ships and submarines (Ships Inertial Navigation System ('SINS')).

Huge research and development programmes have been carried out world-wide involving many billions of dollars expenditure to achieve viable systems. For instance, as mentioned earlier over three orders of magnitude improvement in gyro performance from 15°/hour to 0.01°/hour drift uncertainty was required.

Precision accelerometers had to be developed with bias uncertainties of less than 50μg. The major task of achieving the required computational accuracies had to be solved and in fact the first digital computers operating in real time were developed for IN systems. Once these problems had been solved, however, the INS could provide:

- Accurate position in whatever coordinates are required – eg latitude/longitude etc.

- Ground speed and track angle.
- Euler angles: heading, pitch and roll to very high accuracy.
- Aircraft velocity vector (in conjunction with the air data system).

Accurate velocity vector information together with an accurate vertical reference are essential for accurate weapon aiming and this has led to the INS being installed in military strike aircraft from the early 1960's onwards as a key element of the navigation/weapon aiming system.

The self contained characteristics of an inertial navigation system plus the ability to provide a very accurate attitude and heading reference led to the installation of IN systems in long range civil transport aircraft from the late 1960s. They are now very widely used in airliners operating on long over water routes with triple IN systems installed to ensure availability and the ability to detect failures or degraded performance by cross comparison. They are also a key sensor for the automatic flight control systems (AFCS) in both civil and military aircraft.

RLG based inertial reference systems, with an unaided navigation accuracy of 2 nm/hour, are now being combined with the air data computer and GPS as an integrated 'one box system'. Such systems are now being installed in shorter range civil transport aircraft to enable these aircraft to meet the more stringent navigation accuracy which will be required by the ATC authorities. They also provide a very accurate AHRS.

6.2.2. Basic principles and Schuler tuning

A good question to ask an interviewee who is interested in entering the avionics systems field is 'How can you measure the motion of a vehicle and derive the distance it has travelled without any reference to the outside world?'

The answer is that you can sense the vehicle's acceleration (and also the gravitational vector) with accelerometers. If the vehicle's acceleration components can then be derived along a precisely known set of axes, successive integration of the acceleration components with respect to time will yield the velocities and distances travelled along these axes. This is true provided the initial conditions are known, that is the vehicle velocity and position at the start time.

Figure 6.8(a) and (b) illustrate the basic concepts of deriving the velocity and distance travelled of the vehicle from its acceleration components.

Put like this it seems simple. Any errors, however, in deriving the aircraft acceleration components from the accelerometer outputs will be integrated with time, producing velocity errors which in turn are integrated with time generating position errors.

For example, a constant accelerometer bias error, B, (which can be equated to an initial tilt error) will result in a distance error which is equal to $\iint B \, dt \, dt$, that is $Bt^2/2$. The resulting distance error is thus proportional to the square of the elapsed time.

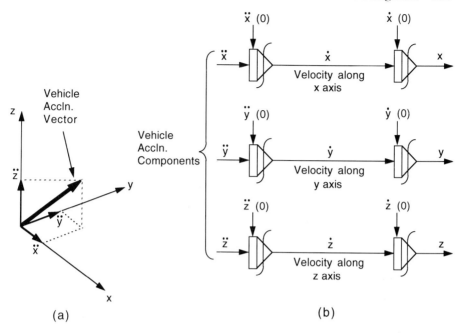

Fig. 6.8 Basic principles of inertial navigation.

It was shown in the preceding chapter that errors in deriving the orientation of the accelerometer input axes with respect to the local vertical, that is tilt errors, introduce gravitational acceleration errors in the acceleration measurements. A constant gyro drift rate, W, will cause the tilt error, $\Delta\theta$, to increase linearly with time ($\Delta\theta = Wt$) resulting in a gravitational acceleration error gWt. This results in a distance error $\iint gWt\,dt\,dt$, that is $gWt^3/6$. The distance error in this case is proportional to the cube of the time.

Clearly these types of error propagation are unacceptable except in applications where the time of flight is short (eg mid-course guidance of short range missiles) or where the system could be frequently corrected by another navigation system. It should be noted that these error propagations approximate closely to the initial error propagation with Schuler tuning over the first few minutes. This is because the correcting action exerted by Schuler tuning on the tilt errors takes place over an 84.4 minute period and so is small over the first few minutes.

It will be shown that Schuler tuning provides an undamped closed loop corrective action to constrain the system tilt errors so that they oscillate about a zero value with an 84.4 minute period. The 'INS vertical' follows the local vertical irrespective of the vehicle accelerations so that the system behaves like a Schuler pendulum. Schuler tuning completely changes the error propagation characteristics from those just described and enables a viable unaided navigation system to be achieved. It is a vital and fundamental ingredient of any IN system.

The concept of a pendulum which would be unaffected by accelerations, and would always define the local vertical was originated by the Austrian physicist Max Schuler in 1924. This arose from studies he carried out on the errors in marine gyro compasses resulting from ship manoeuvres and the effect of ship accelerations on the pendulous gyro compass system. Schuler conceived the idea of a pendulum which would be unaffected by acceleration by considering the behaviour of a simple pendulum whose length was equal to the radius of the earth (see Fig. 6.9). The 'plumb bob' of such a pendulum would always be at the earth's centre and would hence define the local vertical irrespective of the motion, or acceleration, of the point of suspension of the pendulum – any disturbance to the pendulum bob would cause it to osciliate about the local vertical with a period equal to $2\pi\sqrt{\dfrac{R}{g}}$

where R = radius of the earth, g = gravitational acceleration

Using the accepted values of R and g gives the period of the *Earth radius pendulum (or Schuler pendulum)* as 84.4 minutes.

A simple pendulum of length equal to the earth's radius is obviously not feasible. If, however, a pendulous system could be produced with an exact period of 84.4 minutes then it would indicate the local vertical irrespective of the acceleration of the vehicle carrying it.

In 1924, the technology was not available to implement a practical system and in fact it was not until the early 1950s that the technology became available to implement a practical Schuler tuned IN system. The first successful airborne

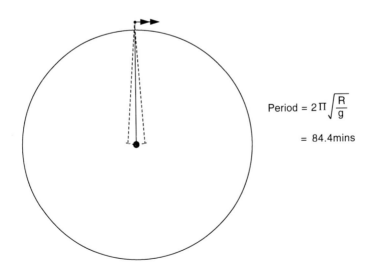

$$\text{Period} = 2\pi\sqrt{\frac{R}{g}}$$

$$= 84.4\text{mins}$$

Fig. 6.9 Schuler or Earth radius pendulum.

demonstration of a Schuler tuned inertial navigation system was achieved by Dr Charles Stark Draper and his team at Massachusetts Institute of Technology (MIT) Boston, USA, around 1952.

A Schuler tuned stable platform INS is shown schematically in Fig. 6.10 (the vertical inertial channel has been omitted for clarity). A stable platform mechanisation of a Schuler tuned INS is used to explain the basic principles as this is easier to visualise initially. It should be noted that a strap-down INS mechanisation maintains an equivalent 'stable platform' within the system computer by computing the horizontal acceleration components along the North and East axes from the body mounted gyro and accelerometer outputs.

The X and Y accelerometers are mounted on the stable platform with their input axes horizontal and the platform is aligned so that the X accelerometer measures the aircraft acceleration component along the East–West axis and the Y accelerometer the acceleration along the North–South axis. The outputs of the X and Y accelerometers, after correcting for the Coriolis and centrifugal acceleration components, are then integrated with respect to time to give the easterly and northerly velocity components of the aircraft, V_E and V_N, (knowing the initial conditions). The computation of the present position latitude and longitude is then as described in Section 6.1.3.

The Schuler tuning is achieved by feeding back the accelerometer derived vehicle rate terms, V_N/R and V_E/R, to torque the vertical gyro so that the platform follows the local vertical as the aircraft moves over the spherical earth.

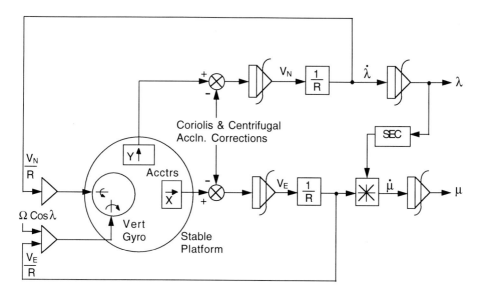

Fig. 6.10 Block diagram of Schuler tuned INS (azimuth gyro and vertical accelerometer omitted from platform for clarity).

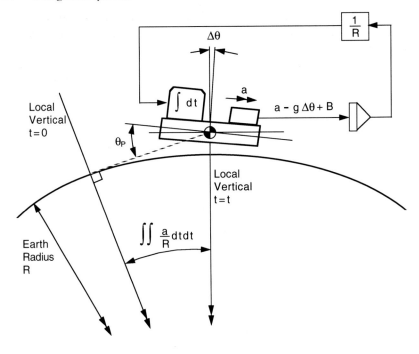

Fig. 6.11 Schuler loop.

The behaviour of a Schuler feedback loop is analysed below, one axis only being considered for simplicity.

Referring to Fig. 6.11:

Accelerometer output $= a - g\Delta\theta + B$

(where B = accelerometer bias)

Angle platform has rotated in space in time $t = \theta_P$

Angle the local vertical has rotated in space in time $t = \displaystyle\int\int \frac{a}{R}\, dt\, dt$

Platform tilt, $\Delta\theta = \theta_P - \displaystyle\int\int \frac{a}{R}\, dt\, dt$ (6.15)

Rate of rotation of platform in space $= \dot{\theta}_P = \displaystyle\int \frac{a - g\Delta\theta + B}{R}\, dt + W$

where W = gyro drift rate

Hence

$$\theta_P = \int\int \frac{(a - g\Delta\theta + B)}{R} dt\, dt + \int W\, dt \qquad (6.16)$$

Substituting for θ_P in equation (6.15) gives

$$\Delta\theta = \int\int \frac{(a - g\Delta\theta + B)}{R} dt\, dt + \int W\, dt - \int\int \frac{a}{R} dt\, dt \qquad (6.17)$$

Differentiating equation (6.17) twice and re-arranging gives

$$\left(D^2 + \frac{g}{R}\right)\Delta\theta = \frac{B}{R} + DW \qquad (6.18)$$

The effects of accelerometer bias and gyro drift are considered separately for the sake of simplicity. The effects on the system of a combination of disturbing inputs of accelerometer bias and gyro drift can then be inferred.

(1) Effect of accelerometer bias
Assume $W = 0$
The accelerometer bias can be equated to an initial tilt $= B/g$ radians.
The solution of equation (6.18) for initial conditions

$$\Delta\theta(0) = B/g \text{ and } \Delta\dot\theta(0) = 0$$

is

$$\Delta\theta = (B/g)\cos\omega_0 t \qquad (6.19)$$

Where $\omega_0 = \sqrt{\dfrac{g}{R}} =$ undamped natural frequency of Schuler loop

The platform thus oscillates about the local vertical with amplitude B/g and a period

$$= 2\pi\sqrt{\frac{R}{g}} = 84.4 \text{ minutes}$$

The acceleration error $= g\Delta\theta$

Velocity error $= \int B\cos\omega_0 t\, dt$

Distance error $= \int\int B\cos\omega_0 t\, dt\, dt$
Hence

Distance error $=$
$$\frac{B}{\omega_0^2}(1 - \cos\omega_0 t) \qquad (6.20)$$

Figure 6.12 shows the effect of an accelerometer bias error equivalent to an initial tilt of one minute of arc ($2.9 \times 10^{-4}g$). This results in an oscillatory error between zero and two nautical miles

(2) Effect of gyro drift
Assume $B = 0$ and gyro drift rate, W, is constant. The solution of equation (6.18) for initial conditions

$$\Delta\theta(0) = 0 \text{ and } \Delta\dot{\theta}(0) = W$$

is
$$\Delta\theta = \frac{W}{\omega_0}\sin\omega_0 t \qquad (6.21)$$

The platform thus oscillates about the vertical as before with an 84.4 minute period. A gyro drift rate of 0.01°/hour results in a very small amplitude oscillation about the local vertical of 0.14 minutes of arc

The velocity error =
$$\int \frac{gW}{\omega_0}\sin\omega_0 t \, dt$$

$$= WR\,(1-\cos\omega_0 t) \qquad (6.22)$$

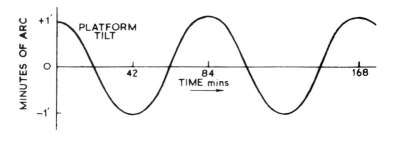

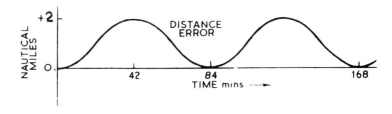

Fig. 6.12 Error propagation due to accelerometer bias of 1 minute of arc.

The velocity error thus oscillates between zero and $2WR$.

The distance error $= \int WR(1-\cos\omega_0 t)\,dt$

$$= WR\left(t-\frac{1}{\omega_0}\sin\omega_0 t\right) \qquad (6.23)$$

The distance error is thus proportional to the time of flight, the small amplitude oscillatory component being swamped after a while.

A constant gyro drift rate of $0.01°$/hour (0.6 minutes of arc per hour) will give an average velocity error of 0.6 knot, that is, a distance error which builds up at a rate of 0.6 NM/hour. It should be noted that this represents the error build up along the North and East axes, so that the radial error will be multiplied by $\sqrt{2}$ and hence will be equal to 1 NM/hour.

Figure 6.13 illustrates the error growth for a constant gyro drift of $0.01°$/hour.

In practice, this is not a constant quantity and the distance error over a long

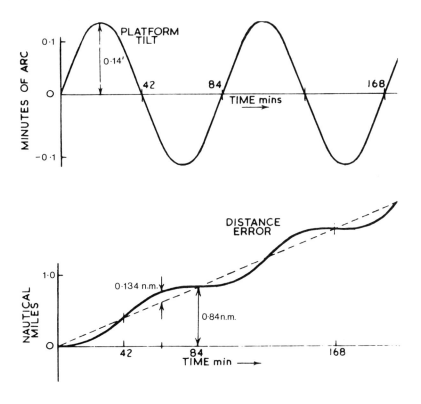

Fig. 6.13 Error propagation due to gyro drift of $0.01°$/hour.

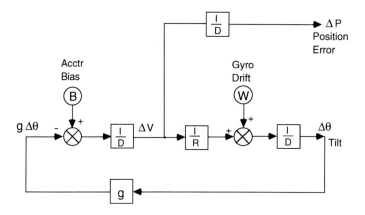

Fig. 6.14 Schuler tuned INS error model.

period is proportional to the square root of the time of flight, if the variation in the gyro drift is random.

The very accurate vertical reference provided by Schuler tuning can be seen – even a gyro drift as high as 0.25°/hour results in a peak vertical error of only 3.5 minutes of arc. Figure 6.14 is a block diagram of the error propagation model of a Schuler tuned IN system.

A physical explanation of Schuler tuning is set out below, as it is important to have a physical appreciation, as well as a mathematical understanding. Referring to Fig. 6.15, suppose the platform is being rotated at the same rate as the local vertical, but is tilted downwards from the local vertical by a small angle, $\Delta\theta$.

At $t = 0$ the platform tilt results in the accelerometers measuring a small negative component of the gravitational acceleration as well as the true acceleration. (Accelerometer input axis tilted downwards results in a negative gravitational acceleration component being measured, as the accelerometer proof mass tends to move in the same direction as when the vehicle is being retarded, i.e. negative acceleration.)

t = 0 to 21 minutes. As the aircraft flies over the earth the negative acceleration error resulting from the platform tilt is integrated with time and causes the computed vehicle rate to become slower than the true vehicle rate. The platform is thus being rotated at a progressively slower rate than the local vertical thereby reducing the tilt error.

At t = 21 minutes, the tilt error is now zero but the platform is still being rotated at a slower rate than the local vertical so that the platform then starts to tilt the other way (i.e. upwards) after 21 minutes.

t = 21 to 42 minutes, the acceleration error due to the platform tilt is now positive so that the platform rate of rotation starts to increase.

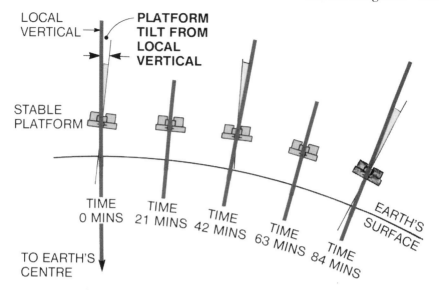

Fig. 6.15 Schuler oscillation.

At t = 42 minutes, the platform is now rotating at the same rate as the local vertical.
t = 42 to 63 minutes, after 42 minutes the platform 'vertical' is now starting to rotate at a faster rate than the true vertical and by 63 minutes the platform tilt is zero again. However, the platform is now being rotated at a faster rate than the local vertical.
t = 63 to 84 minutes, the tilt has changed sign so that the platform rate of rotation now starts to slow down until at 84 minutes it is rotating at the same rate as the local vertical.

The whole cycle is repeated over the next 84 minutes and so on.

6.2.3. Initial alignment and gyro compassing

Inertial navigation can only be as accurate as the initial conditions which are set in. It is, therefore, essential to know the orientation of the accelerometer measuring axes with respect to the gravitational vector, the direction of true North, the initial position and the initial velocity components, to very high accuracy.

The two basic references used to align an inertial system are the Earth's gravitational vector and the Earth's rotation vector.

The alignment process can be regarded as being in two parts, *viz.*, levelling and gyro compassing, although both operations can occur together.

The levelling operation takes place in two stages

(1) A coarse levelling when the accelerometer outputs, which are proportional to the platform tilt, are fed into the appropriate torque motors of the vertical gyro.
(2) This is followed by a fine levelling stage when the accelerometer outputs are filtered before being fed into the vertical gyro torque motors.

The filtering is achieved as follows. The accelerometer outputs are integrated to derive the horizontal velocity components of the aircraft, which should be zero as the aircraft is stationary on the ground. Any platform tilt errors, however, will introduce gravitational acceleration errors which will be integrated with time and so generate horizontal velocity errors. These velocity errors are, therefore, fed back appropriately to correct the tilt errors and level the system, the levelling loops being basically similar to those described in the previous chapter for a strap-down AHRS. The levelling loops are generally third order loops using the integrals of the velocity errors as well as the velocity errors as control terms. The feedback gains can also be varied.

This filtering is necessary to smooth out the spurious accelerations and noise which is being measured by the accelerometers when the aircraft is stationary on the ground. This is due to wind buffet, fuelling, crew and passengers moving about the aircraft etc.

A coarse azimuth alignment with respect to true North is made to within a degree or so using, say a magnetic reference. The fine alignment to achieve the required accuracy is accomplished by the process of gyro compassing. During the gyro compassing phase the platform azimuth is adjusted until the component of the earth's rotation sensed by the vertical gyro about the X axis is zero. The X axis is then aligned due east.

As shown in Chapter 5, the components of the earth's rate of rotation Ω about the North–South, East–West and vertical axes at a latitude of λ are:

North–South axis	$\Omega \cos \lambda$
East–West axis	0
Vertical axis	$\Omega \sin \lambda$

The inset diagram in Fig. 6.16 shows the components of the Earth's rotation rate which are sensed by the vertical gyro when the platform axes are misaligned in the horizontal plane by an amount $\Delta\psi$ from true North. It can be seen that a rotation rate equal to $\Omega \cos \lambda \sin \Delta\psi$ is measured about the X axis of the vertical gyro (nominal East pointing axis), that is, $\Omega \cos \lambda . \Delta\psi$, if $\Delta\psi$ is a small angle.

The North pointing or Y accelerometer will hence be tilting away from the horizontal at a rate $\Omega \cos \lambda . \Delta\psi$ in the absence of any erection or gyro compassing loops. This is assuming the appropriate precession currents are supplied to the vertical and azimuth gyros to compensate for $\Omega \cos \lambda$ and $\Omega \sin \lambda$ respectively.

A closed loop gyro compassing system can be formed as shown in Fig. 6.16 whereby the Y accelerometer output is fed to the azimuth gyro precession amplifier

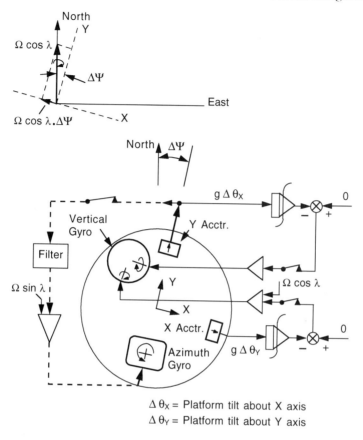

$\Delta \theta_X$ = Platform tilt about X axis
$\Delta \theta_Y$ = Platform tilt about Y axis

Fig. 6.16 Levelling and gyro compassing loops. Levelling loops shown with full lines. Gyro compassing loop shown with dotted lines.

as well as to the vertical gyro X axis precession amplifier. The X axis accelerometer is coupled to the vertical gyro Y axis precession amplifier. In the steady state (assuming perfect gyros and accelerometers) $\Delta \psi$ must come to zero and there will be zero tilt. The allowable vertical gyro drift rate can be determined from the accuracy required of the azimuth alignment. For example, if an accuracy of 0.1° is required for a latitude of 45°, then the component of the earth's rate sensed at this latitude with a misalignment of 0.1° is equal to $\dfrac{0.1}{57.3} \times 15 \sin 45°$ degrees per hour, that is 0.017 degrees per hour.

It can be seen that the magnitude of the component of earth's rate to be sensed decreases with increasing latitude, so that gyro compassing is effectively restricted to latitudes below 80°. It should be noted that the gyro compassing can be carried out within the system computer so that it is not necessary to precess (or torque) the azimuth gyro. Such a system is know as a *free azimuth system*.

The alignment of such a system is carried out by levelling the platform to the local horizontal plane by feeding the appropriate accelerometer outputs into the vertical gyro precession amplifiers. The orientation of the accelerometer input axes with respect to true North can be determined by measuring the accelerometer outputs and automatically adjusting the vertical gyro precession rates until the platform ceases to tilt. The platform azimuth angle can then be computed by comparing the gyro precession rates with the known earth's rotation rate about the North–South axis.

The major factors which affect alignment accuracy and alignment time are:

(a) Initial platform tilt
(b) Aircraft movements e.g. effect of wind gusts etc.
(c) Accelerometer bias errors and gyro drift rates
(d) Change of the above quantities (c) with time as the platform warms up
(e) Accelerometer resolution and gyro threshold.

The loop gains in the levelling and gyro compassing loops are generally controlled by means of a Kalman filter to give an optimal alignment process. Typical alignment times are of the order of seven minutes for full accuracy IN performance.

Reduced alignment times are sometimes used and the system corrected to give full IN accuracy by subsequent position fixes using a position fixing navigation system (e.g. GPS) It should be noted that strap-down IN systems carry out a similar levelling and gyro compassing alignment process within the system computer. This will be discussed in more detail in Section 6.2.7.

6.2.4. Effect of azimuth gyro drift

The effect of azimuth gyro drift is to cause position errors which build up with time and are a function of azimuth error, aircraft velocity and latitude.

From the previous section on gyro compassing it can be seen that an azimuth error $\Delta \psi$, will effectively inject a drift of $\Omega \cos \lambda$. $\Delta \psi$ into the X (or East) axis of the vertical gyro. An error of 0.2° for instance, would generate an effective drift of 0.033°/hour (two minutes of arc/hour approximately.) This is injected into the Schuler loop (Fig. 6.14) as a disturbance $W = 0.033$°/hour and would give a northerly velocity error of 2 NM/hour. An azimuth gyro drift rate will give a distance error which builds up proportionately to the square of the time. This gives a latitude error which affects the accuracy of the other correction terms, e.g. the meridian convergency term $\dfrac{V_E}{R} \tan \lambda$, causing still further error to build up with time. For this reason it is essential to have a good azimuth gyro performance.

6.2.5. Vertical navigation channel

It has been shown in Section 6.2.2 that Schuler tuning effectively constrains the error build up with time in the horizontal channels of the INS. Unfortunately, however, there is no such action in the vertical channel and small errors in computing the vertical acceleration correction terms are integrated with time – in fact the vertical distance error builds up exponentially with time. The principal correction terms for the vertical accelerometer channel are:

(i) Gravity – gravity varies with altitude according to the inverse square law (as explained in Chapter 2). The gravitational acceleration at an altitude H is given by:

$$g = g_0 \frac{R^2}{(R+H)^2}$$

where g_0 = gravitational acceleration at the earth's surface.

Errors in correcting for the variation in gravitational acceleration with height result in a vertical distance error which builds up as cosh function of time. Also, gravity varies over the earth's surface and has anomalous values in certain places.

(ii) Centrifugal acceleration – a vehicle moving over the earth's surface is describing a circular trajectory in space and so will experience a centrifugal acceleration $\dfrac{(V_E^2 + V_N^2)}{R}$ where V_E and V_N are the easterly and northerly velocity components of the vehicle respectively.

(iii) Coriolis acceleration – the component of the earth's rotation rate $\Omega \cos \lambda$ about the North axis is combined with the easterly velocity V_E and generates a Coriolis acceleration along the vertical axis which is equal to $2V_E \Omega \cos \lambda$.

It should be noted that there is also a small centrifugal acceleration due to the earth's rotation which slightly modifies the local value of gravity.

Monitoring of the inertially derived height with some other source of vertical information such as pressure (or barometric) altitude from the air data system, or a radio altimeter is thus essential for long time operation.

The 'Standard Atmosphere' which is used to derive the pressure (or barometric) altitude from the static pressure measurement has already been explained in Chapter 2. Errors can arise in the altitude (or height) derived from the static pressure, however, due to the variations in the lapse rate, the height of the tropopause and in sea level temperature and pressure. Even though a datum is established before take-off, climbing or diving through a non-standard atmosphere (i.e. one which differs from the assumed standard) can result in errors in the height rate of up to 8% with corresponding errors in the height.

An inertial system is accurate over a short period provided it is correctly initialised and it also has an excellent dynamic response. The optimum mixing of inertially derived and barometrically derived height is thus be able to compensate to a very considerable extent for the deficiencies present in the individual sources. The stages in deriving an optimum mixing system are briefly outlined below.

Consider first the simple second order mixing system shown in bold lines in Fig. 6.17(a).

The baro-inertial feedback loop is capable of being opened and the gains K_1 and K_2 which determine the damping and response times respectively are also capable of being varied.

The system response can be shown to be given by

$$H_I = \frac{D^2}{(D^2 + K_1 D + K_2)} \cdot H + \frac{(K_2 + K_1 D)}{(D^2 + K_1 D + K_2)} \cdot H_P + \frac{1}{(D^2 + K_1 D + K_2)} \cdot \delta a_z \quad (6.24)$$

where H = true height, H_P = pressure altitude,
δa_z = uncorrected vertical acceleration error and H_I = baro/inertial mixed height.

The inertially derived component of H_I, that is $\dfrac{D^2}{(D^2 + K_1 D + K_2)} \cdot H$, couples the

inertially measured height through a second order high pass filter transfer function

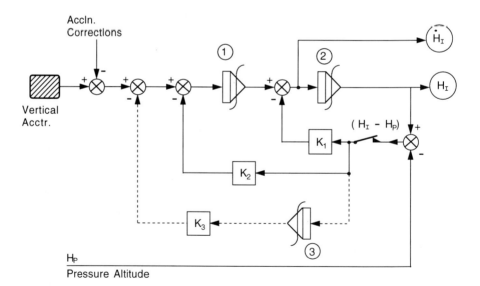

Fig. 6.17(a) Baro-inertial mixing. Simple second order baro-inertial mixing shown in bold lines. Integral term correction for inertial drift shown in dotted lines. (Third order system.)

which responds to dynamic changes in height without lag, but 'DC blocks' the steady state value.

The barometrically derived component of H_I, that is $\dfrac{(K_2 + K_1 D)}{(D^2 + K_1 D + K_2)} \cdot H_P$,

couples the barometrically derived height through a low pass filter which smooths the pressure height signal and attenuates the noise present in the signal. The dynamic response is thus slugged by the low pass filter but the steady state value of the pressure height is not affected. The combination of the two components results in a fast response with a low noise content.

However, there is no drift compensation for the inertial channel. Assuming the system is trimmed before flight, the dominant error would probably be poor 'g' compensation at high altitudes due to the error in the height so that δa_z will increase. This in turn will cause an increasing offset between H_P and H_I. In a closed loop climb, the inertial height, H_I, will be forced to follow changes in pressure height, H_P, with possible resultant height rate errors of up to 8%. An open loop climb, however, would result in the system becoming divergently unstable because of the uncompensated acceleration errors. The transient which would normally occur on closing the loop could be reduced by a 'fast re-set' phase in which the K_1 feedback gain is temporarily increased.

The next stage is to correct the inertial channel drift by the addition of an integral correction term $K_3 \int (H_P - H_I) \, dt$ which is fed to the input of the first integrator (1) as shown in the dotted lines on Figure 6.17(a). This would improve the open loop (i.e. pure inertial) climb performance as the inertial channel would start in a well trimmed state. However, even with a fast re-set the system cannot settle without a lengthy transient on re-closing the loop. This is because to force H_I to equal H_p requires additional drift compensation from integrator (3) to accommodate the new error in the 'g' correction following the open loop climb.

This situation can be improved by adding the fourth integrator (4) as shown in Fig. 6.17(b) to correct the pressure height error as the inertial system will measure the changes in height over a short period accurately if correctly initialised. This eliminates the transient on re-closing the inertial loop and makes an attempt to operate in 'true' height.

Optimum performance over the flight envelope requires the gains K_1, K_2, K_3 and K_4 to be variable. This can be effectively achieved by means of a Kalman filter. With a Kalman filter, the terms in the covariance matrix which compares the errors in the system state variables, reflect the previous history of the system. For example, if a prolonged period of level flight had resulted in a very well trimmed inertial system and hence correspondingly small values of variances associated with it, almost the whole discrepancy arising at the start of a climb would be attributed to an abnormal atmosphere. The Kalman filter can thus be designed to have a capability to learn the structure of the atmosphere and make short term use of the knowledge.

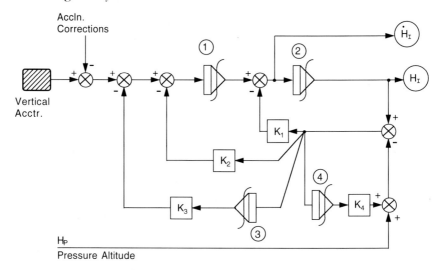

Fig. 6.17(b) Fourth order baro-inertial mixing (with correction for inertial channel drift and pressure altitude error).

Mention has been made earlier of the importance of being able to derive the aircraft's velocity vector from the ground speed vector V_G output of the INS and the baro/inertially derived vertical velocity, $\dot{H}_I$. The magnitude of the velocity vector is equal to $\sqrt{V_G^2 + \dot{H}_I^2}$. The angle the velocity vector makes with the horizontal is equal to $\tan^{-1} \dot{H}_I / V_G$.

6.2.6 Choice of platform axes

The 'local level' set of platform axes are generally used in a stable platform INS. The advantages are:

(a) The gimbal angles provide a direct read-out of the Euler angles of the aircraft, that is the heading, pitch and bank angles.
(b) The X and Y accelerometers measure the horizontal acceleration components directly. This avoids errors in measuring the horizontal accelerations due to inexact compensation of the gravitational acceleration – a significant advantage.

It can hence be seen why a locally level mechanisation is normally used and this leads to a choice of either 'North slaved azimuth' or 'free azimuth'.

The North slaved azimuth system can involve high gyro precession rates of the azimuth gyro at high latitudes due to the 'meridian convergency' term $\dfrac{V_E}{R}\tan\lambda$.

This term becomes infinite at the poles and hence the practical limits of latitude for this system are ±80°. The gyro compassing rate is also limited by the maximum gyro precession rate.

The correction terms for a North slaved locally level platform mechanisation are set out in Table 6.2.

Table 6.2 Correction terms

Precession term	X Axis	Y Axis	Z Axis
Earth's rate	0	$\Omega \cos \lambda$	$\Omega \sin \lambda$
Vehicle rate	V_N/R	V_E/R	$(V_E/R) \cdot \tan \lambda$

Where V_N and V_E are the northerly and easterly velocity components of the aircraft respectively.

As already mentioned, constraining the platform axes to be aligned with the earth's axes introduces additional accelerations, namely Coriolis accelerations as well as centrifugal accelerations. The Coriolis acceleration is equal to $2V\omega$, where V is the vehicle velocity with respect to the rotating reference axes and ω is the rate of rotation of the reference axes in space.

The various acceleration terms involved with a locally level, north slaved platform are set out in Table 6.3.

Table 6.3 Acceleration terms

Acceleration component	X Axis	Y Axis	Z Axis
	(East)	(North)	(Vertical)
Coriolis component	$2V_N \Omega \sin \lambda$	$2V_E \Omega \sin \lambda$	$2V_E \Omega \cos \lambda$
Centrifugal component	$V_N \dfrac{V_E}{R} \tan \lambda$	$\dfrac{V_E^2}{R} \tan \lambda$	$\dfrac{V_E^2 + V_N^2}{R}$
Gravitational component			$g_0 \dfrac{R^2}{(R+H)^2}$

Still further correction must be applied to allow for the fact that the earth is an ellipsoid and not a perfect sphere.

The free azimuth system has the major advantage that the azimuth gyro is not torqued, thereby eliminating the restrictions of being unable to 'follow' North at high latitudes and the limits to the maximum gyro compassing rate. It also avoids gyro torquer errors, scale factor and linearity etc. The aircraft heading with respect to true North is derived by the IN computer.

6.2.7 Choice of navigation co-ordinates

The choice of navigation co-ordinate system reduces practically to either:

(a) Spherical co-ordinates (latitude/longitude)
(b) Direction cosines.

A spherical co-ordinate system has two mathematical 'poles' or singularities where the value of one co-ordinate (the longitude) is indeterminate and near which the high rates of change of that co-ordinate make any system implementation impractical (longitude rate $\mu = \dfrac{V_E}{R} \sec \lambda$ becomes infinite when $\lambda = 90°$). With the latitude/longitude co-ordinate system (also known as the geographical or 'geodetic' system) these poles are the true North and South poles of the Earth. This co-ordinate system is satisfactory provided there is no requirement for navigation in the polar regions. It also provides the latitude/longitude information for display to the flight crew.

When polar navigational capability is required there are three possible ways to overcome the difficultly, namely:

(a) A spherical co-ordinate system with the poles removed to some other regions
(b) A 'unipolar' system
(c) Direction cosines.

The spherical co-ordinate system with transferred poles gives rise to added complications in the ellipsoidal corrections because of the asymmetry which results from the misalignment between the main co-ordinate axis and the axis of the geographic ellipsoid. More seriously, it is a solution of limited usefulness because the singularities still exist.

Space does not permit the 'unipolar' system to be discussed at length. It is a compromise solution which has just one singularity (which may be made the South pole if arctic navigation is required for instance.) Again it is not a general solution and has nothing to recommend it in preference to the third solution, namely the direction cosine system.

The direction cosine system is the system generally used. In this system the co-ordinates of a point are the three direction cosines of its radius vector in an orthogonal, Earth centred, Earth-fixed reference frame (together with its height above the Earth's surface). The attitude of the platform may also be described by the direction cosines of its axes in the same reference frame. This system has no singularities. The basic computations are not significantly more involved than those for latitude/longitude computations and present no problems to the modern microprocessor based system. An additional separate computation is required to derive latitude and longitude for display purposes but this presents no significant problems.

6.2.8 Strap-down IN systems

A strap-down IN system carries out the same functions as a stable platform type. INS and many elements and functional areas are common to both systems. There are two crucial areas in the strap-down mechanisation brought about the elimination of the gimbal system. These are:

(1) Attitude integration whereby the vehicle attitude is derived by an integration process from the body incremental angular rotations measured by the gyros.
(2) Accelerometer resolution whereby the corrected outputs of the body mounted accelerometers are suitably resolved to produce the horizontal and vertical acceleration components of the aircraft.

The derivation of the aircraft attitude from the strap-down gyro outputs by continuously updating the four Euler symmetrical parameters each iteration period has been explained in the preceding chapter (Section 5.3.3.1). Very high accuracy is required in the attitude integration process, the integration period should be as short as possible necessitating a high speed processor and accurate integration algorithms must be used (e.g. Runge-Kutta algorithms). The ortho-normalisation of the transition matrix is essential using the constraint equation for the Euler parameters $e_0^2 + e_1^2 + e_2^2 + e_3^2 = 1$.

The basic computing flow diagram for a strap-down INS is shown in Fig. 6.18. The velocity increments derived from the pulse rate outputs of the three body mounted accelerometers need to be corrected to allow for vehicle rotation during the integration interval when the velocity increments are being accumulated. These corrections are known as Coriolis body rotation corrections. The corrected velocity

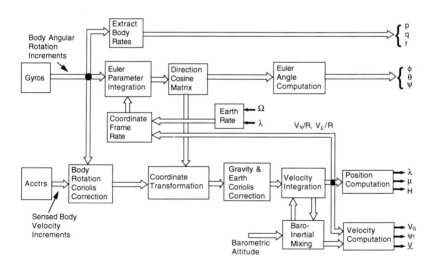

Fig. 6.18 Strap-down IN system.

increments along the vehicle body axes are then transformed by the direction cosine attitude matrix to yield the velocity increments with respect to earth reference axes. (locally level, North slaved or locally level, free azimuth). A set of velocity increments is thus obtained which is identical to that which would be obtained from a set of three accelerometers mounted on a stabilised Earth reference platform. The basic similarities to a stable platform INS mechanisation can be seen in Fig. 6.18 (as well as the differences).

The levelling and gyro compassing alignment of a strap-down INS is basically similar to that for a stable platform INS as described earlier.

The accelerometers for an aircraft strap-down INS are generally mounted along the aircraft's principal axes so that the 'horizontal' accelerometers mounted along the forward and side-slip axes do not sense a large component of gravity. The pitch and bank angles of the aircraft are small as the aircraft is normally fairly level when stationary on the ground. The aircraft attitude integration process, using the incremental body angular rotations measured by the pitch, roll and yaw strap-down gyros,can be initialised by assuming the pitch and bank angles are both zero (if these are not known). The fine levelling is carried out by using the fact that any tilt, $\Delta\theta_X$ and $\Delta\theta_Y$ of the computed horizontal plane from the true horizontal will couple gravitational acceleration components $g\Delta\theta_X$ and $g\Delta\theta_Y$ into the horizontal acceleration components derived from the Y and X accelerometers. The horizontal acceleration components are then integrated with respect to time to produce the horizontal velocity components of the aircraft. These horizontal velocity components should be zero as the aircraft is stationary on the ground. Any resulting horizontal velocity components that are measured are therefore fed back appropriately to correct the tilt and level the system.

The basic gyro compassing loop adjusts the computed heading until the East component of the gyro angular rate measurement in the horizontal plane is nulled, the angular rate of rotation about the East axis being estimated from the summed East axis tilt correction.

6.2.9 Aided IN systems and Kalman filters

The time dependent error growth of an IN system makes it necessary that some form of aided navigation system using an alternative navigation source is introduced to correct the INS error build up on long distance flights. For example, as mentioned earlier, a good quality unaided INS of 1 NM/hour accuracy could have a positional error of 5 NM after a five hour flight.

A variety of navigation aids can be used for this purpose, for example GPS, OMEGA, LORAN, VOR/DME etc.

Consider, firstly, a simple position reset at each position fix as shown in Fig. 6.19. The error growth is limited but follows a saw tooth pattern, the amplitude

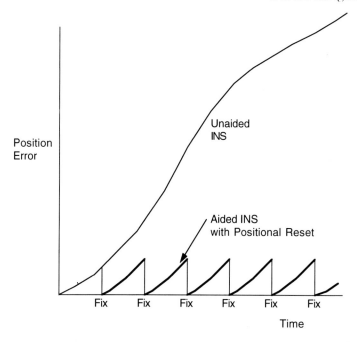

Fig. 6.19 Aided INS with simple positional reset.

depending on the period between updates and the magnitude of the inertial velocity errors and the tilt errors.

Now suppose the errors present in the INS, such as attitude errors, velocity errors, gyro drifts, accelerometer errors etc. could be determined from the positional fix information. This is what a Kalman filter does and corrections can then be applied to the INS as shown in the block diagram in Fig. 6.20. The Kalman filter provides an optimum estimate of the IN system errors taking into account the errors present in the position fixing system. The resulting error propagation using a Kalman filter to estimate and correct the INS errors follows the pattern shown in Fig. 6.21 – a substantial improvement. An accurate velocity reference system, such as a Doppler radar velocity sensor, can also be used in conjunction with a Kalman filter to estimate and correct the INS errors. In fact, a number of navigational aids can all be combined with an INS in an optimal manner by means of a Kalman filter. The dissimilar nature of the error characteristics of an INS and the various position (and velocity) aids is exploited by the Kalman filter to achieve an overall accuracy and performance which is better than the individual systems. The optimum blending of the individual system characteristics which can be achieved can be seen from the brief summary of the various navigation sources below. The complementary filtering of dissimilar systems using simple fixed gain mixing has

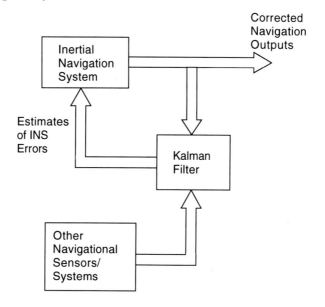

Fig. 6.20 Block diagram of aided INS with Kalman filter.

already been discussed in Chapter 5 and also in the preceding section on baro/inertial mixing. To recap, the various navigation sources comprise:

(1) Position Data

- GPS, OMEGA, LORAN, VOR/DME, TACAN
- Terrain reference navigation systems
- Radar

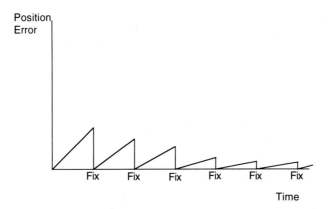

Fig. 6.21 Aided INS with Kalman filter.

- Visual fixes (eg use of helmet mounted sight)
- Astro (Stellar) navigation (using automatic star trackers)

(2) Velocity Data

- Doppler radar
- GPS

(3) Altitude Data

- Barometric altitude from the air data computer
- Radio altimeter

These sources provide good information on the average at low frequency but are subject to high frequency noise due to causes such as instrument noise, atmospheric effects, antenna oscillation, unlevel ground effects etc.

In contrast, IN systems provide good high frequency information content (above the Schuler frequency) despite vehicle motion. The low frequency information, however, is poor due to the inherent long term drift characteristics, as already explained.

It should be stressed at this point that a Kalman filter can be used to provide an optimum estimate of the errors in any measuring system and its use is not confined to navigation systems, although it has been particularly effective in this field. Examples of other applications include:

- Radar and infra-red automatic tracking systems
- Fault detection and in monitoring of multiple (redundant) sensors
- Initial alignment and gyro compassing of an INS.

The Kalman filter was first introduced in 1960 by Dr Richard Kalman (see reference list at the end of the chapter). It is essentially an optimal, recursive data processing algorithm which processes sensor measurements to estimate the quantities of interest (states) of the system using:

(1) A knowledge of the system and the measurement device dynamics
(2) A statistical model of the system model uncertainties, noises, measurement errors
(3) Initial condition information

The recursive nature of the filter, that is using the same equations over and over again make it ideally suited to a digital computer. The filter only requires the last value of the state of the system it be stored and does not need the value of the old observations to be retained. This considerably reduces the amount of computer storage required.

The basic application of a Kalman filter to a mixed navigation system is shown in the flow diagram in Fig. 6.22. The filter contains an error model for all the systems involved, enabling the dynamic behavior of the system errors to be

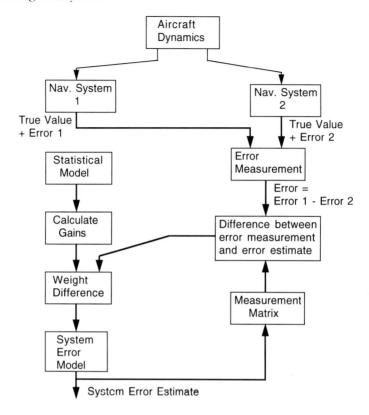

Fig. 6.22 Application of Kalman filter to mixed navigation systems.

modelled. The computer contains a current estimate for each term in the error model and this estimate, which is based on all previous measurements, is periodically updated. At the time of each new measurement, the difference in the outputs of the systems is predicted based on the current estimate of the errors in the systems. This difference between the predicted and actual measurements is then used to update each of the estimates of the errors through a set of weighting coefficients – that is the Kalman gains. The weighting coefficients are variables which are computed periodically in the system computer and are based on the assumed statistical error model for the errors. The configuration takes into account the past history of the system including the effects of previously applied information and of vehicle motions which affect the system errors.

A fundamental feature of the Kalman filter is that the error measurements made of one quantity (or set of similar quantities) can be used to improve the estimates of the other error quantities in the system. For example, the Kalman filtering technique generates an improvement in the INS velocity accuracy by virtue of the strong correlation between position error (the measured quantity) and velocity error, which is essentially the direct integral of the position error.

An introductory overview of Kalman filtering is set out in the following pages with the aim of explaining some of the essential processes and the terms used so that the interested reader can follow up the extensive literature on the subject (references are given at the end of the chapter). It should be noted that applications of Kalman filtering alone can occupy a fairly thick book – for instance AGARDOGRAPH No. 139 'Theory and Application of Kalman Filtering is a soft cover book nearly 2.5cm (1 inch) thick.

An essential element in the Kalman filter is the *System Error Model* which models the dynamic behaviour of the system errors. The system dynamic behaviour can be represented by *n* linear differential equations, where *n* is the number of state variables in the system. In matrix form this becomes

$$\dot{\mathbf{X}} = \mathbf{A}\mathbf{X} + \mathbf{B}\mathbf{U} \qquad (6.25)$$

where **X** is the system state vector comprising *n* state variables, **A** is the coefficient or plant matrix, **B** is the driving matrix and **U** is the input state vector. (Bold letters denote matrices.)

In deriving the Kalman filter, **U** is assumed to be a vector of unbiased, white, Gaussian noise sequences.

The state equations for the errors in one axis of an IN system are derived below as an example. Referring to the simplified block diagram, Fig. 6.14:

Position error = ΔP
Velocity error = ΔV
Tilt error = $\Delta\theta$
Gyro drift error = W
Accelerometer bias error = B

The relationships between these variables are set out below

$$\Delta\dot{P} = \Delta V \qquad (6.26)$$

$$\Delta\dot{V} = -g\Delta\theta + B \qquad (6.27)$$

$$\Delta\dot{\theta} = \frac{1}{R}\Delta V + W \qquad (6.28)$$

These equations can be represented more compactly in matrix form as shown below.

$$\begin{bmatrix} \Delta\dot{P} \\ \Delta\dot{V} \\ \Delta\dot{\theta} \end{bmatrix} = \begin{bmatrix} 0 & 1 & 0 \\ 0 & 0 & -g \\ 0 & \frac{1}{R} & 0 \end{bmatrix} \begin{bmatrix} \Delta P \\ \Delta V \\ \Delta\theta \end{bmatrix} + \begin{bmatrix} 1 & 0 \\ 0 & 1 \end{bmatrix} \begin{bmatrix} B \\ W \end{bmatrix} \quad (6.29)$$

ie

$$\dot{\mathbf{X}} = \mathbf{A}\mathbf{X} + \mathbf{B}\mathbf{U}$$

As explained in Chapter 5, the state transition matrix Φ_n, relates the system state vector at time nT, that is $\mathbf{X}_n$, to the system state vector at time $(n + 1)T$, that is $\mathbf{X}_{(n+1)}$, where T is the iteration period. The relationship is shown below:

$$\mathbf{X}_{(n+1)} = \Phi_n \cdot \mathbf{X}_n \quad (6.30)$$

Assuming a linear system

$$\Phi_n = e^{\mathbf{A}T} = 1 + \mathbf{A}T + \frac{\mathbf{A}T^2}{2} + \cdots \quad (6.31)$$

If the iteration period T is short

$$\Phi_n \approx 1 + \mathbf{A}T + \mathbf{A}^2 \frac{T^2}{2} \quad (6.32)$$

For the above simple example

$$\Phi_n = \begin{bmatrix} 1 & 0 & 0 \\ 0 & 1 & 0 \\ 0 & 0 & 1 \end{bmatrix} + \begin{bmatrix} 0 & 1 & 0 \\ 0 & 0 & -g \\ 0 & \frac{1}{R} & 0 \end{bmatrix} T + \begin{bmatrix} 0 & 1 & 0 \\ 0 & 0 & -g \\ 0 & \frac{1}{R} & 0 \end{bmatrix} \begin{bmatrix} 0 & 1 & 0 \\ 0 & 0 & -g \\ 0 & \frac{1}{R} & 0 \end{bmatrix} \frac{T^2}{2}$$

$$\Phi_n = \begin{bmatrix} 1 & T & -gT^2/2 \\ 0 & (1-gT^2/2R) & -gT \\ 0 & T/R & (1-gT^2/2R) \end{bmatrix} \quad (6.33)$$

hence

$$\Delta P_{(n+1)} = \Delta P_n + T\Delta V_n - gT^2/2 \cdot \Delta\theta_n \quad (6.34)$$

etc.

From initial estimates of the uncertainties in ΔP, ΔV, $\Delta \theta$ at time, $t = 0$, the values at time $t =(n+1)T$ can thus be derived by a step by step integration process using the transition matrix at each iteration.

As stated earlier this is a simplified example to illustrate the process of deriving a dynamic error model.

A typical Kalman filter INS error model consists of 17–22 error states. The 19 state model comprises the following error states: two horizontal position errors, two horizontal velocity errors, three attitude errors, three gyro biases, three gyro scale factor errors, three accelerometer bias errors, three accelerometer scale factor errors. The vertical channel is not modelled since it is unstable, as shown earlier, because of the gravity compensation with altitude which results in the positive feedback of altitude errors. The barometric altitude information from the air data system is combined with the INS vertical channel as shown earlier in a baro/inertial mixing loop. The GPS received error model typically comprises 12 states.

The measurement matrix, $\mathbf{H}$, is used to select the part or the component of the state vector $\mathbf{X}$ which is being measured. For example, suppose the state vector

$$\mathbf{X} = \begin{bmatrix} \Delta P \\ \Delta V \\ \Delta \theta \\ \Delta P_2 \end{bmatrix}$$

where ΔP, ΔV, $\Delta \theta$ are the inertial system errors and ΔP_2 is the error in the position reference system. To extract $(\Delta P - \Delta P_2)$ the measurement matrix $\mathbf{H} =. [1\ 0\ 0\ -1]$

$$\mathbf{H} \cdot \mathbf{X} = \begin{bmatrix} 1 & 0 & 0 & -1 \end{bmatrix} \begin{bmatrix} \Delta P \\ \Delta V \\ \Delta \theta \\ \Delta P_2 \end{bmatrix} = \Delta P - \Delta P_2$$

The covariance matrix of the estimation errors is formed by multiplying the error state vector matrix by its transpose - the transpose of a matrix means interchanging the rows and columns of the matrix.

For example, matrix $\qquad \mathbf{M} = \begin{bmatrix} a & b \\ c & d \end{bmatrix}$

The transpose of matrix $\mathbf{M}$ denoted $\mathbf{M}^T$ is thus

$$\mathbf{M}^T = \begin{bmatrix} a & c \\ b & d \end{bmatrix}$$

As a simple example, the covariance matrix, P, of a state vector $\mathbf{X}$, comprising position, velocity and tilt errors,

that is

$$\mathbf{X} = \begin{bmatrix} \Delta P \\ \Delta V \\ \Delta \theta \end{bmatrix}$$

is thus

$$\mathbf{P} = \mathbf{X}\mathbf{X}^T. \tag{6.35}$$

$$\mathbf{P} = \begin{bmatrix} \Delta P \\ \Delta V \\ \Delta \theta \end{bmatrix} \begin{bmatrix} \Delta P & \Delta V & \Delta \theta \end{bmatrix}$$

$$\mathbf{P} = \begin{bmatrix} \Delta P^2 & \Delta P \cdot \Delta V & \Delta P \cdot \Delta \theta \\ \Delta V \cdot \Delta P & \Delta V^2 & \Delta V \cdot \Delta \theta \\ \Delta \theta \cdot \Delta P & \Delta \theta \cdot \Delta V & \Delta \theta^2 \end{bmatrix}$$

It should be noted that the covariance matrix is symmetrical about a diagonal, with the diagonal elements comprising the mean square position, velocity and tilt errors respectively. The off diagonal terms are cross-correlations between these same three quantities.

The covariance matrix changes with time as the initial errors propagate with time. It is modified with time by means of the transition matrix using the following relationship:

$$\mathbf{P}_{(n+1)} = \boldsymbol{\Phi}_n \cdot \mathbf{P}_n \cdot \boldsymbol{\Phi}_n^T \tag{6.36}$$

($\boldsymbol{\Phi}_n^T$ is the transpose of $\boldsymbol{\Phi}_n$)

The use of the Kalman filter equations is briefly set out below. The purpose of the Kalman filter is to provide an optimum estimate of the system state vector at iteration n. This is denoted by $\hat{\mathbf{X}}_n$; the circumflex indicating a best estimate. The filter may be described in two stages. 'Extrapolation' indicates the period during which the filter simulates the action of the system between measurements. 'Update' occurs when a measurement is made on the system and is incorporated into the filter estimate. Quantities after extrapolation, immediately preceding the nth update are shown in the manner $\hat{\mathbf{X}}_n$ (−), and those immediately succeeding that update as $\hat{\mathbf{X}}_n$ (+). During extrapolation the filter estimate of the system state vector is modified according to the best available knowledge of the system dynamics

$$\hat{\mathbf{X}}_n(-) = \Phi_{n-1} \cdot \hat{\mathbf{X}}_{n-1}(+) \tag{6.37}$$

The error covariance matrix is also extrapolated

$$\mathbf{P}_n(-) = \Phi_{n-1} \cdot \mathbf{P}_{n-1}(+) \cdot \Phi_{n-1}^T + \mathbf{Q} \tag{6.38}$$

where $\mathbf{Q}$ is the covariance matrix of the random system disturbances.

During the filter update procedure, the difference between the actual measurement and a measurement made of the estimated state is weighted and used to modify the estimate state

$$\hat{\mathbf{X}}_n(+) = \hat{\mathbf{X}}_n(-) + \mathbf{K}_n \cdot \left[\mathbf{Z}_n - \mathbf{H}\hat{\mathbf{X}}_n(-)\right] \tag{6.39}$$

$\mathbf{Z}_n$ is the actual measurement made of the state variables and $\mathbf{K}_n$ is the Kalman gain matrix.

The weighting factors, or Kalman gains, are calculated from the current estimate of the error covariance.

$$\mathbf{K}_n = \mathbf{P}_n(-) \cdot \mathbf{H}^T \cdot \left[\mathbf{H} \cdot \mathbf{P}_n(-) \cdot \mathbf{H}^T + \mathbf{R}\right] \tag{6.40}$$

where $\mathbf{R}$ is the measurement noise covariance matrix.

The error covariance matrix is then also updated

$$\mathbf{P}_n(+) = \left[1 - \mathbf{K}_n \mathbf{H}\right] \cdot \mathbf{P}_n(-) \tag{6.41}$$

6.3 GPS – GLOBAL POSITIONING SYSTEM

6.3.1 Introduction

GPS is basically a radio navigation system which derives the user's position from the radio signals transmitted from a number of orbiting satellites. The fundamental difference between GPS and other radio navigation systems (such as LORAN and OMEGA) is simply the geometry of propagation from ground based transmitters compared with space borne transmitters. An orbiting satellite transmitter can provide line of sight propagation over vast areas of the world. This avoids the inevitable trade-offs of less accuracy for greater range which are inherent with systems using ground based transmitters. The satellite signals also penetrate the ionosphere rather than being reflected by it so that the difficulties encountered with sky waves are avoided.

GPS provides a superior navigation capability to either LORAN or OMEGA and although both these systems will be maintained for a number of years, GPS will supersede them. For these reasons and also space constraints, coverage of radio navigation systems has been confined to GPS.

Satellite navigation can be said to have started with the successful launching by the Russians of the world's first orbiting satellite, SPUTNIK 1 in October 1957. The development of the first satellite navigation system TRANSIT 1, was triggered by observations made on the radio signals transmitted from SPUTNIK 1 and was initiated at the end of 1958. TRANSIT 1 resulted in a worldwide navigation system which has been in continuous operation since 1964.

GPS started as a series of preliminary system concept studies and system design studies in the late 1960's. A Phase 1 'Concept and Validation Programme' was carried out from 1973 to 1979 followed by a Phase 2 'Full Scale Development and System Test Programme' from 1979 to 1985. The Phase 3 'Production and Deployment Programme' was initiated in 1985. Twelve development satellites have been used to develop and prove the system and the first production satellite was launched in February 1989. Some delay to the program was incurred by the space shuttle CHALLENGER disaster in 1986, as it had been originally intended to insert all the production standard GPS satellites into orbit using the space shuttle. A DELTA 2 launch vehicle has been used subsequently and deployment of the planned 24 production standard GPS satellites in orbit is nearing completion.

6.3.2. GPS system description

The overall GPS system comprises three segments, namely the space segment, the control segment and the user segment and is shown schematically in Fig. 6.23. The three segments are briefly summarised below.

Space Segment This comprises up to 24 GPS satellites placed in six orbital planes at 55° to the equator in geo-synchronous orbits at 20,000km above the earth. The orbit tracks over the earth, in fact, form an 'egg beater' type pattern.

21 satellites are required for full world wide coverage and three satellites act as orbiting spares.

The GPS satellites use two frequency transmissions, L1 at 1575.42 MHz and L2, at 1227.6 MHz for transmitting the digitally encoded navigation message data at 50 Hz modulation on both the L1 and L2 channels. The navigation message data will be explained in more detail in the next section but basically comprises the satellite

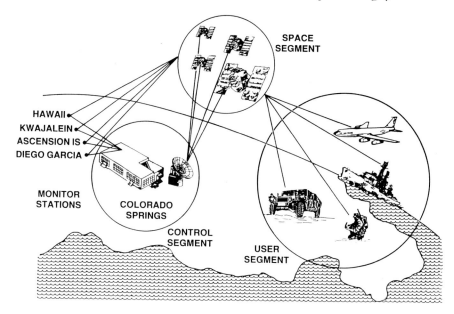

Fig. 6.23 The GPS system.

orbital position parameters, clock correction parameters and health information for itself and the other satellites, and the almanac data for all the satellites.

Spread spectrum techniques are used on both the L1 and L2 frequency channels.

L1 provides the coarse/acquisition or C/A code which is available to all users and which has been deliberately degraded. It also provides the precision or P code which is encrypted (Y code), and is only available to authorised military users.

L2 also provides the precision or P code. The correction for the degradation of accuracy of the C/A code L1 can be made by the military GPS receiver. L2 also enables corrections to be made for ionospheric delay uncertainties, the dual frequency transmission enabling these corrections to be derived.

Control Segment This comprises a Master Control Station at Colorado Springs in the USA and five monitor stations located world wide. The control segment is operated by the United States Department of Defense (DoD). The control segment tracks the satellites and predicts their future orbital position data and the required satellite clock correction parameters, and updates each satellite on the uplink as it goes overhead.

The GPS full system accuracy is only available when the operational control system is functioning properly and navigation messages are uploaded on a daily basis. The GPS satellites are, however, designed to function with the control system inoperable for a period of 180 days with gradually degraded accuracy. This gives the GPS system a high degree of robustness.

User Segment The user segment equipment as mentioned earlier is entirely passive and comprises a GPS receiver. A wide variety of compact, light weight miniaturised GPS receivers is now available, all using the same basic concepts.

The user system operation is very briefly as follows. The operator first enters the estimated present position and the time. The GPS receiver then starts to search for and track satellites. The data coming in identifies the satellite number, locates the satellite in space and establishes the system time. As will be explained in the next section the GPS receiver needs to track the signals from at least four satellites to determine the user's position.

As mentioned in the introduction to this chapter, the user's position is determined to a horizontal 2D accuracy of 100m RMS for civil users. Military users with access to the P code can determine their 3D position to an accuracy of 16m RMS, 3D velocity to 0.1m/s RMS by measuring the Doppler shifts, and time to within 100 ns (1 sigma).

6.3.3 Basic principles of GPS

The basic principle of position determination using the GPS system is to measure the spherical ranges of the user from a minimum of four GPS satellites. The orbital positions of these satellites relative to the earth are known to extremely high accuracy and each satellite transmits its orbital position data.

The spherical range of the user from the individual transmitting satellite can be determined by measuring the time delay for the satellite transmission to reach the user. Multiplying the time delay by the velocity of light then gives the spherical range, R, of the user from the transmitting satellite. The user's position hence lies on the surface of a sphere of radius, R, as shown in Fig. 6.24.

The system depends on precise time measurements and requires atomic clock

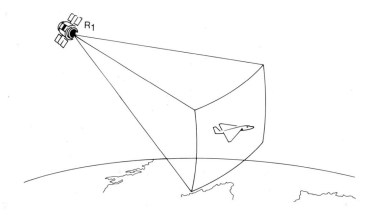

Fig. 6.24 GPS spherical ranging.

reference standards. The need for extremely high accuracy in the time measurement can be seen from the fact that a 10 ns (10^{-8} seconds) time error results in a distance error of 3 metres, as the velocity of light is 3×10^8 m/s.

Each GPS satellite carries an atomic clock which provides the time reference for the satellite data transmission. Assume for the moment that this time is perfect – the corrections required will be explained shortly. Given a perfect time reference in the user equipment, measurement of the spherical ranges of three satellites would be sufficient to determine the user's position. The user's equipment, however, has a crystal clock time reference which introduces a time bias in the measurement of the transit times of the satellite transmissions. The measurement of the time delay is thus made up of two components. The first component is the transit time of the ranging signal and the second component is the time offset between the transmitter clock and the receiver clock due to the non-synchronisation of the clocks. Measuring the spherical ranges from four satellites as shown in Fig. 6.25 enables the user's position to be determined and yields four equations containing the four unknowns, viz the three position co-ordinates of the user and the time bias in the user's clock. The position co-ordinates of the user can thus be determined together with very accurate time information. Figure 6.26 shows the data transmission waveforms and illustrates the user time bias ΔT, and the time delays Δt_1, Δt_2, Δt_3 and Δt_4 for the signals transmitted from the satellites to reach the user.

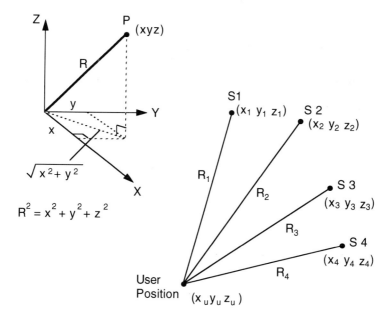

Fig. 6.25 User–satellite geometry.

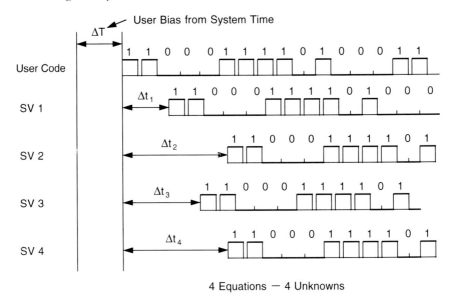

Fig. 6.26 GPS satellite signal waveforms with perfect satellite clocks.

Four pseudo ranges R_{1p}, R_{2p}, R_{3p}, R_{4p} to the four satellites S1, S2, S3, S4 can be determined, *viz* :

$$R_{1p} = c\Delta t_1$$
$$R_{2p} = c\Delta t_2$$
$$R_{3p} = c\Delta t_3 \qquad (6.42)$$
$$R_{4p} = c\Delta t_4$$

Let the range equivalent of the user's clock offset = C_u

ie $C_u = c\Delta T$

Hence, from basic 3D co-ordinate geometry (see Fig. 6.22)

$$R_1 = \left[(X_1 - X_u)^2 + (Y_1 - Y_u)^2 + (Z_1 - Z_u)^2 \right]^{1/2} = R_{1p} + C_u \qquad (6.43)$$

$$R_2 = \left[(X_2 - X_u)^2 + (Y_2 - Y_u)^2 + (Z_2 - Z_u)^2 \right]^{1/2} = R_{2p} + C_u \qquad (6.44)$$

$$R_3 = \left[(X_3 - X_u)^2 + (Y_3 - Y_u)^2 + (Z_3 - Z_u)^2 \right]^{1/2} = R_{3p} + C_u \qquad (6.45)$$

$$R_4 = \left[(X_4 - X_u)^2 + (Y_4 - Y_u)^2 + (Z_4 - Z_u)^2 \right]^{1/2} = R_{4p} + C_u \qquad (6.46)$$

when R_1, R_2, R_3, R_4 are the actual ranges from the user's position to the four satellites S1, S2, S3, S4 and the coordinates of these satellites are $(X_1\ Y_1\ Z_1)$, $(X_2\ Y_2\ Z_2)$, $(X_3\ Y_3\ Z_3)$, $(X_4\ Y_4\ Z_4)$ respectively.

These four equations with four unknowns can thus be solved and yield the user's position coordinates $(X_u\ Y_u\ Z_u)$ and the user's time offset, ΔT.

The assumption of perfect satellite clocks made initially in the discussion, however, is not a valid one and in fact the clocks are slowly but steadily drifting away from each other. The satellite clocks are therefore, mathematically synchronised to a defined GPS Master time which is maintained at the Master Control Station. This GPS Master time is continuously monitored and related to the Universal Time Coordinate (UTC) maintained by the United States Naval Observatory.

Each satellite time is related to GPS time by a mathematical expression and the user corrects the satellite time to the GPS time using the equation

$$t = t_{s/c} - \Delta t_{s/c} \qquad (6.47)$$

where t is the GPS time in seconds, $t_{s/c}$ is the effective satellite time at signal transmission in seconds and $\Delta t_{s/c}$ is the time offset between the satellite and GPS master time.

The time offset is $\Delta t_{s/c}$ computed from the following equation

$$\Delta t_{s/c} = a_0 + a_1 (t - t_{o/c}) + a_2 (t - t_{o/c})^2 + \Delta t_r \qquad (6.48)$$

where a_0, a_1, and a_2 are polynomial coefficients representing the phase offset, frequency offset and ageing term of the satellite clock with respect to the GPS master time and Δt_r is the relativistic term (seconds). The parameter t is the GPS time and $t_{o/c}$ is the *epoch time* at which the polynomial coefficients are referenced and generally $t_{o/c}$ is chosen at the mid point of the fit interval. The polynomial coefficients a_0, a_1, a_2 are estimated by the control segment for each satellite clock and periodically uplinked to the satellite.

These coefficients are transmitted together with the satellite orbital position data, termed the *Ephemeris parameters*, to the navigation user equipment as navigation messages. The clock corrections for the four satellites are designated τ_1, τ_2, τ_3, and τ_4.

The spherical ranges R_1, R_2, R_3 and R_4 are thus given by

$$
\begin{aligned}
R_1 &= c\,(\Delta T_1 + \Delta T - \tau_1) \\
R_2 &= c\,(\Delta T_2 + \Delta T - \tau_2) \\
R_3 &= c\,(\Delta T_3 + \Delta T - \tau_3) \\
R_4 &= c\,(\Delta T_4 + \Delta T - \tau_4)
\end{aligned}
\tag{6.50}
$$

All satellite clocks are mathematically synchronised to the GPS master time by means of these clock correction terms. The error in synchronisation will grow, however, if the polynomial coefficients a_0, a_1 and a_2 are not updated periodically.

The navigation user requires the Ephemeris parameters, that is the instantaneous position data of the GPS satellites which are being used for range measurement, as well as the clock parameters in order to compute the users position. The Ephemeris parameters defining the satellite orbital position data with respect to earth reference axes comprise 16 parameters in all. Figure 6.27 illustrates the satellite–Earth geometry and the definition of the orbit parameters. The control segment processes the tracking data acquired from the monitor stations to generate the orbit estimates for the GPS satellites. The predicted estimates of the satellite position co-ordinates are generated by integrating the equations of motion of the GPS satellites. These Cartesian position co-ordinates are then fitted mathematically over a specified interval of time to compute the Ephemeris parameters.

Both clock and Ephemeris parameters are down linked to the user at 50 bits per second data rate modulated in both C/A and P code (Y code) navigation signals. The navigation message uses a basic format consisting of a 1500 bit long frame made up of five subframes, each subframe being 300 bits long. Subframes 4 and 5 are subcommuted 25 times each so that a compete data message takes a transmission of 25 full frames. Subframe 1 contains the clock parameters and subframes 2 and 3 the Ephemeris parameters for the satellite. Subframes 1, 2 and 3 are repeated every 30 seconds so that it is possible for the user to update the clock and Ephemeris parameters every 30 seconds. Subframes 4 and 5 have each 25 pages so that these subframes are repeated only once in 12.5 minutes.

In addition to the Ephemeris and clock parameters of the satellite, each satellite transmits almanac data of all satellites to the user. This is primarily to facilitate satellite acquisition and to compute what is known as the *geometric dilution of precision* (GDOP) values to assist the selection of satellites to achieve better accuracy. Health data for all the satellites is also transmitted. The almanac data and satellite health data are contained in subframes 4 and 5. Ionospheric data for a single frequency user and the conversion parameters from GPS time to UTC are contained in subframe 4.

The navigation equations are basically non-linear as can be seen in equations (6.43), (6.44), (6.45) and (6.46), but can be linearised about nominal values for their solution by applying Taylor's series approximations. Space constraints limit further treatment of this topic.

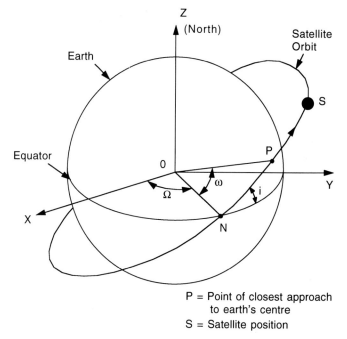

P = Point of closest approach
   to earth's centre
S = Satellite position

Earth Centred, Earth Fixed (ECEF) inertial co-ordinate frame
   Z axis North
   X axis in plane of equator through γ, the first point of Aries
   Y axis in plane of equator at right angles to X axis
   i = inclination of satellite orbit
   Ω = right ascension of ascending mode
   ω = argument of the perigee
   (perigee, P, is point of closest approach)
   Angle NOS = argument of latitude, and defines satellite position, S, at any time.

Fig. 6.27 Satellite–Earth geometry. Ephemeris parameters definition.

6.3.4. Integration of GPS and INS

GPS and INS are wholly complementary and their information can be combined to the mutual benefit of both systems. For example:
• Calibration and correction of INS errors – the GPS enables very accurate calibration and correction of the INS errors in flight by means of a Kalman filter.
• The INS can smooth out the step change in the GPS position output which can occur when switching to another satellite because of the change in inherent errors.

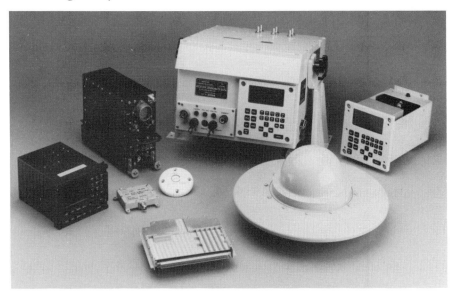

Fig. 6.28(a) GPS hardware.

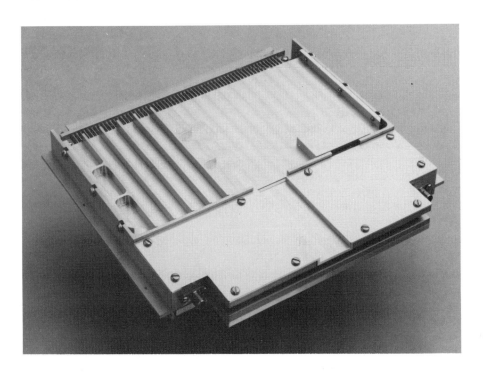

Fig. 6.28(b) GPS module.

• Jamming resistance – like any radio system, GPS can be jammed, albeit over a local area, although it can be given a high degree of resistance to jamming. The INS, having had its errors previously corrected by the Kalman filter, is able to provide accurate navigation information when the aircraft is flying over areas subjected to severe jamming.

• Antenna obscuration – GPS is a line of sight system and it is possible for the GPS antenna to be obstructed by the terrain or aircraft structure during manoeuvres.

• Antenna location corrections – the GPS derived position is valid at the antenna and needs to be corrected for reference to the INS location. The INS provides attitude information which together with the lever arm constants enables this correction to be made.

Figure 6.28(a) and (b) are photographs showing typical GPS hardware (courtesy of GEC-Marconi Defence Systems Ltd).

6.4 TERRAIN REFERENCE NAVIGATION

6.4.1 Introduction

Terrain reference navigation (TRN) is a generic classification which covers any technique of aided navigation which relies on correlating terrain measurement data from a suitable terrain sensor (or sensors) with the data stored in a digital map data base. The system is operated in conjunction with a DR navigation system; the position fixes derived by the TRN system are then used to update and correct the DR system errors by means of a Kalman filter. TRN systems can only be used over land and require a very accurate data base. The latter may need to be derived from satellite data when accurate map data are not available. Terrain reference navigation systems can be divided into three basic types:

(1) Terrain contour navigation (TCN)   In this system, the terrain profile is measured by a radio altimeter and is matched with the stored terrain profile in the vicinity of the aircraft's position as previously estimated by the DR system.

(2) Terrain characteristic matching (TCM)   This type of system can use a variety of sensors – eg radar or radiometric sensors to sense changes in the terrain characteristics below the aircraft. For example, flying over lakes, rivers, roads, woods, buildings etc. The data can then be processed to detect the edges or boundaries where the terrain characteristics change abruptly. The detected edges of the overflown features can then be matched with the stored terrain features in the vicinity of the aircraft's estimated position.

Both TCN and TCM can be used in a complementary manner, TCN providing very good positional fixes where there are reasonable terrain contour variations.

Over very flat terrain the TCN accuracy is degraded; however, the TCM system can then provide very good positional fixes from the detected edges of specific terrain features.

(3) Scene matching area correlation (SMAC) This is also known as digital scene matching area correlation (DSMAC) in the USA. These systems generally use an infra-red imaging sensor to locate and lock on to specific recognisable landmarks or set of features at known positions on the route. An area correlation technique is used to match the processed image data with the stored feature data so that the system will track and lock on to the feature. The aircraft position must be known fairly accurately in the first place using, say, TCN.

The order of magnitude accuracies which can be achieved with these systems are:

TCN   around 50 metres
TCM   around 10 to 20 metres
SMAC   around 1 to 2 metres

It should be noted that while the basic concepts of these TRN systems are relatively straightforward, there is a very considerable mathematical content involved in explaining the detail implementation of these systems. For example, statistics, probability theory, correlation techniques, state estimation methods and Kalman Filters. Space constraints have, therefore, limited the coverage of TRN systems to a brief overview only.

6.4.2 Terrain contour navigation

The basic concepts of terrain contour navigation are illustrated in Fig. 6.29. Three types of measurement data are required by the system.
(1) A sequence of measurements of the ground clearance (ie height above ground level) of the aircraft. This is normally obtained by sampling the output of a radio altimeter, possibly with some additional pre-filtering. A horizontal sampling interval of about 100 metres is typically used, but this is not critical.
(2) Data from a barometric or baro-inertial height system is required to measure any vertical motion of the aircraft between successive ground clearance measurements.
(3) Some form of DR navigation system, eg an INS or Doppler radar plus heading reference is required to measure the relative horizontal positions of the ground clearance measurements.

The essence of TCN is to use this data to reconstruct the terrain profile under the aircraft's track. A digital map of terrain height is then searched to find a matching profile in the vicinity of the aircraft's position as previously estimated. This can then be used as the basis of an update to the estimated position.

There are a considerable number of TCN implementations which have been developed in the US and Europe to provide very accurate navigation systems such

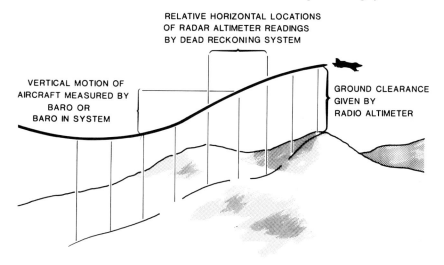

Fig. 6.29 Terrain reference navigation.

as TERCOM, TERPAC, SITAN, CAROTE, TERPROM, SPARTAN. Space constraints limit further coverage and the reader is referred to a number of papers which have been published on TRN systems in the Further Reading.

6.4.3 Terrain characteristic matching

The edge detection of specific terrain features has been mentioned earlier in this section. The basis of the technique is shown in Fig. 6.30 which shows a sequence of 'position likelihood' distributions which indicate how the position estimate improves as feature boundaries are detected.

6.4.4 Civil exploitation of TRN

It should be noted that while TRN systems have been used in exclusively military applications to date, they can clearly be exploited in civil aircraft as a very accurate self contained navigation aid particularly in the terminal area. Accurate radio altimeters are installed in most civil aircraft together with accurate DR navigation systems such as IN systems. The TRN computer and terrain data base is a modest equipment increment with current electronic technology.

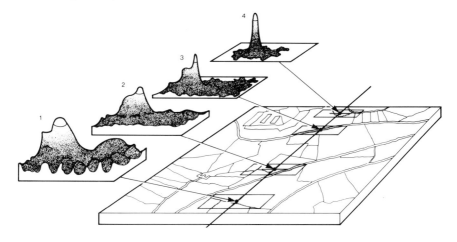

Fig. 6.30 Terrain characteristic matching – edge detection navigation.

Further Reading

AGARDOGRAPH No.139, *Theory and Application of Kalman Filtering.*

Andreas, R.D./, Hostetler, L.D., and Beckmann, C., *Continuous Kalman Upating of an Inertial Navigation System using Terrain Measurements*: NAECON 1978 pp 1263-70

Britting, K.R., *Inertial Navigation System Analysis*: Wiley-Interscience New York 1971

Chen, G., and Chui, C.K., *Kalman Filtering with Real-Time Applications*: Springer Series in Information Sciences Vol.17 Springer-Verlag 1991

Farrell, J.L., *Integrated aircraft navigation*: Academic Press 1976

Forssell, B., *Radio Navigation Systems*: Prentice Hall 1991

Hofmana-Wellanhot, B., Lichtenegger, H., and Collins, J., *GPS Theory and Practice*: Springer-Verlag 1992

Hostetler, L.D., *Optimal Terrain-aided Navigation Systems*: AIAA Guidance and Control Conference August 1978

Kalman, R.E., *A new approach to linear filtering and prediction problems*: Journal of basic engineering transactions of the American Society of Mechanical Engineers March 1960

Kalman, R.E., Bucy, R.S., *New results in linear filtering and prediction theory*: Journal of basic engineering transactions of the American Society of Mechanical Engineers March 1961

O'Donnell, C.F., *Inertial Navigation Analysis and Design*: McGraw Hill, New York 1964

Siouris, G.M., *Aereospace Avionic Systems*: Academic Press 1993

Schlee, F.H., Toda, N.F., Islam, M.A., and Standish, C.J., *Use of an External Cascaded Kalman Filter to improve the Performance of a Global Positioning System (GPS) Inertial Navigator*: Proceedings of the NAECON '88 Conference, Dayton, Ohio 23-27 May 1988 pp 142-147

Skarman, E., *Kalman Filter for Terrain Aided Navigation*: Conference on Remotely Piloted Vehicles, Bristol, U.K., 1979

7

Displays and man–machine interaction

7.1. INTRODUCTION

The cockpit display systems provide a visual presentation of the information and data from the aircraft sensors and systems to the pilot (and crew) to enable the pilot to fly the aircraft safely and carry out the mission. They are thus vital to the operation of any aircraft as they provide the pilot, whether civil or military, with:

Primary flight information
Navigation information
Engine data
Airframe data
Warning information.

The military pilot has also a wide array of additional information to view, such as:

Infra-red imaging sensors
Radar
Tactical mission data
Weapon aiming
Threat warnings.

The pilot is able to rapidly absorb and process substantial amounts of visual information but it is clear that the information must be displayed in a way which can be readily assimilated, and unnecessary information must be eliminated to ease the pilot's task in high work load situations. A number of developments have taken place to improve the pilot–display interaction and this is a continuing activity as new technology and components become available. Examples of these developments are:

- Head up displays
- Helmet mounted displays
- Multi-function colour displays
- Digitally generated colour map displays
- Synthetic pictorial imagery
- Displays management using intelligent knowledge based system (IKBS) technology

- Improved understanding of human factors and involvement of human factors specialists from the initial cockpit design stage.

Equally important and complementary to the cockpit display systems in the 'man machine interaction' are the means provided for the pilot to control the operation of the avionic systems and to enter data. Again, this is a field where continual development is taking place. Multi-function keyboards and multi-function touch panel displays are now widely used. Speech recognition technology has now reached sufficient maturity for 'direct voice input' control to be installed in the next generation of military aircraft. Other methods of data entry which are being evaluated include the use of eye trackers.

It is not possible in the space of one chapter to cover all aspects of this subject which can readily fill several books. Attention has, therefore, been concentrated on providing an overview and explanation of the basic principles involved in the following topics:

Head up displays (7.2)
Helmet mounted displays (7.3)
Discussion of HUDs vs HMDs (7.4)
Head down displays (7.5)
Data fusion (7.6)
Intelligent displays management (7.7)
Display technology (7.8)
Control and data entry (7.9)

7.2 HEAD UP DISPLAYS

7.2.1 Introduction

Without doubt the most important advance in the visual presentation of data to the pilot in the past thirty or so years has been the introduction and progressive development of the Head Up Display or HUD. (The first production HUDs, in fact, went into service in 1962 in the BUCCANEER strike aircraft in the UK.)

The HUD has enabled a major improvement in man–machine interaction (MMI) to be achieved as the pilot is able to view and assimilate the essential flight data generated by the sensors and systems in the aircraft whilst head up and maintaining full visual concentration on the outside world scene.

A head up display basically projects a collimated display in the pilot's head up forward line of sight so that he can view both the display information and the outside world scene at the same time. Because the display is collimated, that is focused at infinity (or a long distance ahead), it overlays the outside world scene. The pilot is thus able to observe both distant outside world objects and display data at the same time without having to change the direction of gaze or re-focus the eyes.

There are no parallax errors and aiming symbols for either a flight path director, or for weapon aiming in the case of a combat aircraft, remain overlaid on a distant 'target' irrespective of the pilot's head movement. (Try sighting on a landmark using a mark on a window. The aiming mark moves off the 'target' if the head is moved sideways – this effect is parallax.)

The advantages of head up presentation of essential flight data such as the artificial horizon, pitch angle, bank angle, flight path vector, height, airspeed and heading can be seen in Fig. 7.1 which shows a typical head up display as viewed by the pilot during the landing phase.

The pilot is thus free to concentrate on the outside world during manoeuvres and does not need to look down at the cockpit instruments or head down displays. It

Fig. 7.1 Head-up presentation of primary flight information (by courtesy of GEC-Marconi Avionics Ltd.).

should be noted that there is a transition time of one second or more to re-focus the eyes from viewing distant objects to viewing near objects a metre or less away, such as the cockpit instruments and displays and adapt to the cockpit light environment. In combat situations, it is essential for survival that the pilot is head up and scanning for possible threats from any direction. The very high accuracy which can be achieved by a HUD and computerised weapon aiming system together with the ability to remain head up in combat have made the HUD an essential system on all modern combat aircraft. They have also been widely retro-fitted to earlier generation fighters and strike aircraft as part of a cost effective avionic system up date. Fig. 7.2 illustrates a typical weapon aiming display.

Using a Forward Looking Infra-Red (FLIR) sensor, an electro-optical image of the scene in front of the aircraft can be overlaid on the real world scene with a raster mode HUD. The TV raster picture generated from the FLIR sensor video is projected on to the HUD and scaled one to one with the outside world enabling the pilot to fly at low level by night in fair weather. This provides a realistic night attack capability to relatively simple day ground attack fighters. A wide field of view HUD is required, however, as will be explained later.

HUDs are now being installed in civil aircraft for reasons such as:

1.   Increased safety in landing the aircraft in conditions of severe wind shear using the HUD to provide a flight path director display which allows for the effects of wind shear. The flight path is computed from the flight path vector derived

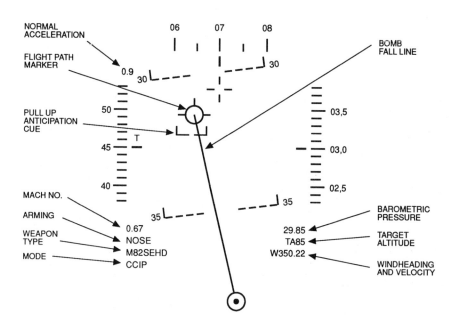

Fig. 7.2 Typical weapon aiming display (by courtesy of GEC-Marconi Avionics Ltd.).

from the INS, airspeed and height from the air data system and the aircraft's aerodynamic characteristics.

2. To display automatic landing guidance to enable the pilot to land the aircraft safely in conditions of very low visibility due to fog, as a back up and monitor for the automatic landing system.

3. Enhanced vision using a raster mode HUD to project a FLIR video picture of the outside world from a FLIR sensor installed in the aircraft, or, a synthetic picture of the outside world generated from a forward looking millimetric radar sensor in the aircraft. These enhanced vision systems are still under development but will enable the pilot to land the aircraft in conditions of very low or zero visibility at airfields not equipped with adequate all weather guidance systems such as ILS (or MLS).

Figure 7.3 illustrates a civil HUD display format for an enhanced vision system (courtesy of GEC-Marconi Avionics Ltd).

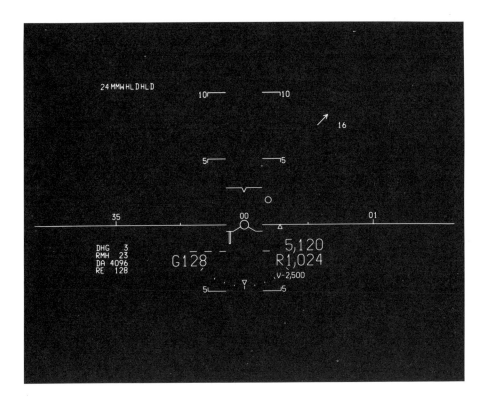

Fig. 7.3 Civil HUD symbology. (by courtesy of GEC-Marconi Avionics Ltd.).

7.2.2. Basic Principles

The basic configuration of a HUD is shown schematically in Fig. 7.4. The pilot views the outside world through the HUD combiner glass (and windscreen). The combiner glass is effectively a 'see through' mirror with a high optical transmission efficiency so that there is little loss of visibility looking through the combiner and windscreen. It is called a combiner as it optically combines the collimated display symbology with the outside world scene viewed through it. Referring to Fig. 7.4, the display symbology generated from the aircraft sensors and systems (such as the INS and air data system) is displayed on the surface of a cathode ray tube (CRT). The display images are then relayed through a relay lens system which magnifies the display and corrects for some of the optical errors which are otherwise present in the system. The relayed display images are then reflected through an angle of near 90° by the fold mirror and thence to the collimating lens which collimates the display images which are then reflected from the combiner glass into the pilot's forward field of view. The virtual images of the display symbology appear to the pilot to be at infinity and overlay the distant world scene, as they are collimated. The function of the fold mirror is to enable a compact optical configuration to be achieved so that the HUD occupies the minimum possible space in the cockpit.

The fundamental importance of collimation to any HUD system merits further explanation for the benefit of readers whose knowledge of optics needs refreshing.

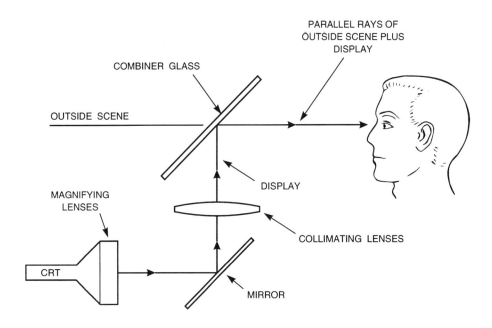

Fig. 7.4 HUD schematic.

A collimator is defined as an optical system of finite focal length with an image source at the focal plane. Rays of light emanating from a particular point on the focal plane exit from the collimating system as a parallel bunch of rays.

Fig. 7.5(a) and (b) shows a simple collimating lens system with the rays traced from a source at the centre, O, and a point, D, on the focal plane respectively. A ray from a point on the focal plane which goes through the centre of the lens is not refracted and is referred to as the 'non-deviated ray'. The other rays emanating from the point are all parallel to the non-deviated ray after exiting the collimator.

It should be noted that the collimating lens, in practice, would be made of several elements to minimise the unacceptable shortcomings of a single element.

It can be seen from Fig. 7.5(a) that an observer looking parallel to the optical axis will see point O at eye positions A, B and C, and the angle of gaze to view O is independent of the displacement of the eye from the optical axis.

Similarly, it can be seen from Fig. 7.5(b) that the angle of gaze from the observer to see point D is the same for eye positions A, B and C and is independent of translation.

The refractive HUD optical system, in fact, is basically similar to the simple optical collimation system shown in Fig. 7.6. The rays are traced for the observer to see points D, O and E on the display with eye positions at points A, B and C. It can

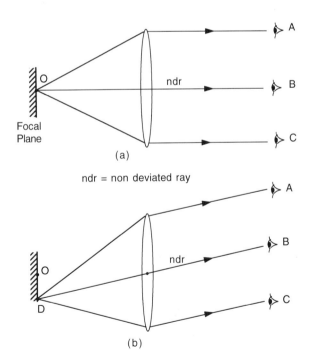

Fig. 7.5 Simple optical collimator.

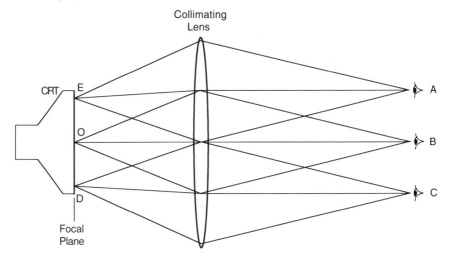

Fig. 7.6 Simple optical collimator ray trace.

be seen that the angles of gaze to see points D, O and E are the same from points A, B or C. The appearance of the collimated display is thus independent of the position (or translation) of the eye and is only dependent on the angle of gaze. Also because of collimation, the display appears to be at infinity as the rays emanating from any point on the display are all parallel after exiting the collimating system.

It should be noted that the display images must be correctly collimated. De-collimation is the term used when the light from a given point on the display does not leave the optical system parallel over the entire lens surface. The light can converge, diverge or otherwise 'misbehave' resulting in 'swimming' of the display images when the pilot's head moves. Sometimes this creates discomfort and in the case of convergence can even cause nausea.

A very important parameter with any HUD is the field of view (FOV), which should be as large as possible within the severe space constraints imposed by the cockpit geometry. A large FOV is particularly important when the HUD forms part of a night vision system and the only visibility the pilot has of the outside world is the FLIR image displayed on the HUD, as mentioned earlier.

It is important to distinguish between the instantaneous field of view (IFOV) and the total field of view (TFOV) of a HUD as the two are not the same in the case of the refractive type of HUD.

The instantaneous field of view is the angular coverage of the imagery which can be seen by the observer at any specific instant and is shown in the simplified diagram in Fig. 7.7(a). It is determined by the diameter of the collimating lens, D, and the distance, L, of the observer's eyes from the collimating lens.

$$\text{IFOV} = 2\tan^{-1} D/2L$$

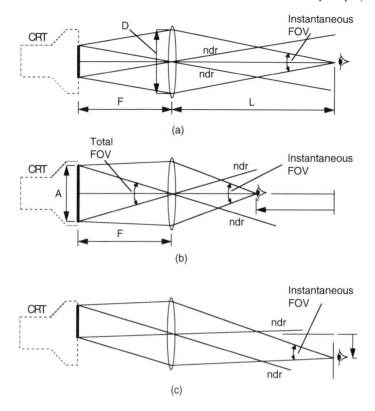

Fig. 7.7 Instantaneous and total FOV.

The total field of view is the total angular coverage of the CRT imagery which can be seen by moving the observer's eye position around. TFOV is determined by the diameter of the display, A, and the effective focal length of the collimating lens, F.

$$\text{TFOV} = 2\tan^{-1} A/2F$$

Reducing the value of L increases the IFOV as can be seen in Fig. 7.7(b) which shows the observer's normal eye position brought sufficiently close to the collimating lens for the IFOV to equal the TFOV. However, this is not practical with the conventional type of HUD using refractive optics. This is because of the cockpit geometry constraints on the pilot's eye position and the space constraints on the diameter of the collimating lens. The IFOV is generally only about two thirds of the TFOV.

It can be seen from Fig. 7.7(c) that by moving the head up or down or side to side the observer can see a different part of the TFOV, although the IFOV is unchanged.

The effect is like looking through and around a porthole formed by the virtual image of the collimating lens as can be seen in Fig. 7.8. The diagram shows the IFOV seen by both eyes (cross hatched), the IFOV seen by the left and right eyes respectively and the TFOV.

The analogy can be made of viewing a football match through a knot hole in the fence and this FOV characteristic of a HUD is often referred to as the 'knot hole effect'.

The constraints involved in the HUD design are briefly outlined below.

For a given TFOV, the major variables are the CRT display diameter and the effective focal length of the collimating lens system. For minimum physical size and weight, a small diameter CRT and short focal length are desired. These parameters are usually balanced against the need for a large diameter collimating lens to give the maximum IFOV and a large focal length which allows maximum accuracy. The diameter of the collimating lens is generally limited to between 75mm and 175mm (3 inches and 7 inches approximately) by cockpit space constraints and practical considerations. Large lenses are very heavy and liable to break under thermal shock.

The HUD combiner occupies the prime cockpit location right in the centre of the pilot's forward line of view at the top of the glare shield. The size of the combiner is determined by the desired FOV and the cockpit geometry, especially the pilot's seating position. The main body of the HUD containing the optics and electronics must be sunk down behind the instrument panel in order to give an unrestricted view down over the nose of the aircraft during high attitude manoeuvres (refer to Fig. 7.8).

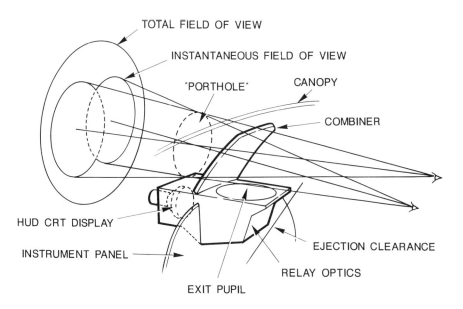

Fig. 7.8 HUD installation constraints and field of view.

The pilot's design eye position for the HUD is determined by pilot comfort and the ability to view the cockpit instruments and head down displays and at the same time achieve the maximum outside world visibility. In the case of a combat aircraft, there is also the ejection line clearance to avoid the pilot being 'kneecapped' by the HUD on ejecting, which further constrains the design eye position.

Typical IFOVs range from about 13° to 18° with a corresponding TFOV of about 20° to 25°. The total vertical FOV of a HUD can be increased to around 18° by the use of a dual combiner configuration rather like a venetian blind. Effectively two overlapping portholes are provided, displaced vertically.

The effect of the cockpit space and geometry constraints is that the HUD design has to be 'tailor made' for each aircraft type and a 'standard HUD' which would be interchangeable across a range of different aircraft types is not a practical proposition.

The conventional combiner glass in a refractive HUD has multi-layer coatings which reflect a proportion of the collimated display imagery and transmit a large proportion of the outside world, so that the loss of visibility is fairly small. A pilot looking through the combiner of such a HUD sees the real world at 70% brightness upon which is superimposed the collimated display at 30% of the CRT brightness (taking typical transmission and reflection efficiencies). The situation is shown in Fig. 7.9 and is essentially a rather lossy system with 30% of the real world

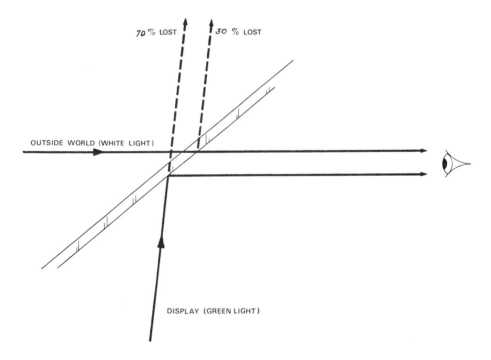

Fig. 7.9 Conventional refractive HUD combiner operation.

brightness thrown away. (equivalent to wearing sunglasses) as is 70% of the CRT display brightness.

In order to achieve an adequate contrast ratio so that the display can be seen against the sky at high altitude or against sunlit cloud it is necessary to achieve a display brightness of 30,000 Cd/m^2 (10,000 ft L) from the CRT. In fact, it is the brightness requirement in particular which assures the use of the CRT as the display source for some considerable time to come, even with the much higher optical efficiencies which can be achieved by exploiting holographic optical elements.

The use of holographically generated optical elements can enable the FOV to be increased by a factor of two or more, with the instantaneous FOV equal to the total FOV. Very much brighter displays together with a negligible loss in outside world visibility can also be achieved, as will be explained in the next section. High optical transmission through the combiner is required so as not to degrade the acquisition of small targets at long distances.

It should be noted, however, that recent development in 'Rugate' dielectric coatings applied to the combiners of conventional refractive HUDs can enable very bright displays with high outside world transmission to be achieved, comparable, in fact, with holographic HUDs. A Rugate dielectric coating is a multi-layer coating having a sinusoidally varying refractive index with thickness which can synthesise a very sharply defined narrow wavelength band reflection coating. The coating exhibits similar high reflection and transmission values to holographic coatings but is not capable of generating optical power.

The IFOV of a refractive HUD using a combiner with a Rugate dielectric coating still suffers from the same limitations and cannot be increased like a holographic HUD. Nevertheless, for applications where a maximum IFOV of up to 20° is acceptable, a refractive HUD configuration exploiting Rugate dielectric coatings can provide a very competitive solution.

7.2.3 Holographic HUDs

The requirement for a large FOV is driven by the use of the HUD to display a collimated TV picture of the FLIR sensor output to enable the pilot to 'see' through the HUD FOV in conditions of poor visibility, particularly night operations. It should be noted that the FLIR sensor can also penetrate through haze and many cloud conditions and provide 'enhanced vision' as the FLIR display is accurately overlaid one to one with the real world. The need for a wide FOV when manoeuvring at night at low level can be seen in Fig 7.10. The wider azimuth FOV is essential for the pilot to see into the turn. (The analogy has been made of trying to drive a car round Hyde Park Corner with a shattered opaque windscreen with your vision restricted to a hole punched through the window.)

In a modern wide FOV holographic HUD, the display collimation is carried out by the combiner which is given optical power (curvature) such that it performs the

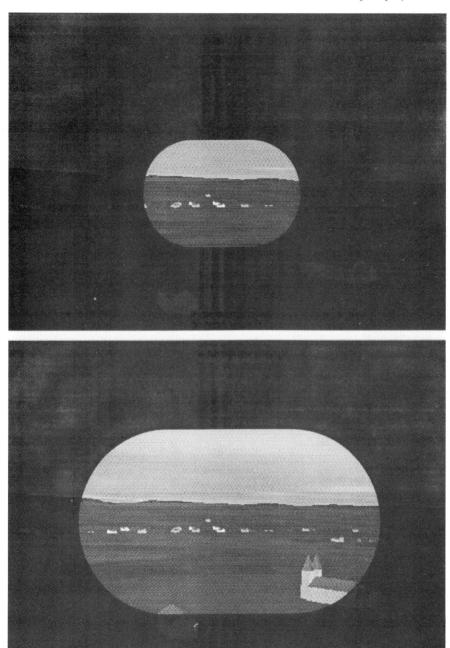

Fig. 7.10 Comparison of instantaneous FOVs of conventional refractive HUD and wide
FOV holographic HUD.

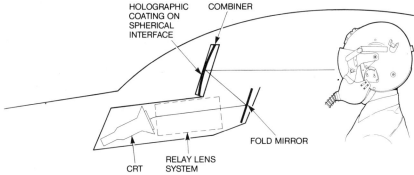

Fig. 7.11 Off-axis holographic combiner HUD configuration.

display image collimation. Figure 7.11 shows the basic configuration of a modern single combiner holographic HUD. The CRT display is focused by the relay lens system to form an intermediate image at the focus of the powered combiner. The intermediate image is then reflected from the fold mirror to the combiner. This acts as a collimator as the tuned holographic coating on the spherical surface of the combiner reflects the green light from the CRT display and forms a collimated display image at the pilot's design eye position.

Figure 7.12 illustrates the collimating action of a spherical reflecting surface.

Because the collimating element is located in the combiner, the porthole is considerably nearer to the pilot than a comparable refractive HUD design. The collimating element can also be made much larger than the collimating lens of a refractive HUD, within the same cockpit constraints. The IFOV can thus be

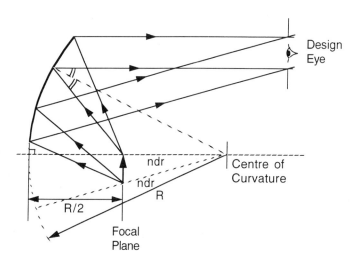

Fig. 7.12 Collimation by spherical surface.

increased by a factor of two or more and the instantaneous and total FOVs are generally the same, as the pilot is effectively inside the viewing porthole.

This type of HUD is sometimes referred to as a 'Projected Porthole HUD' and the image is what is known as pupil forming. The display imagery can, in fact, only be seen within the 'head motion box'. If the eyes or head move outside a three dimensional box set in space around the pilot's normal head position, then the display fades out. It literally becomes a case of 'now you see it – now you don't' at the head motion box limits. Modern holographic HUDs are designed to have a reasonably sized head motion box so that the pilot is not unduly constrained.

The combiner comprises a parallel-faced sandwich of plano-convex and plano-concave glass plates with a holographic coating on the spherical interface between them. The holographic coating is formed on the spherical surface of the plano-convex glass plate and the concave surface glass forms a cover plate so that the holographic coating can be hermetically sealed within the combiner. The holographic coating is sharply tuned so that it will reflect the green light of one particular wavelength from the CRT display with over 80% reflectivity but transmit light at all other wavelengths with around 90% efficiency. (The CRT display phosphors generally used are P43 or P53 phosphors emitting green light with a predominant wavelength of around 540nm, and the hologram is tuned to this wavelength.)

This gives extremely good transmission of the outside world through the combiner. (The outer faces of the combiner glass are parallel so that there is no optical distortion of the outside scene.) The outside world viewed through the combiner appears very slightly pink as the green content of the outside world with a wavelength of around 540nm is not transmitted through the combiner. Holographic HUDs, in fact, are recognisable by the green tinge of the combiner.

The spherical reflecting surface of the combiner collimates the display imagery but there are large optical aberration errors introduced which must be corrected. These aberration errors are due to the large off-axis angle between the pilot's line of sight and the optical axis of the combiner which results from the HUD configuration. Some corrections can be made for these aberrations by the relay lens system but there is a practical limit to the amount of correction which can be achieved with conventional optical elements without resorting to aspherical surfaces. This is where a unique property of holographically generated coatings is used, namely the ability to introduce optical power within the coating so that it can correct the remaining aberration errors. The powered holographic coating produces an effect equivalent to local variations in the curvature of the spherical reflecting surface of the combiner to correct the aberration errors by diffracting the light at the appropriate points. The holographic coating is given optical power so that it behaves like a lens by using an auxiliary optical system to record a complex phase distribution on the combiner surface during the manufacturing process. This will be explained shortly.

A very brief overview of holographic optical elements is set out below to give an appreciation of the basic principles and the technology.

Holography was invented in 1947 by Denis Gabor, a Hungarian scientist working in the UK. Practical applications had to wait until the 1960s, when two American scientists, Emmet Leith and Josph Upatnieks, used coherent light from the newly developed laser to record the first hologram.

Holographic HUDs use reflection holograms which depend for their operation on refractive index variations produced within a thin gelatin film sandwiched between two sheets of glass. This is really a diffraction grating and hence a more accurate name for such HUDs is diffractive HUDs. A holographic reflection coating is formed by exposing a thin film of photo-sensitive dichromated gelatin to two beams of coherent laser light. Due to the coherent nature of the incident beams a series of interference fringes are formed throughout the depth of the gelatin film. During the developing process these fringes are converted into planes of high and low refractive index parallel to the film surface. To a first approximation, the refractive index change between adjacent planes is sinusoidal as opposed to the step function associated with multi-layer coatings. During the developing process the gelatin swells, producing an increase in the tuned wavelength. Re-tuning the hologram is achieved by baking the film which reduces the thickness and hence the spacing between planes of constant refractive index. The designer therefore specifies a construction wavelength at a given angle of incidence after baking. Figure 7.13 illustrates the planes or layers of varying refractive index formed in the holographic coating.

The bandwidth of the angular reflection range is determined by the magnitude of the change in refractive index. This variable can be controlled during the developing process and is specified as the hologram modulation.

At any point on the surface, the coating will only reflect a given wavelength over a small range of incidence angles. Outside this range of angles, the reflectivity drops off very rapidly and light of that wavelength will be transmitted through the

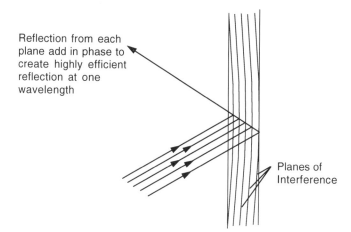

Fig. 7.13 Holographic coating.

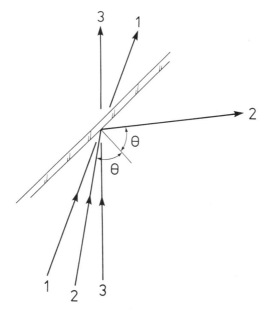

Fig. 7.14(a) Angularly selective reflection of monochromatic rays.

coating. The effect is illustrated in Fig. 7.14 (a) and (b). Rays 1 and 3 are insufficiently close to the reflection angle, θ, for them to be reflected whereas Ray 2 is incident at the design angle and is reflected.

There is another more subtle feature of holograms which gives particular advantages in an optical system. That is the ability to change the tuned wavelength uniformly across the reflector surface. Figure 7.11 shows that the reflection coating must reflect the display wavelength at a different incident angle at the bottom of the combiner from that at the top. It is possible to achieve this effect with a hologram because it can be constructed from the design eye position.

The process for producing the powered holographic combiner is very briefly outlined below.

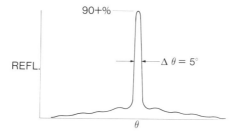

Fig. 7.14(b) Holographic coating performance.

A computer generated hologram (CGH) is first designed and manufactured which will enable the required arbitrary wavefronts to be produced to form the constructor hologram.

The constructor hologram is then used in the auxiliary optical system of the holographic manufacturing facility to produce the interference pattern for the powered holographic combiner coating. The constructor hologram functions in a similar manner to a 'master template' and can be used repeatedly in the manufacturing process.

It should be pointed out that the design of holographic optical systems is a highly computer intensive operation and would not be possible without a very extensive suite of optical design software. Development of the optical design software has required many man years of effort and holographic optical design is very much a task for the experts and professionals. A very close liaison with the holographic optical manufacturing team and facility is essential.

Figure 7.15 is a photograph of a modern wide FOV single combiner holographic HUD (by courtesy of GEC-Marconi Avionics Ltd.)

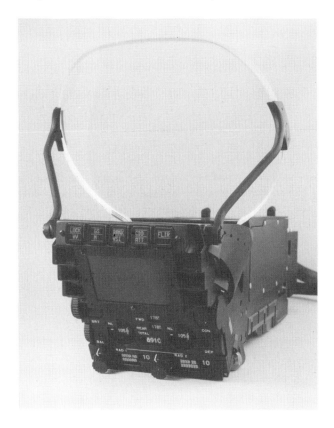

Fig. 7.15 Wide FOV holographic HUD (by courtesy of GEC-Marconi Avionics Ltd.).

Attention has been concentrated on the single combiner off-axis HUD configuration exploiting powered holograms as this is the configuration being adopted for the latest fighters in the USA and Europe which will be entering service from the late 1990s. It is also in service with some existing aircraft.

However, mention should be made of the quasi-axial configurations using a multi-element holographic combiner. These configurations avoid the large aberration errors present in the off-axis configurations and enable a wide FOV (around 30° azimuth) to be achieved with an acceptable head motion box. The holographic elements use simple reflection holograms without optical power. They do, however, require a fairly bulky support structure for the three element combiner which can obstruct the pilot's view of the outside world in certain conditions. Some configurations also involve a small additional loss in outside world transmission as the outside world is seen through two holographic elements in the combiner.

The clean uncluttered appearance of the later generation off-axis single combiner configurations exploiting powered holograms together with the excellent transmission characteristics is resulting in their adoption in preference to quasi-optical configurations.

7.2.4 HUD electronics

The basic functional elements of a modern HUD electronic system are shown in Fig. 7.16. These functional subunits may be packaged so that the complete HUD system

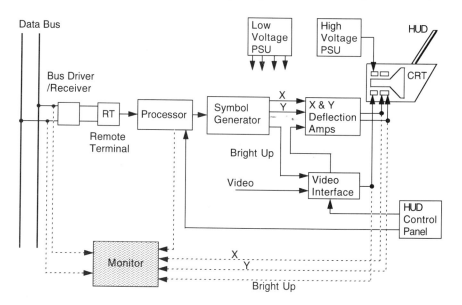

Fig. 7.16 HUD electronics.

is contained in a single unit. This packing configuration is referred to as a 'smart HUD'. Alternatively, the system may be configured as two units, namely the display unit and the electronics unit. The display unit comprises the HUD optical assembly, CRT and display drive electronics including high voltage and low voltage power supplies. The electronics unit carries out the display processing and symbol generation tasks.

There is also a third packaging configuration for the new generation of aircraft installed with integrated modular avionics. In such aircraft the standardised cards/modules carrying out the HUD display processing and symbol generation tasks would be housed in a rack in one of the environmentally controlled cabinets of the integrated avionics system.

The basic functional elements of the HUD electronics are described briefly below.

The data bus interface decodes the serial digital data from the aircraft data bus (typically a MIL STD 1553B data bus) to obtain the appropriate data from the aircraft sub systems and inputs this data to the display processor.

The input data includes the primary flight data from the air data system and the INS, such as height, airspeed, vertical speed, pitch and bank angles, heading, flight path velocity vector. Other data include MLS or ILS guidance signals, stores management and weapon aiming data in the case of a combat aircraft, discrete signals such as commands, warnings etc.

The display processor processes this input data to derive the appropriate display formats, carrying out tasks such as axis conversion, parameter conversion and format management. In addition the processor also controls the following functions:

Self test
Brightness control (especially important at low brightness levels)
Mode sequencing
Calibration
Power supply control.

The symbol generator carries out the display waveform generation task (digitally) to enable the appropriate display symbology (eg lines, circles, alpha-numerics etc.) to be stroke written on the HUD CRT. The symbols are made up of straight line segments joining the appropriate points on the display surface in an analogous manner to a 'join the dots' child's picture book. Fixed symbols such as alpha-numerics, crosses, circles etc. are stored in the symbol generator memory and called up as required. The necessary D to A conversions are carried out in the symbol generator which outputs the appropriate analogue x and y deflection voltage waveforms and 'bright up' waveforms to control the display drive unit of the HUD CRT.

The CRT beam is thus made to trace the outline of the line drawn symbols, the process being known as stroke or cursive writing.

Slow and fast cursive display writing systems are required for a day and night mode operation HUD.

Daytime operation requires a slow cursive writing system to produce a sufficiently bright display which can be viewed against a 30,000 Cd/m^2 (10,000 ftL) white cloud background.

Night time operation requires identical appearance symbology to the daytime symbology to be overlaid on the raster TV FLIR picture displayed on the HUD. The appearance of raster generated symbology can differ significantly from stroke written symbology and pilots wish to see identical symbology night or day. A 'fast cursive' display system is thus used in night mode operations whereby the cursive symbology is drawn at high speed during the raster fly back interval. The display is written at about ten times the slow cursive speed and this makes significant demands on the bandwidth and power requirements of the deflection system. By taking a small number of lines from the top and bottom of the picture, typically around twenty lines total, the whole of the day time symbology can be superimposed over the TV picture for night use.

Because of the high writing speeds, the brightness of the raster and fast cursive display was initially only suitable for night or low ambient brightness use. However, improvements to CRTs, development of more efficient phosphors such as P53, and high efficiency wavelength selective combiners have allowed raster and fast cursive brightness to be increased to around 3,000 Cd/m^2 (1,000 ft L) which is useable in daytime.

The very wide range of ambient brightness levels which a day/night mode HUD must operate over is worthy of note. CRT luminance levels range from 30,000 Cd/m^2 to 0.3 Cd/m^2, a range of 10^5:1.

The display drive unit contains all the display drive electronics for the HUD CRT including the high voltage and low voltage power supply units. High linearity and high bandwidth x and y deflection amplifiers are required to meet the HUD accuracy and cursive writing requirements.

Video interface electronics are incorporated in the display drive electronics for the TV raster mode.

Modern HUD systems are used as the prime display of primary flight data such as height, airspeed, attitude, heading etc. The HUD thus becomes a safety critical system when the pilot is flying head up at low altitude and relying on the HUD. Modern HUD systems for both military and civil aircraft are, therefore, being designed so that the probability of displaying hazardous or misleading information is of the order of less than 1×10^{-9}. The self contained 'SMART HUD' makes it an ideal system configuration for a safety critical HUD. Forward and backward path monitoring techniques can be employed to ensure that what is on the display is what was actually commanded by the aircraft system. The monitoring features incorporated in the HUD electronics are shown in Fig. 7.16.

7.2.5 Worked example on HUD design and display generation

A simple worked example based on typical parameters for a refractive HUD is set out below to show how:-

1   The total and instantaneous FOVs are determined by the HUD CRT, collimating optics and cockpit geometry.
2   The attitude display symbology is generated to be conformal with the outside world.

The objectives are to give an engineering appreciation by deriving realistic numerical values and to reinforce the descriptive content of the preceding sections.

The following data are provided:-

HUD CRT diameter (measured across display face)	50mm (2 ins approx)
Required total FOV	20°
Collimating lens diameter	125mm (5 ins approx)
Collimating lens centre from combiner glass (measured along axis of lens)	100mm (4 ins approx)
Pilot's design eye position from combiner glass (measured along pilot's LOS through centre of combiner)	400mm (16 ins approx)

Assume x and y deflection drive waveform amplitudes of ± 100 units produces full scale deflection of the CRT spot in both positive and negative directions along the OX and OY axes (origin, O, at the centre of the display).

Questions

1   What is the required focal length of the collimating lens?
2   What is the instantaneous FOV?
3   How much up or down, or sideways head movement must the pilot make to see all the total FOV of 20°?
4   How much nearer to the combiner must the pilot's eye position be in order to see the total FOV?
5   Derive the equations for displaying the horizon line in terms of the pitch angle, θ, bank angle, Φ, and the angle θ_0 by which the horizon line is depressed from the centre of the display in straight and level flight.
6   What are the co-ordinates of the end points of the horizon line when the aircraft pitch angle is $-3°$ and bank angle $+10°$, given that the horizon line takes up 50% of the azimuth FOV and that the horizon line is depressed by 4° from the centre of the display in straight and level flight.
7   Sketch the display and show the aircraft velocity vector which is inclined at an angle of 4° below the horizontal.
8   Sketch output waveform of symbol generator to generate the horizon line for the above values.

Solution to Question 1 TFOV = $2\tan^{-1} A/2F$ where A = display diameter and F = effective focal length of collimating lens (refer to Fig. 7.7(b)).

Hence, for a TFOV of 20° and a CRT diameter A of 50mm, the required effective focal length F is given by

$$F = \frac{50}{2\tan 10°}$$

ie $F = 141.8$mm

Solution to Question 2 IFOV = $2\tan^{-1} D/2L$ where D = diameter of collimating lens and L = distance of virtual image of collimating lens from pilot's eyes (refer to Fig. 7.8).

From data given $L = (100 + 400) = 500$mm and $D = 125$mm.
Hence, IFOV = $2\tan^{-1} 125/2\text{x}500$ rads = 14.25°

Solution to Question 3 To see the TFOV of 20°, the required up, down, or sideways head movement of the pilot is given by

$$500\tan\left(\frac{20°-14.25°}{2}\right)$$

that is, 25mm (1 inch approx).

Solution to Question 4 To see the total FOV of 20°, the pilot's eye position must be brought nearer to the combiner by a distance, Z, which is given by

$$\frac{125}{2(500-Z)} = \tan 10°$$

From which $Z = 145$mm (5.7 inches approx).

Solution to Question 5 Referring to Fig. 7.17, the co-ordinates (x_1, y_1) and (x_2, y_2) of the end points of a horizon line of length $2a$, which in straight and level flight is depressed by an amount θ_0 from the centre of the display, are given by

$$\begin{cases} x_1 = -a\cos\Phi \\ y_1 = -(\theta_0 + \theta + a\sin\Phi) \end{cases}$$
$$\begin{cases} x_2 = a\cos\Phi \\ y_2 = -(\theta_0 + \theta - a\sin\Phi) \end{cases}$$

Solution to Questions 6 & 7 Given $\theta_0 = 4°$, $\theta = -3°$ and $\Phi = 10°$
$2a = 0.5$ x azimuth FOV. Hence $a = 5°$

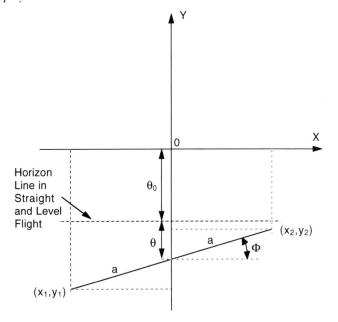

Fig. 7.17 Horizon line.

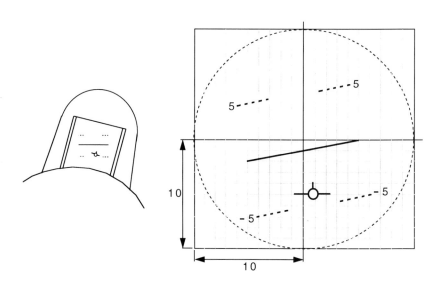

Fig. 7.18 HUD display of horizon line and velocity vector.

From which

$$\begin{cases} x_1 = -5\cos 10° = -4.92 \\ y_1 = -[4 + (-3) + 5\sin 10°] = -1.87 \end{cases}$$

$$\begin{cases} x_2 = 5\cos 10° = 4.92 \\ y_2 = -[4 + (-3) - 5\sin 10°] = -0.13 \end{cases}$$

Horizon line end point co-ordinates are thus (–4.92, –1.87) and (4.92, –0.13) Display. is shown in Fig. 7.18.

Solution to Question 8 The X and Y deflection waveforms are shown in Fig. 7.19.

Full scale deflection of CRT spot is produced when X and Y deflection amplifiers are supplied by signals of ±100 units from the symbol generator. Hence, to draw horizon line for values given, the X axis deflection waveform is a ramp from –49.2 units to +49.2 units, and the Y axis deflection waveform is a ramp from –18.7 units to –1.3 units.

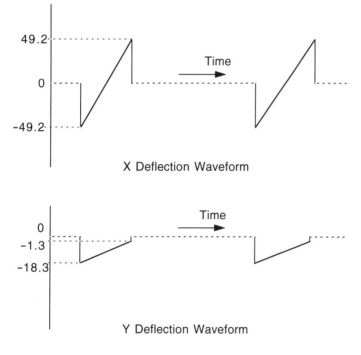

Fig. 7.19 X and Y deflection waveforms.

7.3 HELMET MOUNTED DISPLAYS

7.3.1 Introduction

The advantages of head-up visual presentation of flight and mission data by means of a collimated display have been explained in the preceding section. The HUD, however, only presents the information in the pilot's forward field of view, which even with a wide FOV holographic HUD is limited to about 30° in azimuth and 20° to 25° in elevation. Significant increases in this FOV are not practicable because of the cockpit geometry constraints.

The pilot requires visual information head up when he is looking in any direction and this requirement can only be met by a helmet mounted display (HMD).

In its simplest form the HMD can comprise a simple helmet mounted sighting system which displays a collimated aiming cross or circle and some simple alphanumeric information. An addressable LED matrix array is generally used as the display source, enabling a very light weight and compact HMD to be achieved.

In its most sophisticated form the HMD can provide, in effect, a 'HUD on the helmet'. This can display all the information to the pilot which is normally shown on a HUD but with the pilot able to look in any direction (attitude information is referenced to his line of sight (LOS)). The HMD can also have a wider FOV ranging from 35° to 40° for a fighter aircraft application to up to 55° for a helicopter application. The HMD also enables a very effective night/poor visibility viewing system to be achieved by displaying the TV picture from a gimballed infra-red sensor unit which is slaved to follow the pilot's line of sight. The pilot's LOS with respect to the airframe is measured by a head position sensing system.

Such a helmet can also incorporate night viewing goggles (NVGs) which are integrated into the HMD optical system.

7.3.2 Helmet design factors

It is important that the main functions of the conventional aircrew helmet are appreciated as it is essential that the integration of a helmet mounted display system with the helmet does not degrade these functions in any way. The basic functions are:

1. To provide the pilot with an aural and speech interface with the communications radio equipment. The helmet incorporates a pair of headphones which are coupled to the outputs of the appropriate communications channel selected by the pilot. The helmet and earpieces are also specifically designed to attenuate the cockpit background acoustic noise as much as possible. A speech interface is provided by a throat microphone incorporated in the oxygen mask.

2. To interface with the oxygen breathing system through the oxygen mask attached to the helmet.
3. To protect the pilot's head and eyes from injury when ejecting at high airspeeds. For example, the visor must stay securely locked in the down position when subjected to the blast pressure experienced at indicated airspeeds of 650 knots. The helmet must also function as a crash helmet and protect the pilot's head as much as possible in a crash landing.
4. In addition to the clear protective visor, the helmet must also incorporate a dark visor to attenuate the glare from bright sunlight.
5. The helmet must also be compatible with NBC (nuclear–biological–chemical) protective clothing and enable an NBC mask to be worn.

Integrating an HMD system into the helmet must thus consider the communications and breathing equipment requirement, the protection, comfort and cooling aspects as well as the visual performance of the display.

First generation HMDs were based on modifying and effectively 'bolting on' the HMD optical module, display source and associated electronics to an existing aircrew helmet. This results, inevitably, in a bit of a 'bodge' and the total helmet weight is also too high.

Modern second generation HMDs are totally integrated with the helmet design starting from a 'clean sheet of paper' and optimised to meet all the above requirements for the minimum overall weight. The combined mass of the integrated helmet and HMD system must be as low as possible to reduce the strain on the pilot's neck muscles during manoeuvres. Modern conventional aircrew helmets weigh about 1 kg (2.2 lbs) so that the weight the pilot feels on his head during a 9 g turn is equal to 9 kg (20 lbs approximately), and this is just acceptable. The typical weight for an integrated helmet incorporating a binocular HMD system is around 1.7 to 2.5 kg (3.8 to 5.5 lbs approximately). This is limited by the current technology and lower weight is sought. The helmet CG should be in line with the pivoting point of the head on the spine so that there are minimal out of balance moments exerted on the neck muscles.

The moment of inertia of the integrated helmet system about the yaw and pitch axes of the head should be as low as possible. This is to minimise the inertia torques experienced by the pilot due to the angular acceleration (and deceleration) of the head and helmet in scanning for threats from behind or above. Peak head motion rates of 30°/s can occur in combat.

The helmet must fit securely on the pilot's head and not move about so that the display is visible under all conditions and the pilot's LOS can be inferred from measuring the attitude of the head with the head tracker system.

The HMD optical system must be rigidly fixed to the helmet with adequate mechanical adjustments to accommodate the physical variations within the normal pilot population (interocular distance and eye relief). There should be at least 25mm (1 inch) clearance, or 'eye relief', between the nearest optical surface and the eye.

The display is only fully visible when the eyes are within a circle of a given diameter known as the exit pupil diameter, the centre of the exit pupil being located at the design eye position. The exit pupil diameter should be at least 12mm (0.5 ins approximately). It should be noted that the helmet will be pressed down on the pilot's head during high *g* manoeuvres as the helmet lining is resilient, hence the need for a reasonable exit pupil so that the pilot can still see the display.

7.3.3 Helmet mounted sights

A helmet mounted sight (HMS) in conjunction with a head tracker system provides a very effective means for the pilot to designate a target. The pilot moves his head to look and sight on the target using the collimated aiming cross on the helmet sight. The angular co-ordinates of the target sight line relative to the airframe are then inferred from the measurements made by the head tracker system of the attitude of the pilot's head.

In air to air combat, the angular co-ordinates of the target line of sight (LOS) can be supplied to the missiles carried by the aircraft. The missile seeker heads can then be slewed to the target LOS to enable the seeker heads to acquire and lock on to the target. (A typical seeker head needs to be pointed to within about 2° of the target to achieve automatic lock on.) Missile lock on is displayed to the pilot on the HMS and an audio signal is also given. The pilot can then launch the missiles.

This enables attacks to be carried out at large off-boresight angles, given agile missiles with high *g* manoeuvring capabilities. The pilot no longer needs to turn the aircraft until the target is within the FOV of the HUD before launching the missile. The maximum off-boresight angle for missile launch using the HUD is less than 15° even with a wide FOV HUD, compared with about 120° with an HMS (and highly agile missiles). This gives a major increase in air combat effectiveness – factors of three to four have been quoted in terms of the 'kill' probability, a very effective 'Force Multiplier' in military parlance. Fig. 7.20 illustrates the concept together with a simple HMS.

The helmet mounted sight can also be used to provide visual 'fixes' for the navigation system by designating known landmarks and features and does not require the pilot to overfly them.

The display source on a simple helmet mounted sight typically comprises a LED matrix array. This can range from a simple fixed format addressable array to, say, a 32 × 32 matrix array which can display some limited variable symbology including alpha-numerics. The display image is relayed to the visor by means of a simple optical system and reflected from the visor into the pilot's eye. (Monocular systems are used which normally display the image to the right eye.) The visor, which is of spherical profile, has a reflective coating 'patch' on the portion of the visor in front of the eye which acts as a spherical mirror and collimates the display image. Typical field of view for an HMS is around 7° to 10°. The amount of display

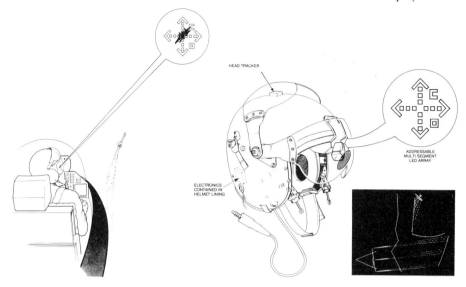

HEAD TRACKER

ADDRESSABLE
MULTI SEGMENT
LED ARRAY

ELECTRONICS
CONTAINED IN
HELMET LINING

Fig. 7.20 Helmet mounted sight (by courtesy of GEC-Marconi Avionics Ltd.).

symbology on a simple sight is generally kept to the minimum: typically an aiming cross (or circle) or possibly cueing arrows (up, down, left, right) to direct the pilot's LOS to a radar acquired target, or a detected threat, and missile status (eg seeker lock).

7.3.4 Helmet mounted displays

As mentioned in the introduction, the HMD can function as a 'HUD on the helmet and provide the display for an integrated night/poor visibility viewing system.

Although monocular HMDs have been built which are capable of displaying all the information normally displayed on a HUD, there can be problems with what is known as 'monocular rivalry'. This is because the brain is trying to process different images from each eye and rivalry can occur between the eye with a display and that without. The problems become more acute at night when the eye without the display sees very little, and the effects have been experienced when night-flying with a monocular system in a helicopter.

It has been shown that a binocular (or biocular) system whereby the same display is presented to both eyes is the only really satisfactory solution. Hence, current HMD designs are binocular systems. Miniature 0.5 inch diameter CRTs are currently used as the display sources.

Night viewing applications require the widest possible FOV. This is particularly important for helicopters operating at night at very low level (eg 30 ft) flying 'Nap of the Earth` missions taking every possible advantage of terrain screening. Binocular HMDs for helicopter applications have minimum FOVs of 40° and FOVs of over 50° are now available.

Fighter/strike aircraft HMDs generally have FOVs of 35° to 40°, trading off FOV for a lower weight optical system. (Helmet weight must be kept to a minimum because of the loads exerted on the pilot`s head during high *g* manoeuvres.)

A number of optical configurations have been developed by the various HMD manufacturers. Fig. 7.21 shows a binocular HMD with a FOV of over 50° for helicopter applications. Collimation of the display images is carried out by the eye piece elements which incorporate a spherical reflecting surface (see inset Fig. 7.21). Very good optical performance and accuracy can be achieved using such configurations with optical errors of less than 1 milliradian – better in fact than the CRT resolution. This design has also excellent stereo viewing performance. The application of the binocular HMD as part of an indirect viewing system for helicopter operations in night/poor visibility conditions is shown in Fig. 7.22. The pan and tilt camera platform, which can mount a low light TV camera and FLIR sensor, is servo controlled to follow the pilot`s LOS (which is measured by the head

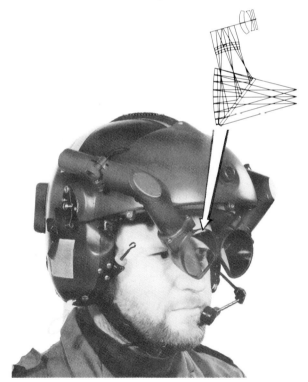

Fig. 7.21 50° FOV binocular HMD for helicopter applications (by courtesy of GEC-Marconi Avionics Ltd.).

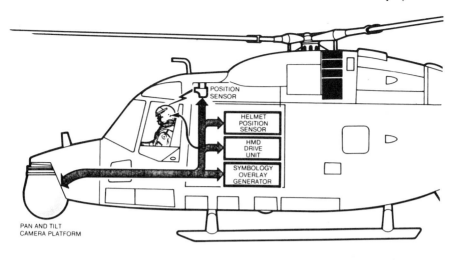

Fig. 7.22 Head steered indirect viewing system.

tracker). Similar systems using a small gimballed infra-red sensor are now installed in a number of modern fighter/strike aircraft.

Figure 7.23 shows a modern integrated helmet with binocular HMD system combined with NVGs. The NVGs are optically combined with the CRT display so that the CRT display overlays the image intensifier display on an accurate one to one basis. This enables the image intensifier display and the infra-red sensor display to be combined in a complementary manner so that the best characteristics of each viewing system are obtained, together with full HUD symbology.

NVGs basically comprise two image intensifiers through which the observer views the scene at night by means of a suitable optical arrangement. Image intensifiers are highly sensitive detectors which operate in the red end of the visible spectrum amplifying light about 40,000 times from less than starlight up to visible level.

It should be noted that special cockpit lighting is necessary as conventional cockpit lighting will saturate the image intensifiers. Special green lighting and complementary filtering are required.

Most night flying currently undertaken by combat aircraft is carried out using NVGs which are mounted on a bracket on the front of a standard aircrew helmet. The weight of the NVGs and forward mounting creates an appreciable out of balance moment on the pilot's head and precludes the pilot from undertaking manoeuvres involving significant *g*. The NVGs must also be removed before ejecting because of the *g* experienced during ejection. (A quick release mechanism is incorporated in the NVG attachment.)

The integrated helmet design shown in Fig. 7.23 avoids these problems as the NVGs are located near the natural pivot point of the head so that they do not create

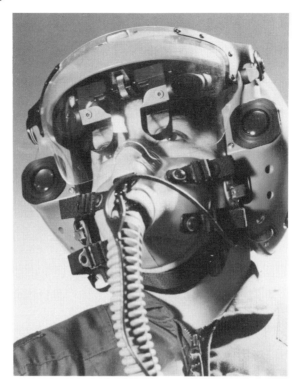

Fig. 7.23 Integrated helmet mounted display system.

significant out of balance moments. (As an added bonus, the NVGs also provide the pilot with night vision after ejecting from the aircraft – for what it is worth.)

A modular approach is likely to be adopted for helmet systems with a common helmet shell and different modules for the different roles. The helmet design shown in Fig. 7.20, in fact, is a modular system. Modules can be simply fixed to the basic helmet shell which uses very strong but light materials and is made in a number of sizes which are precisely form fitted to the pilot by an injection moulding process. The modules can range from a simple visor through a daytime sight to the top of the range system with NVGs and CRTs. The same helmet shell can be used for helicopter applications with the appropriate modules. The expensive optical and electrical parts can be replaced quickly and sent off for repair without the pilot losing his personalised helmet.

7.3.5 Head tracking systems

The need to measure the orientation of the pilot's head to determine the angular co-ordinates of the pilot's line of sight with respect to the airframe axes has already been explained. It should be noted that the problem of measuring the angular orientation of an object which can both translate and rotate can be a fundamental requirement in other applications such as robotics. In fact the solutions for head tracking systems can have general applications.

Space does not permit a detailed review of the various head tracking systems which have been developed. Most of the physical effects have been tried and exploited such as optical, magnetic and acoustic methods.

Optical tracking systems work in a number of ways, for example,

(a) Pattern recognition using a CCD camera
(b) Detection of LEDs mounted on the helmet
(c) Sophisticated measurement of laser generated fringing patterns.

Magnetic tracking systems measure the field strength at the helmet from a magnetic field radiator located at a suitable position in the cockpit. There are two types of magnetic head tracker system:

(a) An AC system using an alternating magnetic field with a frequency of around 10 kHz.
(b) A DC system using a DC magnetic field which is switched at low frequency.

Both systems are sensitive to metal in the vicinity of the helmet sensor. This causes errors in the helmet attitude measurement which are dependent on the cockpit installation. These errors need to be measured and mapped for the particular cockpit and the corrections stored in the tracker computer. The computer then corrects the tracker outputs for the in situ errors. The AC system is more sensitive to these errors than the DC system which is claimed to be 10 times less sensitive to metal than AC systems.

The sensors and radiators for both AC and DC magnetic tracker systems are small and light weight. Typical sensor volume is about 8 cm^3 (0.5 in^3) and weight about 20 gm (0.7 oz). The radiator volume is typically about 16 cm^3 (1 in^3) and weight about 28 gms (1 oz) (excluding installation).

Angular coverage of the magnetic type of head tracker is typically around ±135° in azimuth and ±85° in elevation.

Accuracy on or near boresight is generally within 2 milliradians (0.1° approximately) and around 0.5° at large off-boresight angles.

A reasonable head motion box can be achieved which is operationally acceptable for most applications.

All the systems – optical, magnetic and acoustic – have their individual limitations although the magnetic systems are currently the most successful.

Each system is notable for the complexity of the algorithms needed to sort out the signal from the noise and the man-years of software programming in them.

It should be noted that other errors can be encountered in measuring the pilot's direction of gaze apart from the inherent errors which are present in the head tracker system. This is because the pilot's direction of gaze is inferred from the helmet attitude measurement. The pilot can, in fact, experience great difficulty in keeping the helmet sight on the target when subjected to severe low frequency vibration. This can result from the buffeting which can be encountered at high subsonic speeds in turbulent air conditions at low altitudes. The power spectrum of this buffeting vibration can include significant levels in the 1 Hz to 10 Hz frequency range which the pilot is unable to follow and keep the aiming cross on the target. Tracking errors of over 2° RMS have been measured under replicated buffeting tests carried out on the ground using tape recorded flight data. Some appreciation of the vibration environment can be gained from the fact that a sinusoidal vibration of $0.5g$ amplitude at a frequency of 2 Hz corresponds to an up and down motion of ± 31mm (± 1.2 ins). It has been shown, however, that the pilot's gaze can still be maintained on the target whilst the pilot is experiencing low frequency buffeting type vibration. This is because the control of the eyeball (and hence the direction of gaze) by the eye muscles has a much faster response than the control of the head position by the neck muscles. The 'eyeball servo' has, in fact, a bandwidth of several Hertz in control engineering terms.

The sight line measurement accuracy can hence be greatly improved by the use of an eye tracker system which measures the direction of gaze of the eye with respect to the helmet. The gaze angle relative to the airframe can then be obtained by combining the eye tracker and head tracker outputs.

The eye tracker also offers advantages in high g conditions when head movement is difficult but the eyes can still move to look at a target.

The basic principles of an eye tracker are very briefly explained later in the chapter as an eye tracker can also be used very effectively as a means of data entry – a possible development in future cockpits.

7.3.6 HMDs and the virtual cockpit

The concept of a 'virtual cockpit' where information is presented visually to the pilot by means of computer generated 3D imagery is being very actively researched in a number of establishments both in the USA and the UK.

Only a brief introduction to the topic can be given in this chapter because of space constraints. There is no doubt, however, of its importance in the future and the revolutionary impact it is likely to have on the next generation of aircraft cockpits. The increasing use of remote piloted vehicles (RPVs) and their control from a 'parent' aircraft, or ground station, is another future application for HMDs and virtual cockpit technology. It should be noted that RPVs can include land vehicles or underwater vehicles as well as airborne vehicles.

A correctly designed binocular HMD (BHMD) is a key component in such systems because it is able to present both a display of information at infinity and also stereo images to each eye so that the pilot sees a 3D image. It should be emphasised that the BHMD optical system needs to be accurately designed and manufactured in order to achieve a good stereoscopic capability.

The ability to generate 3D displays opens up entirely new ways of presenting information to the pilot (and crew), the objectives being to present the information so that it can be visually assimilated more easily and in context with the mission. As an example, information related to the outside world can be viewed head up and information related to the mission and the aircraft 'health' can be viewed head down using only one display system, namely the BHMD. Figure 7.24 illustrates the use of a BHMD to implement such a system. When head up, the pilot views the outside world directly, or indirectly by means of a TV display on the HMD from a head steered gimballed electro-optical sensor unit, (eg infra-red imaging sensor, low light TV camera or CCD camera).

When looking down at the instrument panel, the virtual cockpit computer system recognises the pilot's head down sight line from the head tracker output and supplies this information to the display generation system. The display generation system then generates a stereo pair of images of the appropriate instrument display on the panel which corresponds to the pilot's sight line. Thus, looking down into the cockpit at the position normally occupied by a particular instrument display will

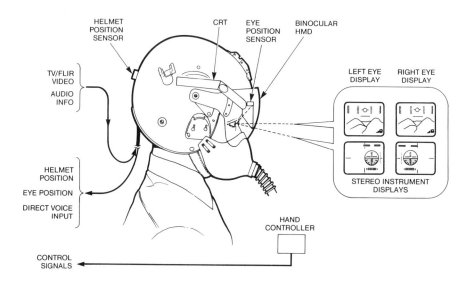

Fig. 7.24 Binocular HMD and virtual cockpit.

result in the pilot seeing a 3D image of that instrument display appearing in the position it normally occupies – ie a virtual instrument panel display. Carried to its logical conclusion, the normal head down displays and instruments would no longer be required. Simple standby instruments of the 'get you home' type would be sufficient to cover the case of a failure in the HMD system. Such a bold step, however, is still several years away from implementation in a production aircraft.

Novel ways of presenting information to the pilot by means of the BHMD include displaying a 3D 'pathway in the sky' as a flight director display, which can be overlaid on the normal outside scene (directly or indirectly viewed) or on a computer generated outside world scene created from a terrain data base. A 'God's eye view' can also be presented: for example, what the pilot would see if he was outside the aircraft and looking down from a height of say 2000 ft above and slightly behind his aircraft. ('Beam me up there, Scotty!'.) The threat situation can be presented on such a 'God's eye view' display – eg enemy aircraft, missile sites, anti-aircraft artillery ('Triple A') sites or enemy radar sites. Threat zones defining the missile (or gun) range, radar field of view etc. can be presented as 3D surfaces so that the pilot can assimilate the threat situation on a single display.

Figure 7.25 shows a 3D 'God's eye view' display which provides the pilot with a stereoscopic view of his helicopter as well as its relationship to its proposed flight path. Further information can be provided by enabling the pilot to change his apparent position outside the aircraft.

With advances in display technology it would seem reasonable to anticipate light weight high resolution colour displays which would be capable of providing a helmet mounted colour display of comparable quality to the current colour head down displays. This, for example, would enable a colour map to be displayed on the HMD.

It is worth sounding a slightly cautionary note on virtual cockpit technology and the need for all the human factors issues to be addressed and satisfactory solutions achieved for any problems that may be encountered.

Some disquiet has, in fact, been expressed recently on the physiological and psychological effects of prolonged viewing of commercial 'virtual reality' systems and the ability of the user to switch back to the real world without delay. Poor quality stereo optics require the visual processing part of the brain to make major adjustments to fuse the resulting stereo images into a 3D picture. To adjust back to normal real world viewing can then take some time.

The cockpit environment, however, is very different and requires frequent viewing of the outside world whether directly, or indirectly, by means of a head steered gimballed camera system. The virtual reality problems referred to above may, therefore, not arise. The stereo quality of the optical system of the BHMD is also far higher than the current commercial virtual reality systems.

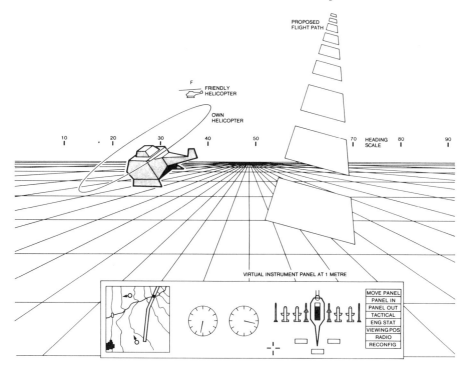

PROPOSED
FLIGHT PATH

F
FRIENDLY
HELICOPTER

OWN
HELICOPTER

10   20   30   40   50   70   HEADING   80   90
                                        SCALE

VIRTUAL INSTRUMENT PANEL AT 1 METRE

MOVE PANEL
PANEL IN
PANEL OUT
TACTICAL
ENG STAT
VIEWING POS
RADIO
RECONFIG

Fig. 7.25 'God's eye view' display concept (by courtesy of GEC-Marconi Avionics Ltd.).

7.4 DISCUSSION OF HUDs *vs* HMDs

7.4.1 Introduction

The current and future roles of the HUD and HMD in military and civil aircraft applications are briefly discussed in the following subsections.

7.4.2 Civil aircraft HUDs

The question of a HUD or an HMD for civil aircraft applications does not arise as civil airline pilots have no need to wear helmets. The pilot's communication requirements are met by a simple push on headset with attached microphone. The acoustic noise levels in the civil aircraft cockpit are much lower than in military aircraft and the head protection requirements do not arise. An oxygen mask is only required in the event of a failure of the cabin pressurisation system whereas the helmet forms part of a special oxygen breathing system for high *g* manoeuvring in a combat aircraft.

It should be noted that the space constraints in a civil cockpit are less severe and roof mounted HUD installations are generally used.

The use of a HUD by civil aviation operators is still a relatively novel practice. In 1992, for example, it was estimated that there were about 100 HUD installations in revenue service in the continental United States. This compares with over 10,000 military HUD installations in service world-wide. Nevertheless, the future for civil HUD systems looks good, although their adoption is likely to be a fairly gradual process.

The main advantages of a HUD in a civil aircraft are:

1. Increased safety by providing a better situational awareness for the pilot to control the aircraft by the head up presentation of the primary flight information symbology so that it is conformal with the outside world scene.

   The problems of severe wind shear conditions have been mentioned earlier in the introduction to this chapter. Figure 7.26 shows how wind shear arises and

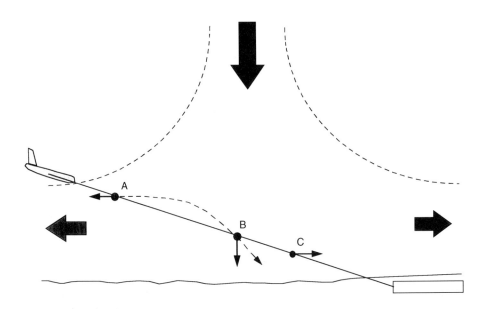

Fig. 7.26 Wind shear. This shows the simplest and probably the most common fatal wind shear scenario. The aircraft enters a downdraught at A and the pilot experiences a rapidly increasing head wind. The indicated airspeed increases and the additional lift causes the aircraft to climb above the glide slope and so the pilot reduces power to recapture the approach. Suddenly, at B, the head wind disappears and, worse, the aircraft is in an intense downdraught. Worse still, at C, a rapidly increasing tail wind further reduces the probability of recovery.

the problems it creates. Over 700 passengers have been killed in the USA alone in accidents in recent years caused by wind shear.

Safety improvements in the course of each stage of a typical flight were predicted in a study published by Flight Safety Foundation (see Further Reading). The study concluded that a HUD would likely have had a positive impact on the outcome of 30% of the fatal jet transport incidents documented over the period 1958 to 1989.

2. Increased revenue earning ability by extending operations in lower weather minima. The HUD is used to monitor the automatic landing system and to enable the pilot to take over on aircraft which are not equipped to full Category III automatic landing standards (automatic landing systems are covered in Chapter 8).

3. Use of the HUD as part of an enhanced vision system to enable operations in lower weather minima at airfields not equipped with automatic landing aids (ILS/MLS). For example, the number of Type II and Type III ILS facilities in the United States is very limited – typically less than 70. The revenue earning potential of enhanced vision systems is thus very considerable. As explained earlier, the HUD displays a video image of the real world ahead of the aircraft derived from a millimetric wavelength radar sensor and a FLIR sensor installed in the aircraft together with the overlaid primary flight information and flight path parameters.

Very encouraging trials have been carried out recently in the United States by the FAA, the trials being completed in 1992. The enhanced vision system used a millimetric wavelength sensor operating at 35 GHz and a 3 to 5 micron infra-red imaging camera (see Further Reading). The infra-red sensor provided useful imaging information in haze and high humidity conditions at a range which substantially exceeded the range of the pilots's eyes. However, descending through typical cloud and fog conditions the range was no better than could be seen by the eye. The 35 GHz radar sensor, however, was able to operate successfully under all weather conditions and enabled a synthetic picture of the runway to be generated from the processed radar image at a range of 2 to 2.5 miles. The pilots were able to identify the runway on which they intended to land at a range of about 1.7 miles. They were routinely able to fly to Category IIIa minima (50 ft decision height/700 ft runway visual range) using the HUD display of the radar generated runway image and the overlaid primary flight symbology as their sole means to conduct the published approach. The display of the aircraft's velocity vector on the HUD was found to be particularly helpful throughout the approach, when used in conjunction with the electronic image of the outside world, and enabled the control of the aircraft's flight path to be precisely managed.

The evaluation pilots reported Cooper-Harper ratings of 3 (satisfactory without improvement and with minimal compensation required by the pilot for desired performance).

7.4.3 Military aircraft HUDs and HMDs

The HUD is likely to remain a key system in military aircraft for the foreseeable future as the prime display of primary flight information and to enable high accuracy weapon aiming with guns, unguided bombs and rockets to be achieved. The boresight errors in a HUD system impact directly on the weapon aiming accuracy and these are typically of the order of 1 milliradian with a HUD compared with 5 to 10 milliradians or more with an HMD. No contest in such an application.

The need for a very wide FOV HUD, however, may be brought into question when the pilot is equipped with an HMD and head steered indirect viewing system. Future developments may well see a combination comprising a binocular HMD and a refractive HUD with an instantaneous FOV of about 20°. The use of a Rugate dielectric coating on the combiner of the refractive HUD would enable a high brightness display to be achieved with very little loss of outside world visibility through the combiner. The display is also bright enough for a raster TV picture to be seen on the HUD in some daylight conditions.

The existence of laser weapons which can cause eye damage has been in the public domain for several years. These can be operated at relatively low power levels to dazzle the pilot of an approaching aircraft so that the pilot becomes disorientated and has to break off the attack. The laser can also be operated at higher power levels which are sufficient to cause permanent damage to the retina of the eye resulting in impaired vision and blindness if sufficiently high. In future conflicts it may, therefore, be necessary to fly with a suitably opaque visor installed on the helmet to protect the pilot's eyes from optical weapons when operating in areas where such a threat exists. The pilot would then fly the aircraft by means of a head steered indirect viewing system.

The potential of the HMD and the exploitation of virtual cockpit concepts has already been briefly discussed and it is likely such concepts will be used in the next generation military aircraft.

Finally, it should be noted that there are civil applications for the HMD as part of an indirect viewing system for night/all weather operation of helicopters (Fig. 7.19). These include air sea rescue operations, servicing North Sea oil rigs, police helicopter surveillance and border patrols using helicopters.

7.5 HEAD DOWN DISPLAYS

7.5.1 Introduction

Head down displays (HDDs) are the oldest type of electronic display, having originated with airborne radar scopes in the second world war, and have subsequently been developed into the workhorse of the cockpit. They comprise the major cockpit displays in a civil aircraft – 'wall to wall' colour displays having transformed the civil flight deck in recent years – from large 'jumbo' jets to small commuter aircraft.

Typically, a HDD contains a CRT with drive circuits, power regulation and, as with the HUD, there is frequently a simple processor and housekeeping software to control the brightness, self test and mode changes. Some units now contain a full processor and symbol generator and data bus interface, but it is more usual for the HDDs to be driven by a central display processor(s).

The size of an HDD is typically defined by an ARINC standard of square format with one inch increments – for example, 5 × 5, 6 × 6 inch etc. This is not ideal for the normal 4:3 aspect ratio video and indeed rectangular format displays are found in both military and civil cockpits.

The high brightness and resolution requirements have determined that the military cockpit is still dominated by monochrome displays – in addition, sensors like FLIR and LLTV are monochrome. It should be noted that green is at the peak of the human visual response. The green CRT display provides both brightness and chromatic contrast against blue sky.

The civil cockpit, which is shaded from direct sunlight to a far greater extent than the military cockpit, saw the introduction of colour CRT displays before the military application. The civil environment is also much more benign from the point of view of vibration, acceleration and shock levels which may be experienced. Colour CRTs, however, are now installed in military cockpits and are essential for video map displays.

The CRT is still the dominant type of display but LCD technology is coming up fast and is likely to displace CRT technology in many head down applications in the not too distant future. A brief overview of display technology is given in Section 7.6.

7.5.2 Civil cockpit head down displays

There has been a major revolution in civil cockpit displays with many of the traditional instrument displays replaced by multi-function colour CRT displays. Some of the traditional instruments have been retained as standby instruments but all the primary displays are now colour multi-function displays. These have transformed the flight deck which now looks far 'cleaner' and less cluttered than the earlier generation civil jet transport aircraft. Figures 1.3 and 1.4 in Chapter 1 show a modern flight deck.

Hybrid raster/stroke techniques in general are used to generate the display formats in civil head down CRT displays. The stroke technique provides precise writing characteristics for the symbology with the colour in-fill being raster generated. The displays are duplicated for the Captain and Second Pilot and being multi-function it is possible to reconfigure the displayed information in the event of the failure of a particular display surface.

The electronic Primary Flight Display (PFD) replaces six electro-mechanical instruments: altimeter, vertical speed indicator, artificial horizon/attitude director indicator, heading/compass indicator and Mach meter. PFD formats follow the classic 'T' layout of the conventional primary flight instruments, as mentioned in Chapter 2. All the primary flight information is shown on the PFD thereby reducing the pilot scan, the use of colour enabling the information to be easily seperated and emphasised where appropriate.

Figure 7.27 shows a representative primary flight display. Airspeed is shown on a scale on the left with pressure altitude and vertical speed on the right hand scales. Aircraft heading information is shown on a 'tape' scale type format below the attitude display. The artificial horizon/attitude display has a blue background above the horizon line representing the sky and a brown background below the horizon line representing the ground. This enables 'which way is up' and the aircraft orientation to be rapidly assimilated by the pilot in recovering from an unusual attitude, for example, as the result of a severe jet upset.

It should be noted that the attitude direction indicator is a key display on any aircraft for recovering from unusual attitudes particularly when the normal outside world visual cues are not present (eg flying in cloud or at night) as the pilot can become disorientated. The normal display of attitude on the HUD is not suitable for

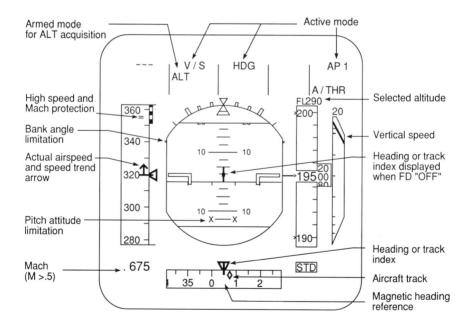

Fig. 7.27 Primary flight display – Airbus A320 (by courtesy of Airbus Industrie).

this purpose, being too fine a scale which consequently moves too rapidly. There are recorded instances of pilots of highly agile fighters becoming disorientated in cloud. In one case the pilot was flying at supersonic speed, upside down and diving towards the ground at an angle of 35° and only became aware of the situation when he emerged from the cloud at about 10,000 ft. Fortunately he was able to recover from the situation in time.

Figure 7.28 shows a typical navigation (or horizontal situation) display. Pilot selectable modes include the traditional compass rose heading display, expanded ILS or VOR formats, a map mode showing the aircraft's position relative to specific waypoints and a North up mode showing the flight plan. Weather radar displays may be superimposed over the map, as shown in the illustration.

Figure 7.29 shows an engine indication and crew alerting messages (EICAS) display. For example, primary caution and warning messages appear in one central location. Also, engine instrument displays use colour changes to clearly show exceedance conditions.

Figure 7.30 shows a synoptic display. Aircraft system schematics (synoptics), display graphics of the fuel system, hydraulics, electrical system and other system configurations can be presented. Monitoring of aircraft systems can be clearly shown in diagrams showing valve positions, current status and other pertinent information.

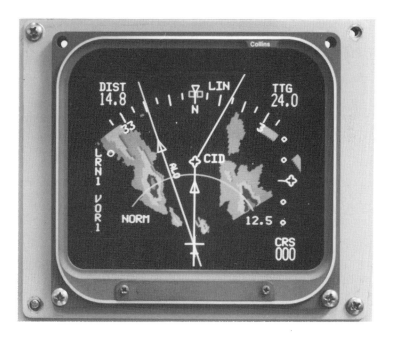

Fig. 7.28 Typical navigation (or horizontal situation) display (by courtesy of Collins Avionics, Rockwell International).

Fig. 7.29 Engine indication and crew alerting messages display (by courtesy of Collins Avionic Systems, Rockwell International).

Fig. 7.30 Synoptic display (by courtesy of Collins Avionic Systems, Rockwell International).

7.5.3 Military head down displays

Video (ie raster) head down displays now include FLIR, LLTV and maps. In the cursive mode, all the HUD functions may be repeated overlaid on the raster pictures. Fuel and engine data, navigation waypoints and a host of 'housekeeping' functions (eg hydraulics, pressurisation) may be displayed. A stores management display is also required showing the weapons carried and their status.

It is usual to have a bezel around the display with keys. Sometimes key functions are dedicated or they may be 'soft keys' where the function is written beside the key on the display. So called tactile surfaces are being introduced using such techniques as infra-red beams across the surface of the display or even surface acoustic waves where the pressure applied can also be measured to give X, Y and Z co-ordinates.

Typical advanced military cockpits are configured with four head down displays:

* HSD (horizontal situation display) 6 × 8 inch colour map display in 'portrait' format with symbol overlay of routeing and threat data.
* VSD (vertical situation display) 8 × 6 inch IR video display showing targeting video at various magnifications. Currently monochrome display in 'landscape' format with symbol overlay. A high resolution colour display will ultimately replace the monochrome display.
* SSD (systems control display) 5 × 5 inch display of system control data. Monochrome symbolic display with tactile data entry overlay.
* SCD (systems control display) 5 × 5 inch display of system control data. Monochrome symbolic display with tactile data entry overlay.

7.5.4 Raster overlay display generation

The generation of symbology in the raster scanning process (eg lines, circles, dials, scales, alpha-numeric characters) to overlay a raster TV video display, to present tabular information, or to produce a map display, is fundamental to such displays. A brief explanation of the basic principles used to generate such overlay displays is, therefore, set out below.

The fundamental advantage of the raster process for generating a video picture should also be appreciated, namely its high bandwidth. The raster scanning process, in fact, made TV practical as it enables every pixel on the display surface to be illuminated at the required intensity at video refresh rates of interlaced 50 Hz. (Alternate raster lines are refreshed at 50 Hz. Modern electronics can now generate a non-interlaced video display at 50 Hz.) As mentioned earlier, however, the raster generated display is inherently of lower brightness than a cursive display.

The first point to be appreciated about a raster overlay generation system is that the position of any pixel to be illuminated on the display is determined by the time interval from the start of drawing the first raster line in the TV frame. For example, consider a 625 line raster display which is refreshed at 50 Hz. The time to generate a complete frame is equal to $10^6/50$, that is 20,000 micro-seconds. (The raster line flyback time has been ignored for simplicity in this example.) A 32 MHz clock enables the position of a point on an individual raster line to be resolved to $1/32 \times 32$, that is 1 in 1024. (It should be noted that the 32 MHz clock is used to generate the raster waveform which in turn is phase locked to the synchronisation pulse of the TV frame.)

Hence to illuminate a pixel, for example, on raster line 121 at 400 pixels from the start of the raster line it is necessary to send a bright up command at

$$(120 \times 32) + \left(\frac{400}{1024} \times 32 \right)$$

that is at 3852.5 microseconds from the start of the TV frame.

The process of drawing a straight line between co-ordinates (x_1, y_1) and (x_2, y_2) is shown in Fig. 7.31. Any figure shape can then be drawn as a set of straight line segment approximations. A suitable display generation programme can thus be implemented by a special fast processor to draw the straight line segments using the process shown in Fig. 7.31.

The raster line steps in the overlaid display line are not too noticeable when the line is near the horizontal or near the vertical. The steps become much more apparent when the line is inclined at an angle to the horizontal (or the vertical), as can be seen in Fig. 7.31. This 'steppiness' in appearance of the line can be very significantly reduced by varying the shape of the CRT bright-up pulse as a function of the slope of the line. The use of such techniques enables reasonable quality raster overlay symbology to be generated in terms of visual appearance, although not as 'crisp' as stroke written symbology.

Special 'task orientated' high speed processors based on discrete computing elements were used for this purpose until the advent of 'graphics chip sets' exploiting VLSI implementation. These are now widely available and provide a very powerful and fast graphics generation capability (including colour).

7.5.5 Digitally generated colour map displays

A map is by far the best way of visually assimilating the aircraft's horizontal situation. That is the position of the aircraft relative to the chosen waypoints and destination, alternative routes, targets and location of specific terrain features such as mountains, hills, lakes, rivers, coast line, towns, cities, railways, roads etc. Colour maps can now be generated digitally from a map data base stored in the computer memory and can be moved with the aircraft so that the aircraft is at the centre of the map display. The map can be oriented 'track-up' or 'North-up' by selecting the appropriate mode. The navigation system provides the basic

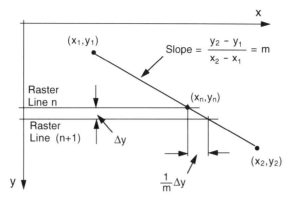

Consider overlay line joining (x_1, y_1) and (x_2, y_2)

Overlay line crosses raster line n when $x = x_n$

Overlay line crosses raster line (n+1) when $x = x_n + \dfrac{1}{m}\Delta y$

Overlay line crosses raster line (n+2) when $x = x_{n+1} + \dfrac{1}{m}\Delta y$

etc.

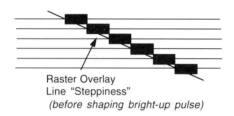

Raster Overlay
Line "Steepiness"
(before shaping bright-up pulse)

Fig. 7.31 Raster overlay line generation.

information to move and rotate the map to maintain the selected orientation (usually track-up).

The system has major advantages in flexibility and the ability to present and overlay information which is not possible with the earlier projected film map displays. Some of the advantages are:

• Scale flexibility. The map can be displayed at any selected scale. The time to re-draw the map at the new selected scale is typically 1 to 2 seconds.

• Look ahead and zoom facility. The pilot can look at the map some distance ahead of the aircraft's present position and 'zoom in' at a large scale to see the terrain details, for example in the target area or destination airfield etc.

• Terrain clearance display. The terrain which is above the pilot's present altitude can be distinctly coloured so that the pilot is aware of the location of potentially dangerous terrain such as hills and mountains relative to the aircraft. Appropriate avoiding action can then be taken in good time.

• Terrain screening display. The areas below the radar LOS of known (or likely) enemy radar sites can be displayed on the map so that advantage can be taken of terrain screening to minimise the probability of detection.

• De-cluttering. Features and information (eg place names) which are not required can be deleted from the map display to de-clutter the map for a particular phase of the mission. Specific features can also be emphasised. The map display can be restored to show the full information whenever required.

• Threat locations. These may be enemy missile sites. 'triple A' sites. fighter airfields or radars. Threat envelopes showing, say, the missile range from a particular site, or sites, can be shown in a distinct colour.

• Known low altitude obstructions. Electricity pylons. power cables. towers etc. can be emphasised on the map, for example by brightening them up. when flying at low altitude.

• 3D map displays. A 3D map can be generated of. say. the target area to assist the pilot in visually assimilating the local terrain and hence assist in target acquisition.

Figure 7.32 illustrates a digitally generated video map display and the high quality which can be achieved. (Unfortunately it is not possible to show this in colour.)

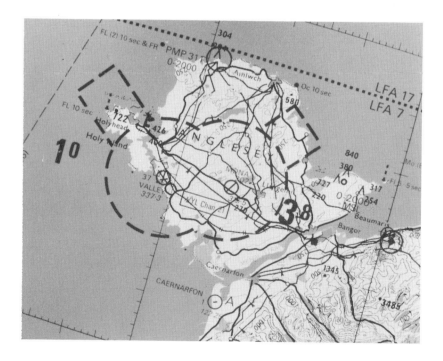

Fig. 7.32 Digitally generated video colour map display (by courtesy of GEC-Marconi Avionics Ltd.).

A brief outline of the principles used to generate a video map display is set out below.

It is important to emphasise that the video map display is raster generated. This is because of the amount of information that has to be displayed, for which a cursive (or stroke) writing system has insufficient bandwidth.

The information to construct the map is stored in a digital data base as sets of specific terrain features such as: coast lines, rivers, lakes, contour lines, town outlines, railways, roads, woods, airfields, electricity pylons etc.

The feature outline is specified as a set of vector co-ordinates which form straight line segment approximations to the feature outline. The colour infill for the feature is also specified in the data.

Map data bases which are stored in this way are known as vector data bases. The method gives an extremely powerful means of data compression to enable all the information depicted on a map to be stored in a practical size of computer memory

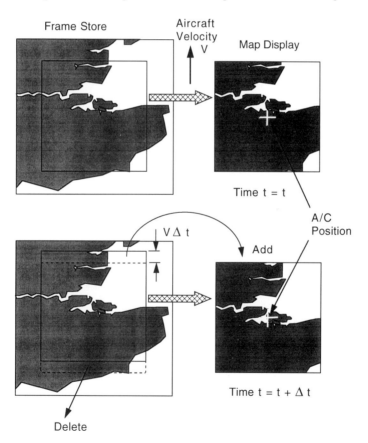

Fig. 7.33 'Scrolling' the map display.

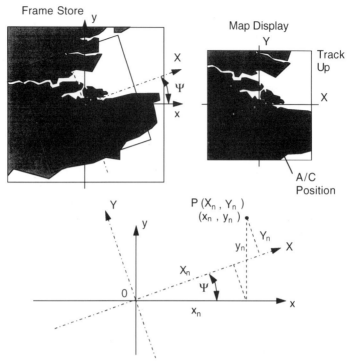

Coordinates of point P with respect to Map Display axes, 0X & 0Y, are (X_n, Y_n), and with respect to Frame Store axes, 0x & 0y, are (x_n, y_n).

Data for point P (x_n, y_n) on Map Display is held in Frame Store coordinates

$$x_n = X_n \cos \Psi - Y_n \sin \Psi$$
$$y_n = Y_n \cos \Psi + X_n \sin \Psi$$

Fig. 7.34 Rotating the map display by rotating the addressing of the frame store.

for airborne applications. The map can easily be de-cluttered by deleting a particular feature set(s) from the display.

Map data bases are available for significant areas of the world but there are very large areas where the data are not yet available. The task of compiling a world-wide mapping data base is a continuing activity by the various mapping agencies around the world, but is clearly a major task. There are vast areas where accurate ground survey data are not available and the map data have to be derived from surveillance satellites.

An alternative approach during the interim period until a world-wide digital data base becomes available, is to extract the map data from paper charts. This can be done by digitising the map feature planes which are used to print the map. The feature data can then be stored as a set of vector co-ordinates. The paper map can also be stored using video compression techniques and then reproduced as a video map display – such a technique is known as a pixel based map. The video map so produced is identical to the paper map in appearance. It should be noted that such a map is only as accurate as the paper chart and in some areas the information on the chart may be of considerably lower accuracy than that from a digital mapping data base.

The task of drawing a map and translating and rotating it in real time with the aircraft's motion is a major processing task but can be accomplished with the modern electronic hardware now available: micro processors, graphics chip sets, semi conductor memory devices etc. Typically, the map data are loaded into a frame store which holds the data for an area whose sides are 1.4 times the sides of the map area being displayed. Scrolling the map display is illustrated in Fig. 7.33. Map rotation is accomplished by rotating the addressing of the frame store as shown in Fig 7.34. (The frame store size factor of 1.4:1 covers the worst case heading changes.)

7.6 DATA FUSION

Data fusion is the name given to the process of combining the data from a number of different sources to provide information which is not present in the individual sources. For example, a synthetic 3D picture can be derived of the terrain in front of the aircraft from an accurate terrain data base and accurate information on the aircraft's position and attitude. The aircraft's position and attitude information are provided by a GPS and an INS, or a TRN system and an INS. The synthetic picture of the terrain can be overlaid one to one with the outside scene derived from a FLIR sensor and displayed on a HUD or a head down display. Ground features which may be hard to detect on the FLIR such as ridge lines can be accentuated together with electricity pylons. This enables the pilot to continue visual flight in conditions of marginal visibility where normally it would be necessary to pull up and fly at a higher altitude. Figure 7.35 illustrates a display enhanced by data fusion.

7.7 INTELLIGENT DISPLAYS MANAGEMENT

The exploitation of intelligent knowledge based systems (IKBS) technology, frequently referred to as 'expert systems', to assist the pilot in carrying out the mission is the subject of a number of very active research programmes, particularly in the United States. One of the US programmes is the 'pilot's associate program' which aims to aid the pilot of a single seat fighter/attack aircraft in a similar way to

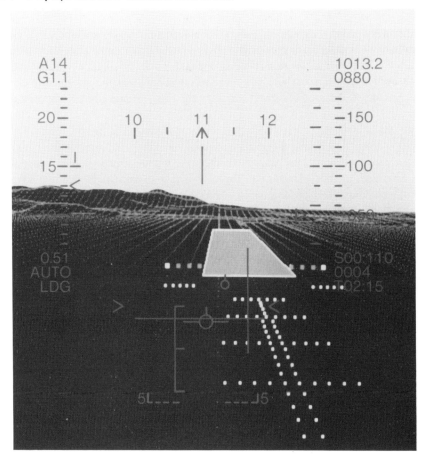

Fig. 7.35 Data fusion enhanced display. Ridge lines generated from terrain data base overlaid on FLIR video picture displayed on HUD together with primary flight information. Enhanced runway image generated from millimetric radar sensor signals (by courtesy of GEC-Marconi Avionics Ltd.).

the way in which the second crew member of a two crew aircraft assists the pilot. The prime aim is to reduce the pilot work load in high work load situations.

Space constraints do not permit more than a very brief introduction to this topic, which is of major importance to the next generation of military aircraft as these will use a single pilot to carry out the tasks which up to now have required a pilot and a navigator/weapons systems officer.

The exploitation of IKBS technology on the civil flight deck will follow as the technology becomes established in military applications.

A subset of all the proposed expert systems on an aircraft is an intelligent displays management system to manage the information which is visually presented to the pilot in high work load situations.

It is the unexpected or uncontrollable that leads to an excessive work load, examples being:

*   The 'bounce' – interception by a counter attacking aircraft with very little warning.
*   Evasion of ground threat – SAM (surface–air missile).
*   Bird strike when flying at low altitude
*   Engine failure
*   Weather abort or weather diversion emergency.

A block diagram illustrating the concepts of an intelligent displays management system is shown in Fig. 7.36. The system comprises an 'aircraft state expert' which deduces 'what is happening' from the aircraft data, pilot inputs and the mission plan by the application of an appropriate set of rules. The aircraft state expert in turn controls the 'displays management expert' which determines the information displayed on the various display surfaces: HUD, map, head down displays or HMD according to an appropriate set of goals and priorities.

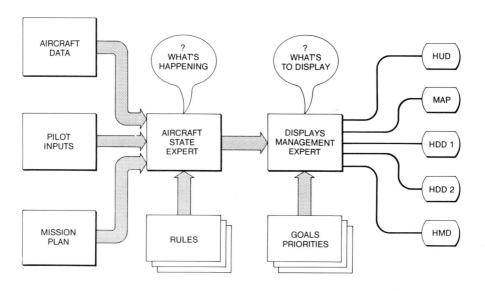

Fig. 7.36 Intelligent display management (by courtesy of GEC-Marconi Avionics Ltd.).

7.8 DISPLAYS TECHNOLOGY

A very brief overview of displays technology is set out below as this is fundamental to the implementation of cockpit electronic displays.

7.8.1 HUD

The display surface for a HUD is provided by a monochrome CRT and this is likely to remain the case for some time as there are no near term technologies which can match the resolution and brightness of the current CRTs. Ever increasing performance, however, is being demanded from the HUD CRT with line brightness of more than 100,000 Cd/m^2 being sought and a raster luminance of 10,000 Cd/m^2. High resolution is demanded in both cursive and raster modes. The spot size, however, is different for the two modes and optimising a CRT for both modes is rarely feasible. CRT phosphors used are P1, P43 and now P53 but, whereas the broad spectrum of P1 gave a gentle line broadening, the spiked response of P43 and P53 can give rise to colour fringing unless the optics are chromatically corrected.

A colour display for a HUD is technically feasible, particularly one based on a Penetron CRT (see later). This offers a limited colour range of two primary colours and shades between. The trade offs, however, in terms of loss of resolution and brightness, additional cost and complexity, and increased weight and volume, make the colour HUD unattractive compared with a monochrome HUD at the present time.

7.8.2 Head down displays

7.8.2.1 *Colour CRTs*

Colour HDDs are currently predominantly based on colour CRTs, although this situation is changing with the advent of the colour LCD. Colour CRTs are available in three major types, namely shadow-mask, beam index and Penetron, and these are briefly described below.

• Shadow-mask CRT. These operate in basically the same way as a domestic colour TV, but are specifically designed to be more rugged and resistant to the effects of vibration. The shadow-mask tube employs a separate electron gun for each primary colour and the three electron beams converge on a thin foil mask with apertures aligned with the appropriate RGB phosphor dot triad. The advent of Invar materials and stretched mask construction has given increased ruggedness, allowing beam currents of up to 1.5 mA without significant loss of colour purity, and dot pitches of 0.2mm (200 microns) to be achieved. Above all the shadow-mask tube is simple to drive and places no excessive demands on video or deflection circuits. The shadow-mask works well in raster, cursive or mixed modes.

• Beam index CRT. The beam index tube has a single electron gun and the beam is scanned in raster format over vertical RGB phosphor stripes. An index phosphor is positioned between each RGB stripe and emission from this is used to detect the beam position and switch the appropriate intensity for that colour. Brightness is potentially much higher than a shadow-mask tube but is, in fact, limited by spot size. Colour purity suffers if the spot exceeds the width of a triad stripe because of scatter as well as ultimately by overlap. The resolution of the tube is also limited by the stripe pitch, but in practice is not markedly inferior to the shadow-mask tube. Perhaps the greatest drawback is the need for complex drive electronics with wide bandwidth video amplifiers and highly linear deflection circuits. The CRT performance is highly dependent upon the electronics. A final problem with the beam index CRT is that the beam cannot be operated at low luminance without loss of indexing. This can result in night operation limitations. The greatest advantage of the beam index CRT is its ruggedness, which is comparable to a monochrome CRT.

• Penetron CRT. The Penetron CRT is a relatively old concept with an established role in a cabin environment where large sizes are usual and low brightness acceptable – for example an ATC centre. The tube works by varying the EHT such that the beam excites different layers of a composite phosphor arranged like onion skin layers. Practically, only two such phosphor layers are suitable and this limits the display to shades of red through yellow to green. The accompanying high voltage switching, however, renders this a poor candidate for cockpit displays. The challenge to the colour CRT is from the LCD.

• LCD shutters and CRT. This permits an intermediate stage between LCD displays and the CRT. With this system a number of LC devices with electrically switched planes of polarisation are placed in front on a monochrome mixed phosphor CRT. The information for each colour is presented frame sequentially with the appropriate shutter switch. The advantage is that the high resolution of the monochrome CRT is retained, but at the cost of a fairly low optical efficiency through the shutters. The incoming video must be frame stored because of the sequential frame operation of the LC shutter and operation is two or three times faster than the normal video rate.

Such displays are operational (eg US F18 fighter) and provide an effective way of achieving a very high resolution colour display.

7.8.2.2 Flat panel displays

Flat panel displays include the following major technologies: light emitting diodes, electro-luminescent, plasma, vacuum fluorescent and liquid crystal. With the exception of the liquid crystal, these displays are all emissive displays.

The characteristics and potential of these technologies are briefly summarised below.

- LEDs. These have potential for data displays but are limited to small areas because of the high power requirements. The display can have good brightness and contrast but resolution is limited to around 64 cells/inch. Colour displays can be produced with red and green LEDs. Applications are in displays for warning, engine monitoring, in programmable switches and in helmet mounted sights.
- Electro-luminescent. These displays have a good viewing angle and give a pleasing display. Matrix displays can be made with a high duty ratio multiplexed drive. Unfortunately, a fairly high drive voltage (200–400 volts AC) is required. This results in fairly expensive drive circuits and has an adverse effect on reliability. A high resolution is possible but colours are limited – blue being especially difficult. Typical applications are control and display units for communications, navigation and maintenance data. Video rate updates are possible and displays of up to 45cm (18 inch) diagonal exist.
- Plasma displays. These are gas discharge displays which can be matrix addressed. High brightness products such as tactical terminals exist with displays of 8 inch square and resolution of 64 pixels per inch.
- Vacuum fluorescent displays. These are limited in size to about 8 inch diameter and are still difficult to make. They can be matrix addressed with a resolution of 60–80 cells per inch. Brightness, colour and life are all adequate but do not indicate much development potential.
- LCDs. The real up and coming technology is LCD. This technology is already in the cockpit in small instruments and panels using super twist displays which operate in a transflective mode. At night, an electro-luminescent back light is available and a good high contrast display can be achieved. Such displays can be matrix addressed, but for high resolution the incorporation of a thin film transistor with every pixel is necessary. This is the so called active matrix LCD (AMLCD).

Active matrix LCDs operate with a back light, usually a fluorescent lamp. Both lamp and LC material suffer from inadequate performance at low temperature and so may require a heater element incorporated into the display. It should be noted that AMLCDs are not flat – they are much less deep than a CRT but still occupy two to three inches over the light source. They are also not low power displays because of the heating and lighting requirements. They are not cheap at the moment and a military environment AMLCD is little cheaper than the equivalent CRT and high voltage power supply unit. However, full colour video rate panels are becoming available and a pixel density of 200 to the inch achieved. Various US military programmes have specified large flat panel displays and clearly it does permit more effective use of the cockpit space.

It is worth noting that any matrix display of limited resolution can give rise to aliasing effects such as Moiré patterns. Special drive circuit techniques are used to minimise these effects. Only LCD and EL can give a resolution of 200 pixels to the inch and this is still really inadequate for a colour display on the traditional 6 inch square HDD.

Fig. 7.37 Active matrix LC display unit with IR touch screen. 10in diagonal, 480x640 RGB
pixels 8 bit primary (by courtesy of Litton Industries, Canada).

Figure 7.37 shows a modern active matrix LC display unit. The grid of LEDs and
infra-red sensors around the bezel which provide the touch panel capability can be
clearly seen.

7.9 CONTROL AND DATA ENTRY

7.9.1 Introduction

As mentioned in the introduction to this chapter, the pilot (and crew) must be able
to control the information being displayed, for example, to switch modes and
information sources at the various phases of the mission, or in the event of
malfunctions, failures, emergencies, threats etc. It is also essential for the pilot to be
able to enter data into the various avionic systems (eg navigation way points).
Control and data entry are thus complementary to the various displays and enable
the pilot (and crew) to interact with the various avionic systems in the aircraft. The
means for effecting control and data entry must be as easy and as natural as

possible, particularly under high work load conditions. This section gives a brief overview of:

Tactile control panels
Direct voice input
Eye trackers

as means of control and data entry.

7.9.2 Tactile control panels

These have already been briefly described in Section 7.5.3. A typical tactile control panel uses a matrix array of infra-red beams across the surface of the display which displays the various function keys. Touching a specific function key on the display surface interrupts the x and y infra-red beams which intersect over the displayed key function and hence signals the operation of that particular key function. Figure 7.38 illustrates the basic principles.

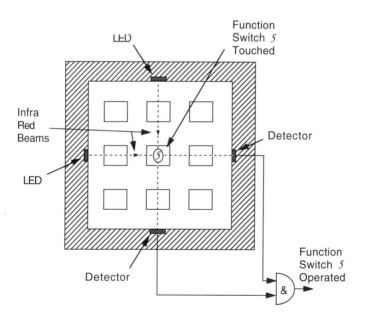

Fig. 7.38 Touch panel concept.

7.9.3 Direct voice input

Direct voice input (DVI) control is a system which enables the pilot to enter data and control the operation of the aircraft's avionic systems by means of speech. The spoken commands and data are recognised by a speech recognition system which compares the spoken utterances with the stored speech templates of the system vocabulary. The recognised commands, or data, are then transmitted to the aircraft sub systems by means of the data bus interconnecting them (eg MIL STD 1553B data bus).

As examples:

(a) To change a communication channel frequency, the pilot says – 'radio' (followed by) 'select frequency three four five decimal six'.
(b) To enter navigation data, the pilot says – 'navigation' (followed by) 'enter waypoint latitude fifty one degrees thirty one minutes eleven seconds North. Longitude zero degrees forty five minutes seventeen seconds West'.

Feedback that the DVI system has recognised the pilot's command correctly is provided either visually on the HUD and HMD (if installed), or aurally by means of a speech synthesizer system. The pilot then confirms the correctly recognised command by saying 'enter' and the action is initiated.

The pilot can thus stay head up and does not have to divert attention from the outside world in order to operate touch panels, switches, push buttons, keyboards etc. DVI can thus reduce the pilot's work load in high work load situations. It should be noted that the vocabulary required is not extensive and pilots (and crew) make good DVI subjects. This is because they are trained to speak clearly and concisely in a strongly structured way when giving commands and information over the communication channels to fellow crew members, other aircraft and ground control.

The main characteristics and requirements for an airborne DVI system are briefly summarised below.

- Fully connected speech. The speech recognition system must be able to recognise normal fully connected speech with no pauses required between words. (Systems which require a pause between each word are known as 'isolated word recognisers'.)
- Must be able to operate in the cockpit noise environment. The background noise level can be very high in a fast jet combat aircraft.
- Vocabulary size. The required vocabulary is around 200 to 300 words.
- Speech template duration. The maximum speech template duration is around 5 seconds.
- Vocabulary duration. The maximum duration of the total vocabulary is around 160 seconds.
- Syntax nodes. The maximum number of syntax nodes required is about 300.

An example of a typical 'syntax tree' is shown below:

RADIO

FREQUENCY	STATION	CODE REFERENCE
315.1	London	Stub 1
346.5	Manchester	Stub 2
etc	etc	etc

- Duration of utterance. There must be no restrictions on the maximum duration of an input utterance.
- Recognition response time. This must be in real time.

Only a very brief outline of speech recognition systems can be given because of space constraints. The basic principles are to extract the key speech features of the spoken utterance and then to match these features with the stored vocabulary templates. Sophisticated algorithms are used to select the best match and to output the recognised words, if the confidence level is sufficiently high. Figure 7.39 illustrates the speech features of an individual word which can be extracted by spectral analysis of the spoken utterances. The distinctive features can be seen in the 3D 'spectragram'.

Very extensive research and development is being carried out world-wide to produce speech recognition systems which are 'speaker independent', that is, they will recognise words spoken clearly by any speaker.

The airborne environment, however, poses particular requirements such as the ability to operate to a very high confidence level in a high background noise level. Recognition accuracies of around 99% are required in the cockpit environment to minimise having to repeat a command as this would defeat the objective of easing pilot work load. The speech recognition system must also recognise commands spoken during the physical stress of manoeuvring. For these reasons, the stored vocabulary templates are currently derived directly from the pilot who thus characterises the system to his particular speech patterns. As already mentioned, the stored vocabulary requirements are not extensive (about 200 to 300 words).

7.9.4 Eye trackers

Eye tracking systems have been used experimentally in airborne applications particularly for evaluating and monitoring the pilot's scanning pattern of the cockpit instruments. They are also being fairly widely used and evaluated in ground simulators for such future applications as:

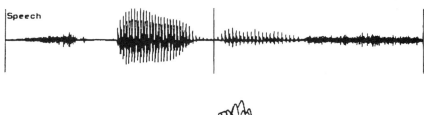

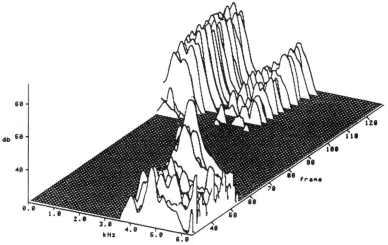

Fig. 7.39 Speech features – spectral analysis – a spectrascape of the word 'spar' (by courtesy of GEC-Marconi Avionics Ltd.).

- Improved target designation accuracy by enabling a more accurate measurement of the pilot's gaze angle to be made in conjunction with a head tracker system.
- Data entry in conjunction with a helmet mounted display. A keyboard can be displayed on the HMD and data can be entered by looking at the appropriate data symbol (eg Function switch and the digits 0 to 9 etc.) and then operating a simple push button. The pilot's gaze angle is measured by the eye tracker and displayed by a simple cross on the HMD so that the pilot can see his LOS is on the chosen symbol. A fast and accurate data entry system can be achieved by this means.

It should be noted that this technique could also be used by disabled people and research is, in fact, being carried out in this area.

A very brief explanation of the operating principles of an eye tracker exploiting corneal reflection is set out below.

Referring to Fig. 7.40, illuminating the eye with parallel light from a collimated light source produces an image or bright spot on the spherical surface of the cornea. The distance of this bright spot from the centre of the pupil is then a direct measure

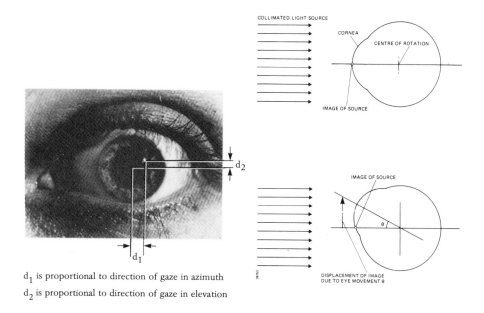

d_1 is proportional to direction of gaze in azimuth

d_2 is proportional to direction of gaze in elevation

Fig. 7.40 Principle of corneal reflection eye tracker (by courtesy of GEC-Marconi Avionics Ltd.).

of the angular rotation of the eyeball. A bright pupil image is also produced by reflection of the light source from the retina. (The 'red eye' image on a camera snap shot when the flash light is used is due to the same cause).

Low levels of illumination from an LED source are used in the eye tracker and the optical system is computer designed so that the levels at the retina are several orders of magnitude below the maximum allowable values, so that eye safety is not an issue. A miniature CCD camera is mounted on the helmet to provide a video image of the eye with the bright spot on the cornea and the bright pupil. Sophisticated image processing techniques and robust algorithms are required to extract the pupil centre, bearing in mind the pupil may be partly obscured by the eye lid and eye lashes. It is also necessary to eliminate any bright spot images on the cornea from unwanted light sources so that only the bright spot image due to the collimated LED source is used in the gaze angle computation.

A calibration procedure is also required to measure and store the individual variations in the eye geometry from the assumed statistical average values, for example, an average eye ball radius of 12mm, corneal spherical radius of 8mm and a distance of 6mm between the corneal centre of curvature and the eyeball centre. In addition, both the eyeball and the corneal surface may not be exactly spherical.

Real time gaze angle computations to an accuracy of the order of 0.5° at video frame rates (50 Hz) can be achieved using the above techniques with modern digital signal processors.

Although integrated helmet mounted eye trackers and HMDs are still in the development stage, they nevertheless have very considerable potential in future systems and hence warrant inclusion in this chapter. 'Watch this space' is perhaps an appropriate phrase for their future use.

Further reading

Bartlett, C.T., *A practical application of computer generated holography to head up display design*: Displays volume 15 number 2 1994 Butterworth Heinemann

Burgess, M.A., and Hayes, R.D., *Synthetic Vision - A View in the Fog*: IEEE Proceedings of 11th Digital Avionics System Conference, Seattle 5-8 October 1992

Flight Safety Foundation study of 543 total loss and 536 partial loss jet transport accidents, including 30 years prior to 1989

Rowntree,T., *The inteligent aircraft*: IEE Review January 1993

'Tomorrow's Cockpit Displays', *Aerospace*, Royal Aeronautical Society, November 1991 pp 12-14

8

Autopilots and flight management systems

8.1 INTRODUCTION

Autopilots and flight management systems (FMS) have been grouped together in one chapter as modern aircraft require a very close degree of integration between the two systems, particularly in their respective control laws.

Much of the background to the understanding of autopilots has been covered in earlier chapters. For instance, the dynamics of aircraft control is covered in Chapter 3 and Fly-by-wire flight control systems in Chapter 4, including pitch rate and roll rate manoeuvre command control systems. Chapter 4 also covers the methods used to achieve the very high integrity and safety required to implement FBW flight control through the use of redundancy. Similar levels of safety and integrity are required in some autopilot modes, for example automatic landing and automatic terrain following (T/F). The digital implementation of flight control systems is also covered in this chapter.

Some of the background to flight management systems has also been covered in previous chapters. For example air data and air data systems in Chapter 2, navigation systems in Chapter 6 and displays and man–machine interaction in Chapter 7.

The basic function of the autopilot is to control the flight of the aircraft and maintain it on a pre-determined path in space without any action being required by the pilot. (Once the pilot has selected the appropriate control mode(s) of the autopilot.) The autopilot can thus relieve the pilot from the fatigue and tedium of having to maintain continuous control of the aircraft's flight path on a long duration flight. The pilot is thus free to concentrate on other tasks and the management of the mission.

A well designed autopilot system which is properly integrated with the aircraft flight control system can achieve a faster response and maintain a more precise flight path than the pilot. Even more important, the autopilot response is always consistent whereas a pilot's response can be affected by fatigue and work load and stress. The autopilot is thus able to provide a very precise control of the aircraft's flight path for such applications as fully automatic landing in very poor, or even zero visibility conditions. In the case of a military strike aircraft, the autopilot in conjunction with a T/F guidance system can provide an all weather automatic terrain following capability. This enables the aircraft to fly at high speed (around 600 knots) at very low altitude (200 ft or less) automatically following the terrain profile to stay below the radar horizon of enemy radars. Maximum advantage of terrain screening can be taken to minimise the risk of detection and alerting the enemy's defences.

The basic autopilot modes are covered in the next section. These include such facilities as automatic coupling to the various radio navigation systems such as VOR and the approach aids at the airport or airfield such as ILS and MLS. The autopilot then steers the aircraft to stay on the path defined by the radio navigation aid. The autopilot can also be coupled to the flight management system which then provides the steering commands to the autopilot to fly the aircraft on the optimum flight path determined by the FMS from the flight plan input by the pilot.

The autopilot is thus an essential equipment for most military and civil aircraft, including helicopters. The advent of the micro-processor has also enabled relatively sophisticated and affordable autopilots to be installed in large numbers of general aviation type aircraft.

The prime role of the flight management system is to assist the pilot in managing the flight in an optimum manner by automating as many of the tasks as appropriate to reduce the pilot workload. The FMS thus performs a number of functions, such as:

Automatic navigation and guidance including '4D' navigation.
Presentation of information.
Management of aircraft systems.
Efficient management of fuel.
Reduction of operating costs.

The broad concepts and operation of an FMS are covered in Section 8.3. It should be appreciated that the detailed implementation of an FMS is a complex subject and can involve over 100 man-years of software engineering effort and very extensive (and expensive) flight trials before certification of the system can be obtained from the regulatory authorities. It is only possible, therefore, because of space constraints to give an overview of the subject.

It should also be pointed out than an FMS has an equally important role in a military aircraft. Accurate adherence to an optimum flight path and the ability to keep a rendezvous at a particular position and time for, say, flight refuelling or to join up with other co-operating aircraft are clearly very important requirements.

8.2 AUTOPILOTS

8.2.1 Basic principles

The basic loop through which the autopilot controls the aircraft's flight path is shown in the block diagram in Fig. 8.1. The autopilot exercises a guidance function in the outer loop and generates commands to the inner flight control loop. These

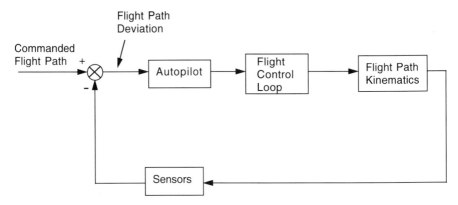

Fig. 8.1 Autopilot loop.

commands are generally attitude commands which operate the aircraft's control surfaces through a closed loop control system so that the aircraft rotates about the pitch and roll axes until the measured pitch and bank angles are equal to the commanded angles. The changes in the aircraft's pitch and bank angles then cause the aircraft flight path to change through the flight path kinematics.

For example, to correct a vertical deviation from the desired flight path, the aircraft's pitch attitude is controlled to increase or decrease the angular inclination of the flight path vector to the horizontal. The resulting vertical velocity component thus causes the aircraft to climb or dive so as to correct the vertical displacement from the desired flight path.

To correct a lateral displacement from the desired flight path requires the aircraft to bank in order to turn and produce a controlled change in the heading so as to correct the error.

The pitch attitude control loop and the heading control loop, with its inner loop commanding the aircraft bank angle, are thus fundamental inner loops in most autopilot control modes.

The outer autopilot loop is thus essentially a slower, longer period control loop compared with the inner flight control loops which are faster, shorter period loops.

8.2.2 Height control

Height is controlled by altering the pitch attitude of the aircraft, as just explained. The basic height control autopilot loop is shown in Fig. 8.2.

The pitch rate command inner loop provided by the pitch rate gyro feedback enables a fast and well damped response to be achieved by the pitch attitude command autopilot loop. As mentioned earlier, a FBW pitch rate command flight control system greatly assists the autopilot integration.

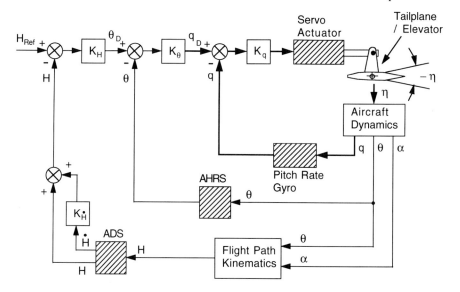

Fig. 8.2 Height control loop (heavy lines show FBW pitch rate command loop). Pitch rate gyro and AHRS are shown separately for clarity. In a modern aircraft, a strap-down AHRS/INS provides both θ and q outputs.

The pitch attitude command loop response is much faster than the height loop response – fractions of a second compared with several seconds. The open loop transfer function of the height loop in fact approaches that of a simple integrator at frequencies appreciably below the bandwidth of the pitch attitude command loop. The height error gain, K_H (or gearing) is chosen so that the frequency where the open loop gain is equal to unity (0 dB) is well below the bandwidth of the pitch attitude loop to ensure a stable and well damped height loop response. (The design of autopilot loops is explained in more detail in the worked example in the next section.)

The pitch attitude loop bandwidth thus determines the bandwidth of the height loop so that the importance of achieving a fast pitch attitude response can be seen.

The transfer function of the flight path kinematics is derived as follows:

Vertical component of the aircraft's velocity vector

$$= V_T \sin \theta_F = \dot{H}$$

where θ_F = flight path angle, that is the inclination of the velocity vector V_T to the horizontal.

$$\theta_F = \theta - \alpha$$

where θ = aircraft pitch angle and α = angle of incidence,
$V_T \approx U$, aircraft forward velocity and θ_F is assumed to be a small angle.

Hence
$$\dot{H} \approx U(\theta - \alpha)$$
(8.1)

Thus
$$H = \int U(\theta - \alpha)dt$$
(8.2)

8.2.3 Heading control autopilot

The function of the heading control mode of the autopilot is to automatically steer
the aircraft along a particular set direction. As explained in Chapter 3, Section 3.5,
aircraft bank to turn, and assuming a perfectly co-ordinated turn (i.e. no sideslip)
Centripetal acceleration = $g \tan \Phi = U\dot{\Psi}$
where Φ = aircraft bank angle, U = aircraft forward velocity and $\dot{\Psi}$ = rate of change
of heading.
   Hence, for small bank angles

$$\dot{\Psi} = \frac{g}{U}\Phi$$
(8.3)

   The basic control law used to control the aircraft heading is thus to demand a
bank angle, Φ_D, which is proportional to the heading error, $\Psi_E = (\Psi_{COM} - \Psi)$,
where Ψ_{COM} is the commanded heading angle and Ψ the heading angle

$$\Phi_D = K_\Psi \cdot \Psi_E$$
(8.4)

where K_Ψ = heading error gain.
   The heading control loop is shown in more detail in the block diagram in
Fig. 8.3. The inner bank angle command loop should have as high a bandwidth as
practical as well as being well damped in order to achieve a 'tight' heading control
with good stability margins. This will be brought out in the worked example which
follows.
   Roll rate command FBW enables a fast, well damped bank angle control system
to be achieved.
   As explained earlier in Chapter 3, the dynamic behaviour of the lateral control
system of an aircraft is complicated by the inherent cross coupling of rolling and
yawing rates and sideslip velocity on motion about the roll and yaw axes.
   The dynamic behaviour of the aircraft is thus represented by a set of matrix
equations as shown in Chapter 3 (Section 3.4.5)
   However, a good appreciation of the aircraft and autopilot behaviour can be
obtained by assuming pure rolling motion and neglecting the effects of cross

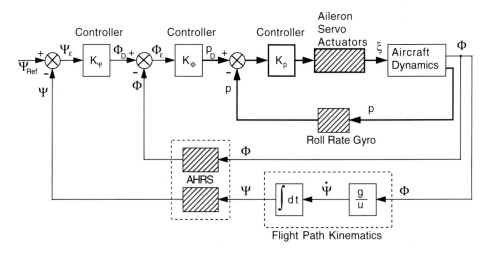

Fig. 8.3 Heading control loop (heavy lines denote FBW roll rate command loop).Roll rate gyro and AHRS are shown separately for clarity. In a modern aircraft, a strap-down INS/AHRS provides both p and Φ outputs.

coupling between the roll and yaw axes. This can also be justified to some extent by assuming effective auto-stabilisation loops which minimise the cross coupling effects.

Considering the bank angle demand loop, the aerodynamic transfer function relating roll rate, p, to aileron angle, ξ, is derived in Chapter 3, Section 3.6.1 which shows

$$\frac{p}{\xi} = \frac{L_\xi}{L_p} \cdot \frac{1}{1 + T_R D} \tag{8.5}$$

where T_R = roll rate response time constant ($= I_x / L_p$) and L_ξ = rolling moment derivative due to aileron angle, ξ and L_p = rolling moment derivative due to roll rate, p and I_x = moment of inertia of aircraft about the roll axis.

Referring to equation (5.20) in Chapter 5, *viz*

$$\dot{\Phi} = p + q \sin\Phi \tan\theta + r \cos\Phi \tan\theta$$

If Φ and θ are small angles

$$\dot{\Phi} = p$$

ie

$$\Phi = \int p \, dt \tag{8.6}$$

These assumptions are used in the worked example which follows in order to simplify the aircraft dynamics. It is also assumed that there is no loss of height when the aircraft banks to correct heading, appropriate changes to the pitch attitude and hence incidence being effected by the height hold autopilot control.

8.2.3.1 Worked example of heading control autopilot

The aircraft in this example has a roll rate response time constant $T_R = 0.5$ seconds and $L_p/L_\xi = 20°$/second per degree aileron deflection.

The response of the aileron servo actuators approximates to that of a simple first order filter with a time constant of 0.08 seconds. The aircraft's forward velocity, $U = 250$m/sec (approximately 500 knots) and is assumed to be constant.

Determine suitable values for the autopilot gains (or gearings) K_p, K_ϕ and K_ψ.

The overall loop for the heading control is as shown in Fig. 8.3. The block diagram for the bank angle demand loop is shown in Fig. 8.4.

It should be noted that the Laplace operator, s, has been substituted for operator D (d/dt) to conform with general control engineering practice. The use of Laplace transforms has many advantages particularly in dealing with initial conditions. Readers not familiar with Laplace transforms will find them covered in most engineering mathematics text books.

• *Determination of roll rate error gain, K_p*

The open loop transfer function of the innermost loop, namely, the roll rate feedback loop is:

$$\frac{p}{p_E} = \frac{K_p 20}{(1+0.08s)(1+0.5s)} = KG(s) \qquad (8.7)$$

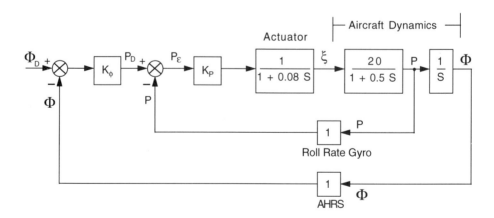

Fig. 8.4 Block diagram of bank angle demand loop.

where $p_E = p_D - p$

The closed loop transfer function p/p_D is given by

$$\frac{p}{p_D} = \frac{KG(s)}{1 + KG(s)} \tag{8.8}$$

The stability of the closed loop is determined by the roots of the characteristic equation

$$1 + KG(s) = 0 \tag{8.9}$$

ie

$$(1 + 0.08s)\,(1 + 0.5s) + 20K_p = 0$$

$$s^2 + 14.5s + 25\left(1 + 20K_p\right) = 0 \tag{8.10}$$

Comparing with the standard form of second order equation

$$2\zeta\omega_0 = 14.5$$

$$\omega_0^2 = 25(1 + 20\,K_p)$$

Choosing a damping ratio, $\zeta = 0.7$, requires the undamped natural frequency, $\omega_0 = 10.4$ rad/sec (14.5/2 x 0.7). From which $K_p = 0.165°$ aileron angle per degree/second roll rate error.

$$K_p = 0.165 \text{ seconds.}$$

from which

$$\frac{p}{p_D} = \frac{0.767}{1 + \dfrac{2\zeta}{\omega_0}s + \dfrac{1}{\omega_0^2}s^2} \tag{8.11}$$

where $\omega_0 = 10.4$ and $\zeta = 0.7$.

The roll rate response with and without roll rate feedback is shown in Fig. 8.5. (The step input to the system is adjusted to give the same steady state rate of roll in each case.) The improved speed of response with roll rate feedback can be seen. The actuator lag limits the loop gain to 20 × 0.165, that is 3.3, in order to achieve satisfactory damping without recourse to shaping (eg phase advance).

- *Determination of bank angle error gain K_ϕ*

Consider now the bank angle demand loop. The open loop transfer function is:

Roll Rate (p)

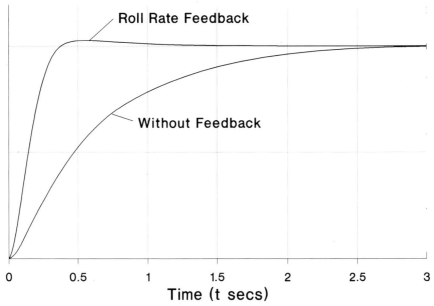

Fig. 8.5 Roll rate response.

$$\frac{\Phi}{\Phi_E} = \frac{0.767 K_\phi}{s\left(1 + \dfrac{2\zeta}{\omega_0}s + \dfrac{1}{\omega_0^2}s^2\right)} \tag{8.12}$$

where $\omega_0 = 10.4$ and $\zeta = 0.7$.

The value of K_ϕ is determined using frequency response methods, as follows. The open loop frequency response can be obtained by substituting $j\omega$ for s.

$$\frac{\Phi}{\Phi_E}(j\omega) = \frac{K}{j\omega\left(1 - \left(\dfrac{\omega}{\omega_0}\right)^2 + j2\zeta\dfrac{\omega}{\omega_0}\right)} \tag{8.13}$$

$$= K F_1(j\omega).F_2(j\omega) \tag{8.14}$$

where $K = 0.767 K_\phi$

$$F_1(j\omega) = \frac{1}{j\omega} \tag{8.15}$$

$$F_2(j\omega) = \frac{1}{1 - \left(\dfrac{\omega}{\omega_0}\right)^2 + j2\,\zeta\,\dfrac{\omega}{\omega_0}} \tag{8.16}$$

$F_1(j\omega)$ and $F_2(j\omega)$ can be expressed in terms of their respective moduli and phase angles (or arguments)

$$F_1(j\omega) = \left|\frac{1}{\omega}\right| \quad , \quad \angle - 90° \tag{8.17}$$

It is convenient to use a non dimensional frequency, $u = \dfrac{\omega}{\omega_0}$, in $F_2(j\omega)$

$$F_2(j\omega) = \frac{1}{1 - u^2 + j2\,\zeta\,u} \tag{8.18}$$

$$F_2(j\omega) = \left|\frac{1}{\sqrt{(1-u^2)^2 + 4\zeta^2 u^2}}\right| \quad , \quad \angle \tan^{-1} \frac{2\zeta u}{1 - u^2} \tag{8.19}$$

The evaluation of $KF_1(j\omega).F_2(j\omega)$ at specific values of ω is carried out using the logarithms of the moduli of the respective elements of the transfer function.

$$20\log_{10}\left|KF_1(j\omega) \cdot F_2(j\omega)\right|$$

$$= 20\log_{10} K + 20\log_{10}\left|F_1(j\omega)\right| + 20\log_{10}\left|F_2(j\omega)\right| \tag{8.20}$$

The log modulus of the overall transfer function is thus the sum of the logs of the moduli of the individual elements.

The overall phase angle is similarly the sum of the individual phase angles of the respective elements.

$$\angle F_1(j\omega) \cdot F_2(j\omega) \quad = \quad \angle F_1(j\omega) \quad + \quad \angle F_2(j\omega) \tag{8.21}$$

The overall open loop frequency response in terms of log gain and phase has been calculated from the above formulae and is set out in Table 8.1 for $K = 1$.

Table 8.1 Frequency response

ω	$F_1 (j\omega)$		$F_2 (j\omega)$		$F_1 (j\omega) \cdot F_2 (j\omega)$	
	20 log M	Phase	20 log M	Phase	20 log M	Phase
0.5	6dB	−90°	0dB	−3.9°	6dB	−94°
1	0dB	−90°	0dB	−7.8°	0dB	−98°
2	−6dB	−90°	0dB	−15.6°	-6dB	−106°
3	−9.5dB	−90°	0dB	−23.8°	−9.5dB	−114°
5	−14dB	−90°	−0.2dB	−41.2°	−14.2dB	−131°
7	−16.9dB	−90°	−0.7dB	−59.9°	−17.6dB	−150°
10.4	−20.3dB	−90°	−2.9dB	−90°	−23.2dB	−180°
15	−23.5dB	−90°	−7dB	−118.1°	−30.5dB	−208°
20	−26dB	−90°	−11.6dB	−135.1°	−37.6dB	−225°
30	−29.5dB	−90°	−18.4dB	−151.1°	−47.9dB	−241°

M = Modulus

It should be noted that there are standard tables and graphs of the frequency responses of the common transfer function elements. There are also computer programs for evaluating the overall frequency response of a transfer function comprising a number of elements.

The open loop frequency response is plotted on Fig. 8.6 which shows the log gain and phase at the specific values of ω chosen in the table. Such a graph is known as a 'Nichol's chart'. Lines of constant gain and phase loci for converting the open loop frequency response into the closed loop frequency response are normally superimposed on such a chart. The lines corresponding to −3 dB and −90° are shown on the figure, but the rest have been omitted for clarity.

The Nichol's chart has the following advantages:

1. The logarithmic gain scale enables the response at all the frequencies of interest to be shown on the graph.
2. The effect of shaping the open loop response can be readily seen. For example, increasing the loop gain, or introducing phase advance or one plus integral of error control, or the effect of additional lags in the system.
3. The closed loop frequency response can be obtained using the constant gain and phase loci.
4. The gain and phase margins can be read off directly.

The value for the open loop scalar gain, $K = 1$, gives a very stable loop, with a gain margin of 23.2 dB and a phase margin of 82°. The bandwidth, however, is relatively low. The closed loop response is 3 dB down at $\omega = 1.2$ approximately (the intersection of the −3 dB loci with the open loop response). The phase lag is 90° at $\omega = 2.8$ approximately (the intersection of the −90° loci with the open loop response).

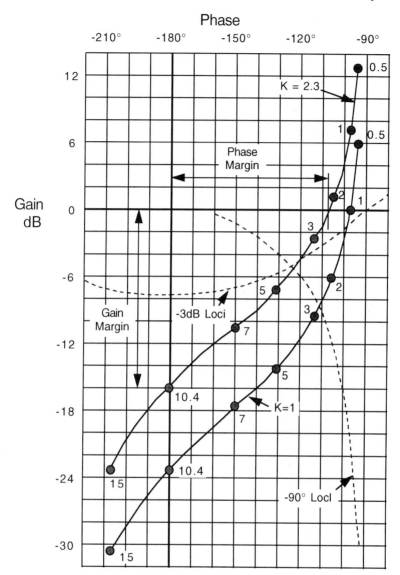

Fig. 8.6 Bank angle open loop frequency response (Nichol's chart).

Increasing the open loop scalar gain K to 2.3, that is by 7.2 dB, improves the closed loop bandwidth considerably. The closed loop response is 3 dB down at $\omega = 3.8$ approximately, and there is 90° phase lag at $\omega = 4.2$ approximately. The gain and phase margins are reduced to 16 dB and 71° respectively, but are still very adequate.

The open loop transfer function is

$$\frac{\phi}{\phi_E} = \frac{2.3}{s(1+0.1346s+0.0092456s^2)} \qquad (8.22)$$

The closed loop stability and damping can be confirmed by finding the roots of the characteristic equation

$$s(1+0.1346s+0.0092456s^2)+2.3 = 0$$

ie
$$s^3 + 14.56s^2 + 108.16s + 246.77 = 0 \qquad (8.23)$$

This factorises into $(s+3.63)(s^2+10.93s+68.5) = 0 \qquad (8.24)$

The $(s + 3.63)$ factor produces an exponential subsidence component in the response with a time constant of $1/3.63$ seconds, that is 0.28 seconds. The quadratic factor $(s^2 + 10.93s + 68.5)$ produces an exponentially damped sinusoidal component with an undamped natural frequency $\omega_0 = \sqrt{68.5}$, that is 8.28 rad/s (1.32 Hz) and with a damping ratio $\zeta = 0.66$, confirming a reasonably well damped closed loop response.

Hence
$$\frac{\Phi}{\Phi_D} = \frac{1}{(1+0.28s)(1+0.16s+0.0146s^2)} \qquad (8.25)$$

The required value of K_ϕ is obtained from $K = 0.767 K_\phi = 2.3$

$$K_\phi = 3 \qquad (8.26)$$

• *Determination of heading error gain, K_ψ*
Referring to Fig. 8.3 it can be seen that the open loop transfer function of the heading control loop is

$$\frac{\psi}{\psi_E} = \frac{K_\psi g / U}{s(1+0.28s)(1+0.16s+0.0146s^2)} \qquad (8.27)$$

Hence
$$\frac{\psi}{\psi_E}(j\omega) = \frac{K}{j\omega\left(1+j\dfrac{\omega}{3.63}\right)\left[1-\left(\dfrac{\omega}{8.28}\right)^2 + j\,0.16\omega\right]} \qquad (8.28)$$

The factors in the denominator have been expressed in a form which enables the cut off frequencies to be readily seen. The scalar gain $K = K_\psi g/U$. It can be seen from the above expression that at low frequencies (considerably lower than the first cut off frequency of 3.63 rad/s), both the first and second order filter components of the transfer function approach unity. Hence at low frequencies

$$\frac{\psi}{\psi_E}(j\omega) \to \frac{K}{j\omega}$$

and

$$\left|\frac{\psi}{\psi_E}(j\omega)\right| \to \frac{K}{\omega}$$

The open loop gain thus falls at –6 dB/octave at low frequencies.

The value of the frequency at which the open loop gain is equal to unity (that is 0 dB) is known as the '0 dB cross-over frequency'. The value of K is chosen to give a 0 dB cross-over frequency such that the rate of change of open loop gain with frequency is –6 dB/octave for well over an octave in frequency either side of the 0 dB cross-over frequency. This satisfies Bode's stability criterion.

The 0 dB cross-over frequency is chosen to be 0.7 rad/s.

Hence

$$\frac{K}{0.7} = 1$$

ie

$$K = 0.7 = K_\psi 9.81/250$$

$K_\psi = 18$ degrees bank angle per degree heading error.

From the open loop transfer function it can be seen that the gain starts to fall off at –12 dB/octave $(1/\omega^2)$ between $\omega = 3.63$ and $\omega = 8.28$ and then at –24 dB /octave $(1/\omega^4)$ above $\omega = 8.28$ rad/s. The phase changes very rapidly over this frequency range, for example –115° at 1 rad/s, –138° at 2 rad/s, –180° at 4.1 rad/s, –246° at 8.28 rad/s, –310° at 15 rad/s and approaches –360° as the frequency increases still higher.

The gain and phase margins with $K_\psi = 18$ are approximately 19 dB and 73° respectively. The closed loop bandwidth is about 1.4 rad/s when the phase lag is –90°. The 3dB down frequency is just over 1 rad/s.

The loop gain clearly varies with forward velocity, U, and a generous gain margin has been allowed to cope with changes in speed. Gain scheduling with speed is required, however, to cover the whole of the aircraft speed envelope.

A bank angle limiter is also required to prevent too large a bank angle being demanded when there is a significant initial heading error on engaging the autopilot. Typical limits would be $\pm 12°$ ($\pm 0.2g$). As stated earlier, the aircraft dynamics have been simplified in this example to give an insight into the design of the inner and outer autopilot loops.

In practice, a full representation of the aircraft dynamics including the yaw/roll cross-coupling effects would be used to establish the autopilot design.

8.2.4 ILS/MLS coupled autopilot control

8.2.4.1 Approach guidance systems

ILS is a radio based approach guidance system installed at major airports and airfields where the runway length exceeds 1800m which provides guidance in poor visibility conditions during the approach to the runway.

A small number of major airports are also now equipped with MLS – microwave landing system. MLS is a later and more accurate system which is superior in all aspects to ILS. ILS, however, is a very widely used system and it will be a long time before it is completely replaced. It will thus be supported and maintained for many years to come.

The runway approach guidance signals from the ILS (or MLS) receivers in the aircraft can be coupled into the autopilot which then automatically steers the aircraft during the approach so that it is positioned along the centre line of the runway and on the descent path defined by the ILS (or MLS) beams. The autopilot control loops are basically the same for ILS or MLS coupling apart from some signal preconditioning.

Space does not permit a detailed description of either system and a very brief outline of the ILS system only is given as this is the most widely used system. The ILS system basically comprises a localizer transmitter and a glide slope transmitter located by the airport runway together with two or three radio marker beacons located at set distances along the approach to the runway. The airborne equipment in the aircraft comprises receivers and antennas for the localizer, glide slope and marker transmissions. The guidance geometry of the localizer and glide slope beams is shown in Fig. 8.7.

The localizer transmission, at VHF frequencies (108–122 MHz), provides information to the aircraft as to whether it is flying to the left or right of the centre line of the runway it is approaching. The localizer receiver output is proportional to the angular deviation γ_L, of the aircraft from the localizer beam centre line which in turn corresponds with the centre line of the runway.

The glide slope (or glide path) transmission is at UHF frequencies (329.3–335 MHz) and provides information to the aircraft as to whether it is flying above or below the defined descent path of nominally $2.5°$, for the airport concerned. The glide slope receiver output is proportional to the angular deviation γ_V, of the aircraft from the centre of the glide slope beam which in turn corresponds with the preferred descent path. (The sign of the γ_L and γ_V signals is dependent on whether

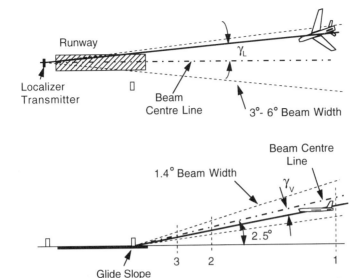

Fig. 8.7 ILS localizer and glide slope geometry.

1 Outer Marker
2 Middle Marker
3 Inner Marker

the aircraft is to the left or the right of the runway centre line, or above or below the defined glide slope.)

The marker beacon transmissions are at 75 MHz. The middle marker beacon is located at a distance of between 1,000 and 2,000m from the runway threshold and the outer marker beacon is situated at a distance of between 4,500 and 7,500m from the middle marker. The inner marker beacon is only installed with an airport ILS system which is certified to Category III landing information standards and is located at a distance of 305m (1,000 ft) from the runway threshold.

It should be noted that ILS does not provide sufficiently accurate vertical guidance information down to touchdown.

The height limits and visibility conditions in which the autopilot can be used to carry out a glide slope coupled approach to the runway depends on the visibility category to which the autopilot system is certified for operation, the ILS ground installation standard, the runway lighting installation and the airport's runway traffic control capability.

Visibility conditions are divided into three categories, namely Category I, Category II and Category III, depending on the vertical visibility ceiling and the runway visual range (RVR). These will be explained in more detail in Section 8.2.5, but briefly the visibility conditions deteriorate as the Category number increases;

'Cat III' includes zero visibility conditions. The required autopilot system capabilities in terms of safety and integrity for operation in these visibility categories are also discussed in Section 8.2.5.

An automatic glide slope coupled approach is permitted down to a height of 30m (100 ft) above the ground, but only if the following conditions are met:-

1. There is sufficient vertical visibility at a height of 100 ft with a runway visual range of at least 400m for the pilot to carry out a safe landing under manual control (Category II visibility conditions). This minimum permitted ceiling for vertical visibility for the landing to proceed is known as the decision height (DH).
2. The autopilot system is certified for Cat. II operation. This will be explained in more detail in Section 8.2.5, but briefly a fail passive autopilot system is required. This is so that the pilot can take over smoothly in the event of a failure in the autopilot system.

Hence, when the decision height is reached the pilot carries out the landing under manual or automatic control. Alternatively, the pilot may execute a go-around manoeuvre to either attempt to land a second time or divert to an alternative airport/airfield.

A very high integrity autopilot system is required for fully automatic landing below a DH of 100 ft – Cat. III conditions. This is covered in Section 8.2.5.

8.2.4.2 Flight path kinematics

The mathematical relationships between the flight path velocity vector and the angular deviations of the aircraft from the guidance beam, or 'beam errors' (γ_L and γ_V) are basically the same for the lateral and vertical planes. These relationships are derived below as they are fundamental to both the localizer and glide slope control loops. The particular parameters for each loop can be readily substituted in the general case which is shown in Fig. 8.8.

θ_F is the angle between the flight path velocity vector V_T and a chosen spatial reference axis. In the vertical guidance case, the reference axis is the horizontal axis. In the lateral case, the reference axis is the runway centre line.

θ_B is the angle between the centre line of the guidance beam and the chosen reference axis. In the vertical case $\theta_B = 2.5°$ nominally and in the lateral case $\theta_B = 0$.

From Fig. 8.8, it can be seen that the component of the flight path velocity vector normal to the sight line is equal to $V_T \sin(\theta_B - \theta_F)$, that is $U(\theta_B - \theta_F)$ as $V_T \approx U$ and $(\theta_B - \theta_F)$ is a small angle.

The rate of rotation of the sight line is thus equal to $U(\theta_B - \theta_F)/R$ where R is the slant range of the aircraft from the transmitter.

θ_B is fixed, so that $\dot{\theta}_B = 0$

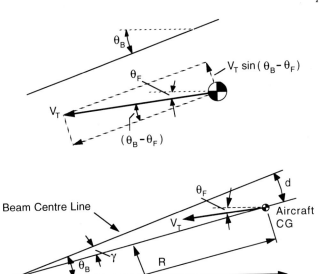

Fig. 8.8 Guidance geometry.

Hence

$$\dot{\gamma} = U(\theta_B - \theta_F) / R \qquad (8.29)$$

and

$$\gamma = \int \frac{U}{R}(\theta_B - \theta_F)dt \qquad (8.30)$$

The beam error $\gamma = d/R$, where d = displacement of the aircraft's CG from the beam centre line. Hence as R decreases, a given displacement produces an increasing beam error. For example, a displacement of 5m at a range of 1,500m gives a beam error of 3.3 milliradians, that is 0.2°. At a range of 300m, the same offset produces a beam error of 1°. The guidance sensitivity thus increases as the range decreases, as can also be seen from equation (8.30).

8.2.4.3 ILS localizer coupling loop

The ILS localizer coupling loop of the autopilot is shown in the block diagram in Fig. 8.9. It should be noted that the VOR coupling loop is basically similar. The range of the ILS localizer is much lower, however, and is usually less than 24 km compared with up to 150 km for VOR. The heading control loop is the same as that described in Section 8.2.3, the heading angle ψ, is referenced relative to the runway centre line bearing.

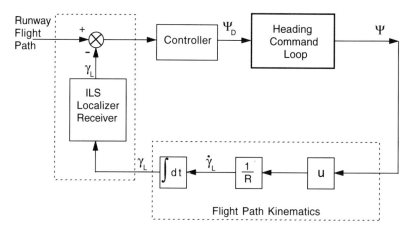

Fig. 8.9 Localizer coupling loop. (ψ is aircraft heading relative to runway centre line bearing).

The flight path kinematics have been derived in the preceding section. Referring to equation (8 29) and substituting the appropriate values of θ_B and θ_F, that is $\theta_B = 0$ and $\theta_F = \psi$

$$\dot{\gamma}_L = U\psi / R \qquad (8.31)$$

$$\gamma_L = \int \frac{U}{R} \psi dt \qquad (8.32)$$

The relationship is shown in the block diagram in Fig. 8.9.

It can be seen that the loop gain increases as R decreases and will reach a point where the loop becomes unstable. Gain scheduling with range is thus required.

The localizer controller in the autopilot provides a proportional plus integral of error control and generally a phase advance term. It should be noted that some filtering of the beam error signal, γ_L, is required to remove the 90 Hz and 150 Hz modulation components inherent in the ILS system and specially to attenuate the noise present. This filtering inevitably introduces some lags.

8.2.4.4 ILS glide slope coupling loop

The ILS glide slope coupling loop is shown in the block diagram in Fig. 8.10. The pitch attitude command loop which controls the inclination of the flight path velocity vector is the same as that described in Section 8.2.2.

The flight path kinematics have been derived in Section 8.2.4.2.

$\theta_F = (\theta - \alpha)$ where θ is the aircraft pitch angle and α is the angle of incidence.

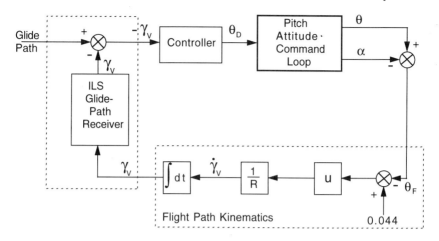

Fig. 8.10 Glide path coupling loop.

Substituting the appropriate values, $\theta_B = 0.044$ radians (2.5°) and $\theta_F = (\theta - \alpha)$ in equation 8.29 yields :

$$\dot{\gamma}_V = U(0.044 - \theta + \alpha)/R \qquad (8.33)$$

Hence

$$\gamma_V = \int \frac{U}{R}(0.044 - \theta + \alpha)\,dt \qquad (8.34)$$

As in the localizer coupling loop, the loop gain increases as the range decreases and will ultimately cause instability. Gain scheduling with range is thus required.

The airspeed, U, is controlled by an auto-throttle system, as described later on in Section 8.2.6, and is progressively reduced during the approach according to a defined speed schedule.

The glide slope controller generally comprises a proportional plus integral and phase advance control terms with a transfer function of the form

$$K_C\left(1 + \frac{1}{T_1 D}\right)\left(\frac{1 + T_2 D}{1 + \dfrac{T_2}{n}D}\right)$$

where K_C = controller scalar gain, T_1 = integral term time constant, T_2 = phase advance time constant and n = phase advance gain.

As with the localizer coupling loop, the filtering of the beam error signal γ_V, introduces lags. The noise present also limits the phase advance gain, n.

The mixing of inertially derived position and velocity information from the INS (if fitted) with the ILS information in a suitable complementary filtering system using a Kalman filter can greatly improve the dynamic quality of the beam error control signals and virtually eliminate the noise. The loop performance and stability is thus improved. It can also smooth the spatial noise and some of the 'kinks' in the beam.

8.2.5 Automatic landing

8.2.5.1 Introduction

The previous sections have described the automatic coupled approach phase of an automatic landing using the guidance signals from the ILS (or MLS) system in Cat.I or Cat.II visibility conditions. The pilot, however, takes over control from the autopilot when the decision height is reached and lands the aircraft under manual control.

Attempting to land an aircraft under manual control with decision heights of less than 100 ft, as in Cat.III conditions, is very demanding because of the lack of adequate visual cues and the resulting disorientation which can be experienced. There are only two alternatives for effecting a safe landing in such conditions:

a) A fully automatic landing system with the autopilot controlling the landing to touch-down. A very high integrity autopilot system is required with failure survival capability provided by redundancy such that the probability of a catastrophic failure is less than 10^{-7}/hour. High integrity autopilot systems capable of carrying out a fully automatic landing in Cat.III conditions are now at a mature stage of development and large numbers of civil jet aircraft operated by major airlines world-wide are now equipped with such systems.

b) The use of an enhanced vision system with a HUD as described in Chapter 7, using a millimetric wavelength radar sensor in the aircraft to derive a synthetic runway image. This is presented on the HUD together with the primary flight data, including the flight path velocity vector, and provides sufficient information for the pilot to land the aircraft safely under manual control. These systems, however, are at a relatively early stage of development.

The development of automatic landing systems has been a world-wide activity, with many companies and airlines, research establishments and regulatory authorities involved. It is appropriate, however, to mention the leading role of UK organisations and companies in the initial development of high integrity automatic blind landing systems. The UK had a distinct economic incentive to develop automatic landing systems because of the frequency of fogs and poor visibility conditions, especially in London, which necessitated frequent diversions. The first major contribution was made by the Blind Landing Experimental Unit, BLEU, of the former Royal Aerospace Establishment (now part of the UK Defence Research Agency, DRA). BLEU was formed in 1946 and by 1958 had developed the system which has subsequently formed the basis of all the automatic landing systems now in operation. The BLEU system is briefly described in Section 8.2.5.3. Another essential contribution was the formulation of the safety and certification requirements by the Air Registration Board around 1960 and the total system approach to safety and regulations including ground guidance equipment and airport facilities initiated by the UK Civil Aviation Authority (CAA). The next major contribution was made by the successful development and certification of two fully automatic landing systems by two British companies in the late 1960s. A monitored duplicate autopilot system for the VC10 airliner was developed by Elliott Brothers (London) Ltd., now part of GEC-Marconi Avionics Ltd., and a triplex autopilot system was developed for the Trident III airliner by Smiths Industries Ltd. The Trident III automatic landing system was in fact the first 'failure-survival fully automatic landing' system to be certified for Cat.III operation in the world, and is cleared to a decision height of 12 ft and a runway visual range of 100m. The contributions made to the development of automatic landing systems during the 1960s by companies and organisations in the United States and France should also be acknowledged. At that time, efforts in both countries were directed towards the development of simpler fail passive automatic landing systems. The Sud Aviation SE 210 Caravelle airliner was in fact certified for Cat.III operation with a decision height of 50 ft around the same time as the Trident III. The experience and the design methodologies developed on these pioneering systems, such as redundancy management and techniques such as failure modes and effects analyses have been of great value in subsequent programmes and have been disseminated world-wide.

It is also noteworthy that much of the technology developed for high integrity autopilot systems has provided the base for the subsequent development of Fly-by-wire flight control systems.

It is appropriate at this point to describe the visibility categories in greater detail and the autopilot capabilities for operation in these categories.

8.2.5.2 Visibility categories and autopilot requirements

As already explained, the two basic parameters used to define the visibility category are the decision height, that is the minimum vertical visibility for the landing to proceed, and the runway visual range. Table 8.2 shows the various visibility categories.

The safety and integrity requirements for the autopilot system to be qualified for operation in the various visibility categories and the limits on its operation are shown in Table 8.3. The flight path guidance system must also meet the appropriate category standards and accuracy. For example, a Cat.II ILS system must provide accurate glide path guidance down to a height of 100 ft above the ground.

Table 8.2 Visibility categories

Category	Minimum Visibility Ceiling	Runway Visual Range
I	200 ft	800m
II	100 ft	400m
IIIa	12–35 ft	100–300m
	Depending on aircraft type and size	
IIIb	12 ft	100m
IIIc	0 ft	0m

Table 8.3 Safety and integrity requirements

Category	Autopilot requirements and operational limits
I	Simplex autopilot system acceptable.
	Pilot takes over the landing at a DH of 200 ft.
II	Fail passive autopilot system required.
	Pilot takes over the landing at a DH of 100 ft.
IIIa	Full automatic landing system with automatic flare.
	Failure survival autopilot system with a probability of catastrophic failure of less than 10^{-7} per hour required.
	Pilot assumes control at touchdown.
IIIb	Same as IIIa as regards autopilot system capability and safety and integrity requirements, but with automatic roll out control after touchdown incorporated. Runway guidance system required.
	Pilot assumes control at some distance along the runway.
IIIc	Same as IIIb as regards autopilot system capability and safety and integrity requirements, but with automatic taxi-ing control incorporated. Runway guidance required to taxi point. No system yet certified for Cat.IIIc operation.

8.2.5.3 *The BLEU automatic landing system.*

The BLEU automatic landing system is shown in Fig. 8.11 and is divided into four phases from the time the outer marker radio beacon is reached, about 8000m from the threshold. These phases are briefly described below.

1. *Final approach.* This phase covers the approach from the outer marker beacon to the inner marker beacon. At the inner marker beacon the aircraft flight path should be aligned with the defined glide path at a height of 100 ft above the ground and also aligned with the centre line of the runway. During this phase, the autopilot controls the aircraft flight path using the guidance signals from the ILS system.

The aircraft height above the ground is measured by very accurate radio altimeters.

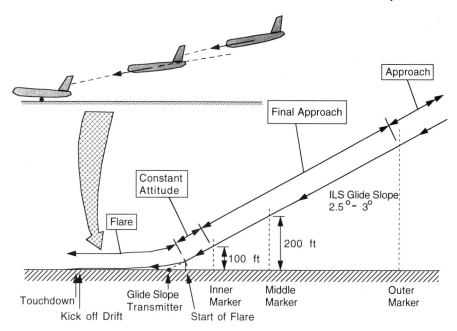

Fig. 8.11 BLEU automatic landing system.

2. *Constant attitude*. The guidance signals from the ILS are disconnected from the autopilot when the aircraft reaches a height of 100 ft above the ground. The autopilot then controls the aircraft to maintain the pitch attitude and heading at the values set up during the approach until the height is reached at which the flare-out is initiated.

3. *Flare*. The aircraft pitch attitude is controlled by the feedback of the radio altimeter derived height to produce an exponential flare trajectory. The flare is initiated at a height of around 50ft where the aircraft is over or very near the runway threshold. The aircraft is progressively rotated in pitch during the flare so that the flight path angle changes from the $-2.5°$ to $-3°$ value at the start of the flare to the positive value specified for touch-down. The vertical velocity is reduced from typically about 10ft/s at the start of the flare manoeuvre to around 1 to 2ft/s at touch-down.

4. *Kick off drift*. Just prior to touch-down a 'kick off drift' manoeuvre is initiated through the rudder control so that the aircraft is rotated about the yaw axis to align it with the runway. This ensures the undercarriage wheels are parallel to the runway centre line so that no sideways velocity is experienced by the wheels when they make contact with the runway.

8.2.5.4 Automatic flare control

The phases of the automatic landing have just been described. The automatic control loops for pitch attitude and heading control during the 'constant attitude' phase have been described earlier. The automatic flare control system, however, merits further explanation.

The mathematical law describing the exponential flare manoeuvre is:

$$\dot{H} = -KH \tag{8.35}$$

where H is the aircraft height above the ground and K is a constant.

The solution of equation (8.35) is:

$$H = H_0 e^{-t/T} \tag{8.36}$$

where H_0 = aircraft height above the ground at the start of the flare manoeuvre, T = time constant = $1/K$. (Time, t, is measured from the start of the manoeuvre.) The horizontal velocity component of the aircraft is effectively equal to U as the flight path angle is a small angle. Assuming U is constant, the trajectory will thus be exponential.

The control law used for the autoflare is:-

$$H + T\dot{H} = H_{REF} \tag{8.37}$$

H_{REF} is a small negative height, or bias, which ensures there is still a small downwards velocity at touchdown. This avoids the long exponential 'tail' to reach zero velocity and enables a reasonably precise touchdown to be achieved.

As mentioned earlier, the auto-flare is initiated at a height of around 50 ft where the aircraft is over or very near the runway threshold so that the radio altimeter is measuring the height of the aircraft above the runway. Low range radio altimeters are used to ensure accuracy. Safety and integrity considerations generally dictate a triplex or even quadruplex configuration of totally independent radio altimeters (rad.alt.).

The block diagram for the automatic flare control loop is shown in Fig. 8.12 which also indicates the redundancy necessary to meet the safety and integrity requirements in an automatic landing system.

The required control law response can be obtained by feeding back the rate of change of height suitably scaled by the required time constant, that is $T\dot{H}$, together with the height measured by the rad.alt. This is because the response of a closed loop system approaches the inverse of the feedback path transfer function if the gain of the forward loop is sufficiently high. The aircraft height response thus approaches that of a simple first order system with a transfer function of $1/(1+TD)$ at low frequencies where the forward loop gain is high.

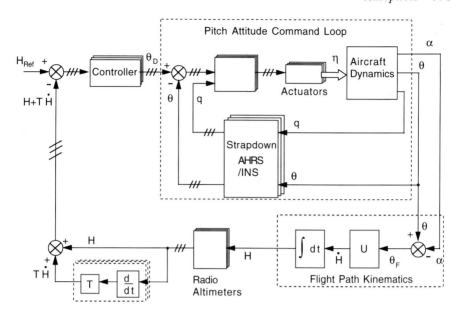

Fig. 8.12 Automatic flare control loop.

The $\dot{H}$ feedback term can be derived by differentiating the suitably smoothed rad.alt. output. A filter is required to smooth the noise present on the rad.alt. output and the differentiation process amplifies any high frequency noise components which are present. The $\dot{H}$ signal so obtained thus has lags present in its response because of the smoothing filters required.

An alternative and superior source of $\dot{H}$ can be derived from inertial mixing of the INS derived vertical velocity. This is assuming there is adequate redundancy, for example a triplex INS installation. The inertial mixing enables an $\dot{H}$ output to be obtained with an excellent dynamic response and low noise content as explained earlier in Chapter 6.

The auto-flare loop is a high order system; apart from the lags present in the filtered rad.alt. signals there are also the lags present in the response of the pitch attitude command loop. This loop controls $\dot{H}$ and its response is significantly slower at the low speeds during the approach.

A proportional plus integral control term is used in the auto-flare controller to ensure accuracy and some phase advance is generally provided to compensate for the lags in the loop and hence improve the loop stability and damping.

The approach speed is typically around 65m/s. (130 knots) so that the vertical velocity at the start of the flare is around 65 sin 2.5°, that is 2.84m/s or 9.3 ft/s. The flare time constant is typically around 5 seconds so that the vertical velocity is reduced exponentially from around 2.84m/s at the start of the flare to about 0.6m/s at touchdown. The corresponding time to touchdown is around 7.7 seconds. Hence, assuming the approach speed stays constant, the touchdown point would be around 500m from the runway threshold.

8.2.6 Speed control and auto-throttle systems

Control of the aircraft speed is essential for many tasks related to the control of the aircraft flight path, for example the position of the aircraft relative to some reference point.

The aircraft speed is controlled by changing the engine thrust by altering the quantity of fuel flowing to the engines by operating the engine throttles. Automatic control of the aircraft's airspeed can be achieved by a closed loop control system whereby the measured airspeed error is used to control throttle servo actuators which operate the engine throttles. The engine thrust is thus automatically increased or decreased to bring the airspeed error to near zero and minimise the error excursions resulting from disturbances. A typical airspeed control system is shown in the block diagram in Fig. 8.13.

In any closed loop system, the lags in the individual elements in the loop resulting from energy storage processes (eg accelerating inertias) exert a destabilising effect and limit the loop gain and hence the performance of the automatic control system. The dynamic behaviour of the engines over the range of flight conditions, the throttle actuator response and the aircraft dynamics must thus be taken into account in the design of the speed control system. The response of the jet engine thrust to throttle angle movement is not instantaneous and approximates to that of a simple first order filter with a time constant which is typically in the range 0.3 to 1.5 seconds, depending on the thrust setting and flight condition. Clearly, the lag in the throttle servo actuator response should be small compared

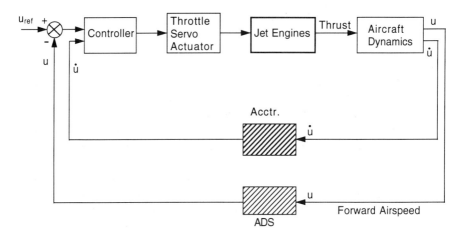

Fig. 8.13 Airspeed control system.

with the jet engine response. The aircraft dynamics introduces further lags as a change in thrust produces an acceleration (or deceleration) so that an integration is inherent in the process of changing the airspeed. The derivation of airspeed from the air data system can also involve a lag.

The rate of change of forward speed, $\dot{U}$, derived from a body mounted accelerometer with its input axis aligned with the aircraft's forward axis, can provide a suitable stabilising term for the control loop. (The $\dot{U}$ term could also be provided by a strap-down AHRS/INS.) A proportional plus integral of error control is usually provided to eliminate steady state airspeed errors.

A duplicate configuration is generally used so that the system fails passive. The throttle actuator is de-clutched in the event of a failure and the pilot then assumes control of the engine throttles.

8.3 FLIGHT MANAGEMENT SYSTEMS

8.3.1 Introduction

The FMS has become one of the key avionics systems because of the major reduction in pilot work load which is achieved by its use. In the case of military aircraft, it has enabled single crew operation of advanced strike aircraft such as the Lockheed F117A Stealth Fighter. The overall block diagram of a flight management system is shown in Fig. 8.14.

It should be pointed out that although flight management systems are now in very wide scale use, ranging from relatively basic systems in commuter type aircraft to 'all singing, all dancing' systems in long range wide body jet airliners, they are nevertheless a fairly recent development. In fact, it was not until the mid-1980s that they became widely installed. A number of factors have led to their installation:

• *Quantifiable economic benefits* – provision of automatic navigation and flight path guidance to optimise the aircraft's performance and hence minimise flight costs.

• *Air traffic* – growth of air traffic density and consequently more stringent ATC requirements, particularly the importance of 4D navigation.

• *Accurate navigation sources* – availability of accurate navigation sources. For example, combined GPS/INS, OMEGA together with the radio navigation aids, such as VOR, DME, VORTAC, TACAN and the approach and landing aids ILS and MLS.

• *Computing power* – availability and affordability of very powerful and reliable computing systems with the progressive development of the micro-processor and low cost solid state memory devices capable of storing very large amounts of data.

• *Data bus systems* – ability to interconnect the various sub systems providing data to the FMS by an efficient data bus system.

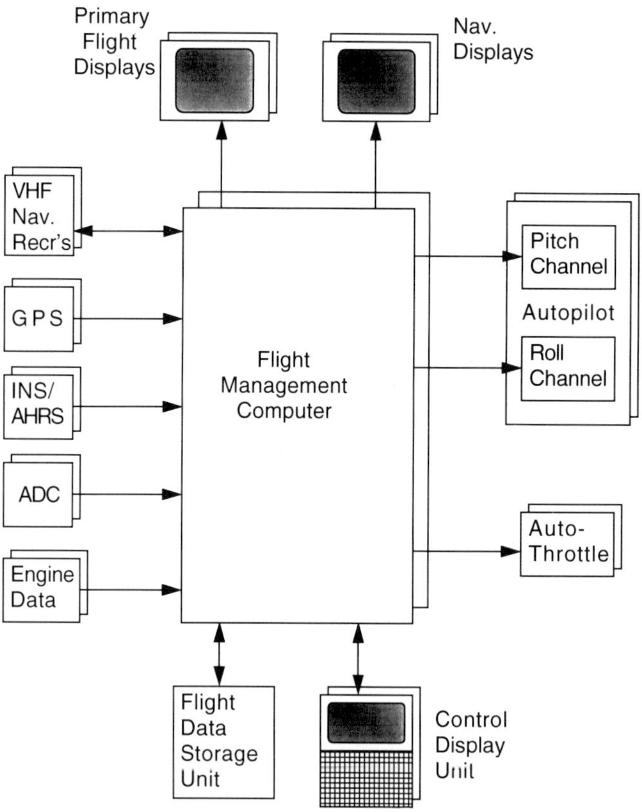

Fig. 8.14 Flight management system block diagram.

The FMS thus carries out the following tasks:

1. Provision of flight guidance and lateral and vertical control of the aircraft flight path.
2. Monitoring the aircraft flight envelope and computing the optimum speed for each phase of the flight and ensuring safe margins are maintained with respect to the minimum and maximum speeds over the flight envelope.
3. Automatic control of the engine thrust to control the aircraft speed.

In addition the FMS plays a major role in the flight planning task, provides a computerised flight planning aid to the pilot and enables major revisions to the flight plan to be made in flight, if necessary, to cope with changes in circumstances.

The FMS architecture for a modern airliner is shown in Fig. 8.15 which shows the Airbus A330/340 Automatic Flight System. Referring to Fig. 8.15, the flight management guidance and envelope computer, FMGEC, takes data and instructions from the multi-function control display units and displays textual information on

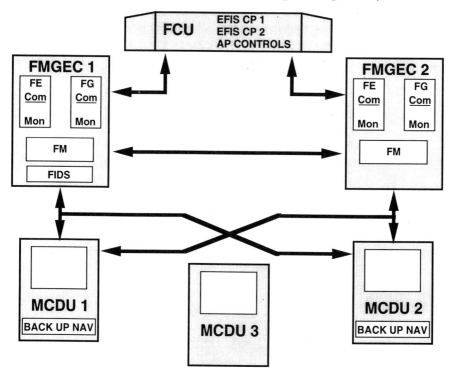

MCDU 3 switchable for FM function in case of MCDU 1 or 2 failure

Fig. 8.15 General architecture (by courtesy of Airbus Industrie).

the multi purpose control display units (MCDU). A redundant bus/structure allows for failure of one or both FMGECs, or one MCDU failure – there are three MCDUs in the central console.

A typical MCDU is shown in Fig. 8.16. This features a colour display for textual information with pilot instructions and data entry via keys. The display can show 14 lines of 24 characters (upper case, two sizes) with highlighting and flashing of text. Multi-level menus are selected by keys at the side of the display with the display prompting for options.

8.3.2 Flight planning

The FMS contains a database of:

- *Radio Navaids* – VOR, DME, VORTAC, TACAN, NDB, comprising identification, latitude/longitude, altitude, frequency, magnetic variation, class, airline figure of merit.

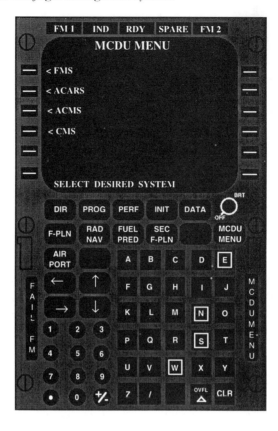

Fig. 8.16 Multipurpose control display unit (by courtesy of Airbus Industrie).

- *Waypoints* – usually beacons.
- *Airways* – identifier, sequence number, waypoints, magnetic course.
- *Airports* – identifier, latitude, longitude, elevation, alternative airport.
- *Runways* – length, heading, elevation, latitude, longitude.
- *Airport procedures* – ICAO code, type, SID, STAR, ILS, profile descent.
- *Company routes* – original airport, destination airport, route number, type, cruise altitudes, cost index.

The navigation data base is updated every 28 days, according to the ICAO AiRAC cycle, and is held in non-volatile memory. Diskette loading is used for acquisition and updating of the data. It is clearly essential to maintain the recency and quality of the data base and the operator is responsible for the detail contents of the data base which is to ARINC 424 format.

Figure 8.17 illustrates a typical airline route section and shows the information content. This can be displayed on a multi-function display surface.

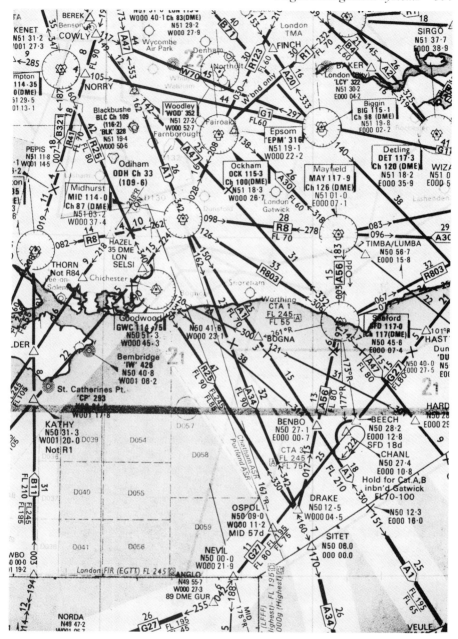

Fig. 8.17 Section of radio navigation chart (by courtesy of British Airways AERAD).

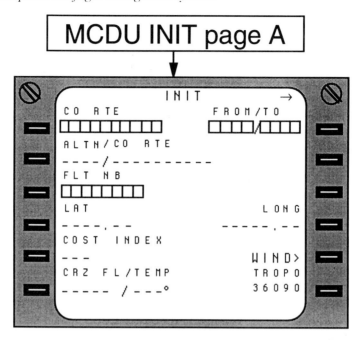

Fig. 8.18 Flight plan assembly (by courtesy of Airbus Industrie).

Figure 8.18 shows the use of the MCDU in the flight plan assembly. The flight plan assembly can be done automatically from the data base by the means of a company route number/city pair or manually by a waypoint by waypoint definition.

The resulting flight plan is composed of active and alternative flight plans, including departure and arrival procedures and missed approach. It is possible to alter the initial flight plan by means of lateral and vertical revisions. The secondary flight plan can also be copied from the active flight plan and modified to the same conditions as the primary flight plan.

A great deal of flexibility is built into the flight planning facilities in the FMS enabling revisions to the flight plan to be made easily. For example, the flight plan page can be called up to insert a waypoint or clear a waypoint or overfly a waypoint. 'Direct to page' can be used to modify the active leg, for instance for a radial interception. A temporary flight plan can be made to insert or modify depart/arrival procedures, insert a waypoint, insert a latitude/longitude crossing, change destination, insert a VIA/GO TO, insert an airways intersection, insert an offset, insert a holding pattern, or select or enable an alternative flight plan.

8.3.3 Navigation and guidance

The FMS combines the data from all the navigation sources to derive the best position estimate. The navigation sources include the IN systems, air data/heading reference, GPS, and the radio navigation aids VOR, DME, TACAN, VORTAC, and the approach and landing systems ILS/MLS. The FMS selects and automatically tunes the navaids specified in the flight plan and carries out the navigation computation. It also processes the data for the displays and generates the guidance information. The FMS computes ground speed, track, wind direction and velocity. (It should be noted that the air data system provides the height information for vertical navigation.) As stated earlier, the FMS provides both lateral and vertical guidance signals to the autopilot to control the aircraft flight path. In the lateral case, the FMS computes the aircraft position relative to the flight plan and the lateral guidance signals to capture and track the flight path specified by the flight plan.

Three dimensional vertical guidance is provided to control the vertical flight profile including the time dimension as will be explained in more detail later. This is of particular benefit during the descent and approach.

8.3.4 Flight path optimisation and performance prediction

The FMS is able to optimise specific aspects of the flight plan from a knowledge of the aircraft type, weight, engines and performance characteristics, information on the wind and air temperature and the aircraft state – airspeed, Mach number, height etc.

The FMS continually monitors the aircraft envelope and ensures that the speed envelope restrictions are not breached. It also computes the optimum speeds for the various phases of the flight profile. This is carried out taking into account factors, such as

- Aircraft weight – computed from a knowledge of the take-off weight and the fuel consumed (measured by the engine flow meters). It should be noted that fuel can account for over 50% of the aircraft weight at take off.
- CG position – computed from known aircraft loading and fuel consumed.
- Flight level and flight plan constraints.
- Wind and temperature models.
- Company route cost index.

The recommended cruise altitude and the maximum altitude are also computed from the above information.

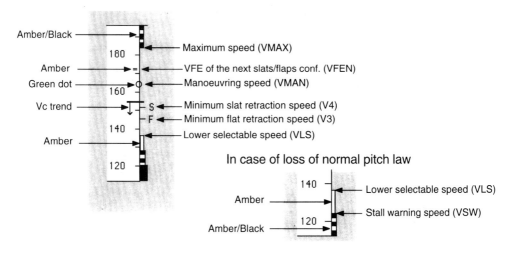

Fig. 8.19 Speed computation (PFD scale) (by courtesy of Airbus Industrie).

Figure 8.19 shows the speed scale displays on the primary flying displays of the A330/A340 airliners. Flight envelope protection is achieved by generating maximum and minimum selectable speed, windshear warning, alpha floor signal and stall warning. The manoeuvring speed and the flap and slat retraction speeds are also computed.

Figure 8.20 shows a display of speed optimisation information on the A330/A340. The FMS also carries out extensive performance predictions and provides the following information:

- Time, altitude, fuel, wind and temperature are predicted at each waypoint.
- Engine out performance.
- Climb and descent computation.
- Altitude and time markers.

8.3.5 Control of the vertical flight path profile

The FMS selects the speeds, altitudes and engine power settings during climbs, cruises and descents taking into account the flight plan, the prevailing conditions and the optimisation of the operation of the aircraft. The vertical definition of a typical flight plan is shown in Fig. 8.21.

The tasks which can be carried out and the facilities provided by the FMS during the various phases of the flight are briefly summarised below:

• TAKE OFF – The critical speeds V1, VR, and V2 are inserted by the crew and displayed on the primary flight displays.

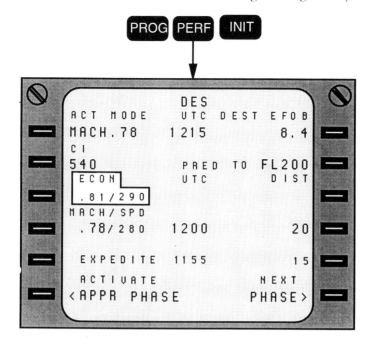

Fig. 8.20 Speed optimisation (by courtesy of Airbus Industrie).

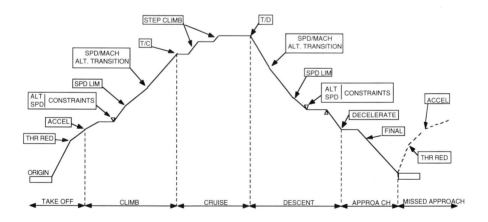

Fig. 8.21 Flight plan – vertical definition (by courtesy of Airbus Industrie).

• CLIMB – The FMS uses the manually input speed, the ATC constraint speed or the economical speed. It determines the start of the climb during take-off and predicts the end of the climb and the optimum cruising flight level.

• CRUISE – Five flight levels can be defined manually in the FMS. (Two flight levels can be stored for every route in the navigation data base. During the cruise, ATC or the crew may change the cruise altitude and the FMS can perform a 'step' climb at economical speed or a 'step' descent at 1000 ft/min at an economical speed. These events are also displayed symbolically on the navigation display.

• DESCENT – The FMS uses the manually input speed, the ATC constraint speed or the economic speed. The altitude and speed during the descent are computed as a function of the distance to the destination and a geometric profile is formed. The flight path is then computed backwards to satisfy the constraints.

·• APPROACH – The FMS can be coupled to the autopilot or alternatively provide guidance information to the pilot for manual control of the aircraft. Speed is critical during this phase and the approach speed is computed with respect to V_{REF} and the landing configuration (flaps, slats etc.) and the wind at the destination.

The approach mode is entered at the end of the descent and the approach ends either with landing or go around. Lateral guidance is provided by the FMS from the computed aircraft position as described (Section 8.3.3) and vertical guidance from barometric altitude when an RNAV approach has been selected. The FMS also provides speed control.

At the end of an RNAV approach the crew takes control to carry out the landing using visual references. When an ILS approach has been selected, the FMS tunes the ILS frequency and selects the runway heading as required for the runway selected by the crew. The approach and landing guidance is carried out by the autopilot using the ILS localizer references for horizontal guidance and the ILS glide slope references for the vertical guidance until the glide extension and flare phases, unless the crew elect to carry out an automatic go around or elect to take over control.

• GO AROUND – This is always assumed. The FMS manages the climb to the accelerating altitude or a selected altitude and provides track guidance from the outbound track defined in the flight plan.

8.3.6 Operational modes.

The FMS provides a number of very useful operational modes which are shown in the little sketches in Fig. 8.22 (a) (b) (c) and (d).

• Tangential go direct to mode – This is shown in Fig. 8.22(a) and provides navigation from the current position to any waypoint in the flight plan or entered during the flight.

• Turn anticipation – This is shown in Fig. 8.22(b) and avoids overshooting waypoints. It reduces the distance flown and reduces off-track manoeuvring.

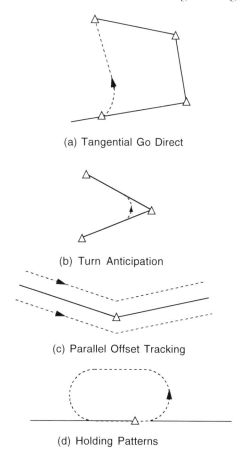

(a) Tangential Go Direct

(b) Turn Anticipation

(c) Parallel Offset Tracking

(d) Holding Patterns

Fig. 8.22 Operational modes.

• Parallel offset tracking – This is illustrated in Fig. 8.22(c). The lateral offset allows ATC to increase traffic flows in certain cases.
• Holding pattern – This is illustrated in Fig. 8.22(d). The FMS produces a precision holding pattern based on published ICAO entry procedure to reduce the pilot work load.

8.3.7 4D flight management

4D navigation has already been briefly referred to a number of times; in fact '4D flight management' is a better description of the process. 4D flight management covers the optimisation of the aircraft's flight along the most fuel conservative 3D path through climb, cruise and descent within the constraints of the air traffic control environment. Most importantly, the arrival time of the aircraft is also controlled so that it fits into the air traffic flow without incurring or causing delays. This is achieved by the automatic closed loop control exercised by the FMS through the autopilot and auto-throttle systems. These control the aircraft's flight path so that its 3D position at any time corresponds closely with the optimum time referenced flight path generated by the FMS computer.

Flexibility is essential for 4D flight management and there must be several minutes of arrival time variation possible so that the arrival time of the aircraft will coincide with the arrival time required by the air traffic controller. Some flexibility is also needed to compensate for differences between the actual and predicted wind and for performance modelling discrepancy with the actual aircraft in the FMS computations, although these are typically quite small. The air traffic control system also specifies the arrival time of the aircraft to the nearest integral minute, so the aircraft must have at least 30 seconds of arrival time flexibility to fit. Delays caused by traffic or weather also require some flexibility. Besides arrival time flexibility, there must also be some path flexibility. For example, air traffic control may require the aircraft to vector off course because of traffic. The 4D guidance and control exercised by the FMS then automatically re-engages the descent and computes the new course to the position specified by the ground ATC and makes good the original or newly specified arrival time. Occasionally aircraft are required to level-off in their descent, for example when the high altitude air traffic controller hands the aircraft off to the low altitude controller. When instructed to resume the descent the FMS then re-engages and guides the aircraft down to the ATC specified position at the pre-established time.

The process of automatic 4D descent control is very briefly described below. An ideal 4D trajectory which is defined by a table of time, range and altitude is pre-computed as briefly explained earlier.

The altitude and range errors from this ideal time referenced trajectory are shown in Fig. 8.23.

For the current aircraft position (current range) the required altitude is computed from the ideal trajectory and a combination of pitch attitude and direct drag or engine thrust is used to control both the altitude error and the speed. Pitch attitude is controlled through the pitch channel of the autopilot by the FMS to maintain the aircraft on the ideal trajectory or to return to it, while drag or thrust is used to control the speed as necessary.

For the current time, the required range is computed and speed changes are effected to control the range error and hence the required time.

The FMS exercises speed control for the 'too low' case by engine thrust control through the auto-throttle system. The 'too high/too fast' case is controlled by direct drag control by the spoilers/speed brakes. For safety/comfort reasons, the speed brake extension is generally controlled by the crew in response to a PFD message and not directly by the FMS/autopilot.

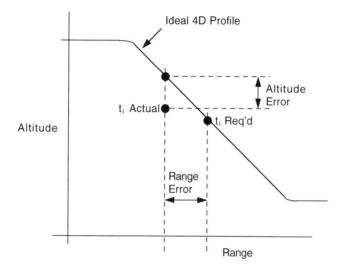

Fig. 8.23 4D descent control variables.

The effectiveness of automatic closed loop 4D guidance and control in reducing the arrival time uncertainty can be seen from the following figures which assume the ATC related errors are zero.

The arrival time uncertainty with closed loop 4D guidance and control is typically less than ± 8 seconds for 95% of all arrivals.

The corresponding arrival time uncertainty for performing an open loop high profile descent without time control is estimated to be of the order of 40 seconds.

The major reduction in arrival time uncertainty together with the ability to provide arrival time flexibility to accommodate delays caused by traffic or weather etc. coupled with flight path flexibility to meet ATC requirements make 4D flight management a very important advance.

Further reading

Combs, S.R., Sanchez-Chew, A.P., Tauke, G.J., *Flight Management System Integration on the F-117A*: American Institute of Aeronautics and Astronautics AIAA-92-1077, 1991

Lee, H.P., and Leffler, M.F., *Development of the L1011 Four-Dimensional Flight Management System*: National Aeronautics and Space Administration Report No. NASA CR 3700, 1 February 1984

McLean, D., *Automatic Flight Control Systems*: Prentice Hall

9

Avionics systems integration

9.1 INTRODUCTION AND BACKGROUND

Major avionic systems generally comprise a number of smaller sub-systems which are combined to form an overall system. The combination, interconnection and control of the individual sub-systems so that the overall system can carry out its tasks effectively is referred to as 'systems integration'. The number of sub-systems which need to be integrated to form a major system can be appreciated from the previous chapter on flight management systems.

It is instructive to review the development of avionic systems and their integration into overall systems in the light of the technology available and the circumstances prevailing at the time. The object is to put the development of today's advanced systems and the even more advanced systems currently under development in perspective. In many cases the current concepts and philosophy are not new – often the originators of particular system developments in the past were far sighted in their concepts, but, as always, were limited by the technology available at the time.

World War 2 (WW2) resulted in a major growth in the electronic equipment installed in aircraft and the birth of avionics, with the very rapid development of airborne radar systems and associated displays, radar warning systems and ECM, and more advanced autopilot systems exploiting electronics. Installation of the electronic equipment (or 'black boxes'), however, was very much on an ad hoc basis due to the very rapid developments and time scale pressures in war time. Some very limited degree of integration between sub-systems was introduced, for example coupling the bomb sight to the autopilot – as readers who have seen the film 'Memphis Belle' may have noted. In general, however, the systems were 'stand alone' systems and their integration into an overall system was carried out by specific crew members such as the navigator, bomb aimer or radar operator.

The 1950s period saw the emergence of a number of avionic sub-systems (some of which were initiated during WW2) which have since undergone continual development and now form part of the avionic equipment suite of most civil and military jet aircraft and helicopters. For example, auto stabilisers (or stability augmentation systems), ILS, VOR/DME, TACAN, Doppler, air data computers, attitude heading reference systems, inertial navigation systems.

The first major step towards integrating avionic systems was taken in the mid-1950s with the establishment of the 'weapon system' concept. These concepts were incorporated in the 1960s generation of aircraft, many of which are still in service. The concept requires a total system approach to the task of carrying out the mission effectively with a high probability of success. The aircraft, weapons and the avionic systems required by the crew to carry out the mission effectively must thus be considered as an integrated combination. It should be appreciated, of course, that the total system approach is applicable to any project, military or civil. As with many methodologies, however, military applications provided the spur and the initial funding. The very widely used 'programme evaluation review technique', or 'PERT' networks, and 'critical path analysis', for example, were originally developed on the POLARIS missile program.

As an example of the overall system approach, consider the requirements for a naval strike aircraft. The aircraft must be able to operate from an aircraft carrier in all weathers and be able to find the target and attack it with a suitable weapon (or weapons) with a high probability of success. Operational analysis shows that to minimise the probability of detection and alerting the enemy's defences, the aircraft needs to approach the target at high sub-sonic speeds (550–600 knots) at very low level at a height of 100 ft or so above the sea so as to stay below the radar horizon of the target as long as possible. The avionic sub-system specifications can then be determined from the overall system requirements with an aircraft crew comprising pilot and observer/navigator. Hence in the above example, the avionic equipment fit would comprise:

- Radar – target acquisition in all weather conditions.
- Doppler – accurate (±4 knots) velocity sensor for DR navigation. (Note - IN systems capable of accurate initial alignment at sea on a moving carrier were still under development in the late 1950s.)
- Attitude heading reference system (or master reference gyro system – UK terminology) – attitude and heading information for pilot's displays, navigation computer, weapon aiming computer, autopilot.
- Air data computer – height, calibrated airspeed, true airspeed, Mach number information for pilot's displays, weapon aiming, reversionary DR navigation, autopilot.
- Radio altimeter – very low level flight profile during attack phase and all weather operation.
- Navigation computer – essential for mission.
- Autopilot – essential for reduction of pilot work load.
- Weapon aiming computer – essential for mission.
- HUD – all the advantages of the HUD plus weapon aiming for low level attack; for example, 'toss' bombing.
- Stores management system – control and release of the weapons.
- Electronic warfare (EW) systems – radar warning receivers, radar jamming equipment. Essential for survivability in hostile environment.
- Identification system (identification friend or foe – 'IFF') – essential to avoid attack by friendly forces.

- Radio navigation aids – location of parent ship on return from mission.
- Communications radio suite – essential for communicating to parent ship, co-operating aircraft etc.

A significant degree of integration was required between the avionic sub-systems. For example, the weapon aiming system required the integration of the HUD, weapon aiming computer, AHRS, air data computer and the radar system.

The basic avionic systems specified for a naval strike aircraft in the mid- to late 1950s, for service introduction in the 1960s, for example the Royal Navy Buccaneer aircraft, would also be required in a 1990s naval strike aircraft. The systems would need to be of higher performance, however, and there would also be additional systems in a 1990s aircraft which did not exist in the late 1950s or early 1960s. For example, laser gyro INS, GPS, FLIR, 'smart weapons' etc. and a much greater level of systems integration would be required. The point being made is the essential continuity of the role of avionic systems and their development, although their implementation changes as new technologies become available.

It should be stressed that although the above example is a military one, the same principles apply to civil avionic systems in terms of their role in enabling the mission to be carried out safely and effectively with the minimum flight crew. Most civil avionic sub-systems have also been directly developed from military avionic systems (eg air data computer, INS, GPS, OMEGA etc.) and, until fairly recently, military aircraft generally required a larger number of avionic sub-systems with a higher degree of integration.

A major step towards facilitating the integration of avionic sub-systems in civil aircraft was taken in the early 1950s with the adoption of ARINC specifications for avionic systems and equipment. ARINC is a non-profit-making organisation in the USA which is run by the civil airlines with industry and establishment representation, which defines systems and equipment specifications in terms of functional requirements, performance and accuracy, input and output interfaces, environmental requirements and physical dimensions and electrical interfaces. For example, air data computers, attitude heading reference systems, INS, communication radio equipment, data bus systems etc. Equipment made to an ARINC specification by one manufacturer should thus be completely interchangeable with equipment made by another manufacturer to the same ARINC specification. The electronic implementation of the two systems can be totally different provided they conform to the ARINC specification in terms of form–fit–function. For example, the LRUs (line replaceable units) must conform to the ARINC specifications dimensionally with the specified rack fixing arrangements, connectors and the pin allocations in the connectors. The systems must also meet the ARINC performance and accuracy specifications and the environmental requirements in terms of temperature range, acceleration, shock and vibration and EMC. The use of avionic equipment qualified to ARINC specifications thus ensures a competitive situation enabling procurement to be made from manufacturers on a world-wide basis.

The late 1950s and early 1960s saw the development of the first real time airborne digital computers, and these were progressively introduced in military aircraft from the mid-1960s for tasks such as navigation, mission management, weapon aiming, radar processing and displays processing.

These first generation airborne digital computers were very expensive and it is interesting to note that a number of proposals were made during the 1960s to promote the concept of carrying out as many of the avionic sub-system computing tasks as possible with a powerful central digital computer. These proposals were not taken up for a number of reasons, such as:

1. Vulnerability – 'all the eggs in one basket', whereby a failure in the central digital computer affected all the sub-systems sharing its computing facilities.
2. Inflexibility – changes in an individual sub-system could involve changes in the main computer software with possible ramifications and 'knock on' effects on all the other sub-systems sharing the computing facility. This was particularly relevant with the computing speeds achievable at the time and the high cost and limited capacity of the non volatile memory technology which depended on magnetic core stores. (Typical store sizes were 8K to 16K at that time.)
3. Cost and weight of redundant central computer configurations was unacceptable.

The availability today, however, of affordable, very powerful processors with large memory capacity and high speed data buses has radically changed the situation. The concept of sharing avionic sub-system processing tasks between a number of processors with spare capacity to take over particular tasks in the event of a processor failure is now economically attractive. The new integrated modular avionics architectures currently under development for the next generation of both military and civil aircraft entering service in the late 1990s and early 2000s, in fact, exploit such concepts as will be discussed later. As stated earlier, many of today's concepts are not new. It is the technology available at the time which limits their economic exploitation.

By the latter half of the 1960s, the development of affordable 'task orientated processors', that is processors which are sized to carry out one (or possibly two) avionic sub-system computing tasks, had become viable with the development of integrated circuit technology. A number of avionic sub-systems had thus been developed and were entering service in both military and civil aircraft by the end of the 1960s which contained their own internal digital computers; for example, digital waveform generators in HUDs, air data computers and inertial navigation systems.

Many of the aircraft sub-systems up to the early 1970s, however, were still largely analogue in their implementation with synchro and potentiometer outputs/inputs requiring point to point wiring to interconnect them. The interface units carrying out the necessary analogue to digital (A-D) and digital to analogue (D-A) conversions to enable the sub-systems to communicate with each other were inevitably bulky and could exceed the size of the digital computer. The complexity

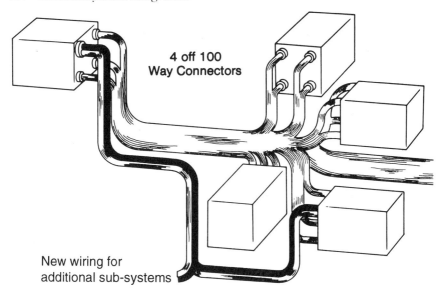

4 off 100 Way Connectors

New wiring for additional sub-systems

Fig. 9.1 Interconnection of avionic sub-systems before introduction of multiplexed data buses.

of the inter-unit cabling looms and wiring harnesses can be seen in Fig. 9.1 which is typical of a late 1960s/early 1970s aircraft installation. It should be noted that many of these aircraft are still being operated. The weight of the cable looms can be very significant and can exceed several hundred kilograms. The large number of multi-way plugs and connectors also inevitably degrades the overall system reliability and can introduce intermittent faults.

The first development to reduce the number of interconnecting wires was the use of time division multiplexing (TDM) of the signals to be transmitted between two units.

TDM enables information from several signal channels to be transmitted through one communication system by staggering the different channel samples in time to form a composite pulse train. Each signal channel is transmitted as a serial digital pulse train at a given time slot in a clock cycle. Thus given the clock time information and the address of the signal (suitably pulse coded), the receiver can then decode and distribute the individual signals. Hence if there are, say, 30 different signals to be transmitted between two units, only two wires are required in principle when the data is multiplexed compared with 60 wires. This type of single-source–single-sink communications link between two units (say A and B) is frequently referred to as an 'A to B' link. Early systems, in fact, used three screened twisted pairs of wires to transmit the signal, address and clock data. This was then reduced to two screened twisted pairs of wires. Later single- and multi-source–multi-sink systems, however, operating at higher clock rates encode the signal data and address with the clock data using Manchester II bi-phase encoding (covered in the next section). A single screened twisted pair of wires only is required with such an encoding system to transmit information data rates of up to 1 to 2 Mbits/s. Higher data rates of up to 50 Mbits/s can be transmitted using a suitably screened coaxial electrical cable and information at even higher data rates can be transmitted as coded light pulses using a fibre optic cable. This will be covered in Section 9.2.

By the mid-1970s, it became possible to implement many more avionic sub-systems digitally, exploiting task orientated processors and the newly developed micro-processors and so eliminate the analogue computing elements and the analogue input/output components such as synchros and potentiometers etc. It thus became possible to interconnect the individual avionic sub-systems by means of a digital data bus system. This enabled the systems to communicate with each other and transfer serial digital information using time division multiplexing in conjunction with a suitable protocol system to control the data transmission to and from each individual sub-system. This will be covered in the next section but the dramatic reduction in inter-unit wiring can be seen in Fig. 9.2 which illustrates the interconnection of the sub-systems using a MIL STD 1553 data bus system.

An essential parallel development in the late 1970s and early 1980s was the implementation of the complex circuitry required by the MIL STD 1553 data bus system terminals in LSI (and later by VLSI). This enabled the system complexity to be encapsulated in highly reliable IC chips with a 'chip set' costing less than some multi-way connectors.

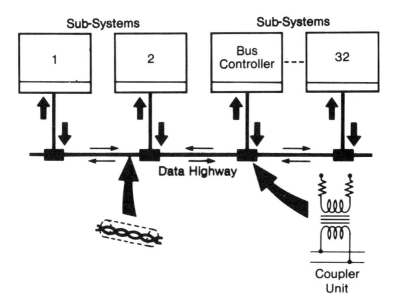

Fig. 9.2 Interconnection of avionic sub-systems by multiplexed data bus.

The avionic system architectures which had become established by the mid-1970s are still the standard architectures in current aircraft in service and are referred to as 'federated architectures'. Federated architectures essentially consist of a number of interconnected but functionally independent sub-systems with some degree of central computer control of overall operating modes. The advance of digital electronics through the late 1970s and 1980s enabled the development of very sophisticated sub-systems which, although still addressing self contained functional areas, shared information via data links such as the 1553 bus.

Federated architectures, built up from individual proprietary designs and available from a wide and experienced supplier base, became firmly established as the low risk avionics system design approach. Also, because of the well defined system boundaries and the clear division of areas of responsibility, federated systems were universally accepted as the low risk approach commercially and contractually as well as technically.

In a federated architecture, however, there is inevitably some duplication of functions in each of the individual sub-systems which makes the overall system heavier and more costly than necessary. LRUs are each designed and optimised separately resulting in many different implementations of similar functions. Items such as processors, memory, software, interfaces and power supplies are individually optimised for each sub-system and the design and support of these unique proprietary parts drastically increases the overall system development and support costs. The sub-system design approach also limits the introduction of fault tolerance because the proprietary sub-systems cannot easily share resources and the cost of building redundancy into every sub-system is unacceptable. The use of proprietary sub-systems also creates additional problems whenever an optimised proprietary sub-system interface must be accessed as part of an upgrade.

The maintenance of federated systems is constrained by the proprietary boundaries of the architecture and requires the complete removal of sub-system black boxes from the aircraft for repair in second, third and fourth line maintenance facilities. Flexible deployment of such systems in military applications is seriously hampered by the need to deploy sophisticated second line, or 'avionics intermediate shop' support equipment. With the reducing numbers of new military designs the need for mid-life upgrades has focused attention on the high cost of upgrading federated architectures. Essentially, whole systems may have to be removed and replaced and a small improvement in system functionality can result in very large costs.

Since the mid-80s, avionics R&D programmes in the USA and Europe have been directed towards finding a better way to build avionics. Over that period the concept of modular systems built from a small number of standard module types, housed in standard racking, and communicating over standard data networks has been developed and has become known as *modular avionics*. It is generally accepted that modular avionics has significant potential for savings in system weight, power, space and cost. However, to achieve the full benefits from modular systems it is necessary to dissolve the sub-system boundaries which exist in conventional avionics so that a common pool of spare resources, shared between sub-systems, can be used to improve system availability. Hence, integrated modular avionics, or IMA, offers additional benefits of fault tolerance and application flexibility through reconfiguration.

The IMA concept has been adopted as the way forward for both civil and military avionics and systems based on IMA concepts are now in the final stages of EM & D (engineering manufacture and development).

The topic of IMA is covered at an overview level in Section 9.3.

9.2 DATA BUS SYSTEMS

Data bus systems are the essential enabling technologies of avionic systems integration in both federated and integrated modular avionics architectures.

They can be broadly divided into electrical data bus systems where the data are transmitted as electrical pulses by wires, and optical data bus systems where the data are transmitted as light pulses by optical fibres. These are discussed and the basic principles outlined in Sections 9.2.1 and 9.2.2 respectively.

9.2.1 Electrical data bus systems

There are several electrical data bus systems in use in avionics systems. Space constraints have restricted the main coverage to the MIL STD 1553B data bus system because of its very wide usage. It is also a relatively sophisticated data bus system and an understanding of its operation reads across to the other systems in many areas, such as ARINC 429 which is a system of lower capabilities used in civil avionic systems and the new ARINC 629 data bus system. The ARINC 629 data bus system will have its first service introduction on the new Boeing 777 airliner which enters airline service in May 1995.

9.2.1.1 MIL STD 1553 bus system

MIL STD 1553B is a US military standard which defines a TDM multiple-source–multiple-sink data bus system which is in very wide scale use in military aircraft in many countries. It is also used in naval surface ships, submarines, and land vehicles such as main battlefield tanks. The system is a half duplex system, that is operation of a data transfer can take place in either direction over a single line, but not in both directions on that line simultaneously.

The system was initially developed at Wright Patterson Air Force Base in the early 1970s and, as MIL STD 1553A, was first introduced in service on the F15 fighter programme in 1975. The standard was upgraded to MIL STD 1553B in 1978 to incorporate additional modes and facilities. This specification has since been progressively refined and amendments incorporated in the light of user experience by the SAE 2K Committee under the auspices of the Society of Automotive Engineers (SAE) in the USA. The standard has also been adopted as a NATO standard and has been given the NATO codification STANAG 3838.

The basic bus configuration is shown in Fig. 9.3; the system is a command-response system with all data transmissions being carried out under the control of the bus controller. Each sub-system is connected to the bus through a unit called a remote terminal (RT). Data can only be transmitted from one RT and received by another RT (or RTs as there may be more than one sub-system requiring the same data) following a command from the bus controller (BC) to each RT. (The operation of the data bus system such that information transmitted by the bus controller or a remote terminal is addressed to more than one of the terminals connected to the data bus is known as the 'broadcast' mode.)

The protocol exercised by the bus controller hence ensures that there are no data clashes on the bus as only one RT is transmitting at any time. The bus controller thus initiates all data transfers and monitors the status of all transfers. It is generally incorporated in one of the sub-systems – usually the one generating the most traffic.

The bus is formed as a single twisted cable pair with one layer of shielding and jacketing and with a maximum length of 100m (328 ft). Although direct coupling to the bus is allowed, this is generally not used in order to avoid the risk of one terminal shorting out the bus. The bus connection is typically via a transformer coupled stub so that shorting of the stub is isolated from the bus. The maximum stub length allowed is 20 ft.

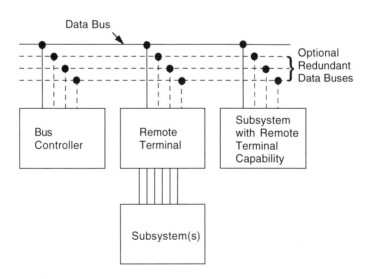

Fig. 9.3 Typical multiplex data bus system architecture.

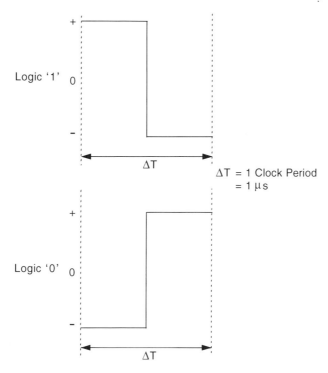

Fig. 9.4 Logic '1' and logic '0'.

The data are transmitted at 1 Mbit/s. The data word size is 20 bits so that the maximum data transmission rate is 50,000 words/s. A maximum of 31 terminals can be connected to the bus. The bus operation is asynchronous, each terminal having an independent clock source for transmission. Decoding is achieved in receiving terminals using clock information derived from the messages.

The technique adopted for data encoding is known as 'Manchester bi-phase' encoding where there must be an active transition for every bit, ie for '0' and '1' signals. This is shown in Fig. 9.4. Apart from the SYNC bits all data bits must conform to these requirements. This eliminates 'stuck high' or 'stuck low' faults as there must be a transition during one clock period. Fig. 9.5 shows the data encoded waveform. The standard requires the transmission rate to be 1 Mbit/s with a combined accuracy and long term stability of 0.01% (ie ±100 Hz). The short term stability (ie stability over a 1.0 second period) is required to be at least 0.001% (ie ±10 Hz) The word size is 16 bits plus the SYNC waveform and the parity bit for a total of 20 bit times. There are three types of words transferred: command words, status words and data words. The formats for these words are illustrated in Fig. 9.6.

A command word comprises six separate fields. These are briefly explained below:

]

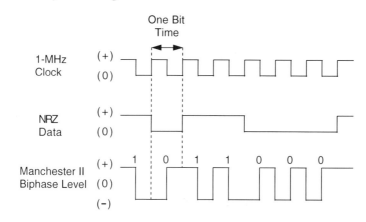

Fig. 9.5 Data encoding.

- The SYNC signal field is an invalid Manchester waveform so that it cannot be 'confused' with any data bits.
- The RT address field occupies 5 bits, each RT being assigned a unique 5 bit address. Decimal address 31(11111) is not assigned as a unique address and is a broadcast address.

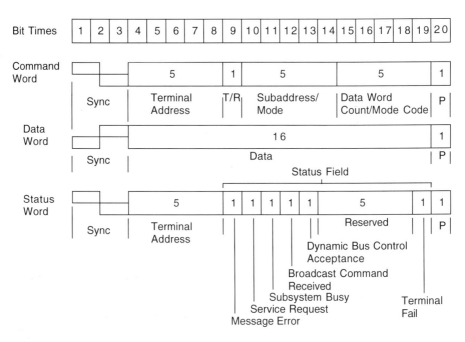

Fig. 9.6 Word formats.

- The T/R bit = 0 if the RT is to receive, and = 1 if the RT is to transmit.
- The sub-address/mode field, comprising 5 bits, is used for either an RT sub-address or mode control. The sub-address is used to route data to and from a location in the RT. A code of all zeros (00000) in the sub-address/mode field indicates that the contents of the word counts/mode field are to be decoded as a five bit mode command.
- The data word count/mode code field, comprising 5 bits, is generally used for data transfers. The word count field indicates the number of data words to be transferred in any one message block, the maximum number being 32 (indicated by all zeros).
- The parity bit = 1 if there is an odd number of bits in fields 1–19.

A status word is the first word of a response by an RT to a BC command. It provides:

(a) a summary of the status/health of the RT
(b) the word count of the data words to be transmitted in response to a command.

The fields are briefly described below:

- The SYNC signal field is the same as with a command word.
- The RT address field (5 bits) confirms the correct RT is responding.
- The status field comprises 11 bits. The message error bit is set if the previous command was not correctly understood. The instrumentation bit = 0 to distinguish the word from a command word.
- The parity bit is set by the RT in the same sense as a command word.

Data words contain the actual data transmitted between stations. The data field is 16 bits. For commands which imply a data content, data words are transmitted corresponding to the word count in the command or status word.

The SYNC signal is the inverse of the command and status word syncs. The most significant bit of the data is transmitted after the SYNC bits.

There are ten possible transfer formats, but the three most commonly used formats are:

- BC to RT
- RT to BC
- RT to RT

These are shown in Fig. 9.7. An intermission gap time of at least 2µs is provided by the bus controller between messages. A status word gap time of at least 2µs but not more than 10µs is provided by the RT before transmitting a status word.

A high degree of data checking and monitoring is built into the MIL STD 1553B system. For example:

- *Message data validation* – the terminal is designed to detect improperly coded signals, data drop-outs or excessively noisy signals.

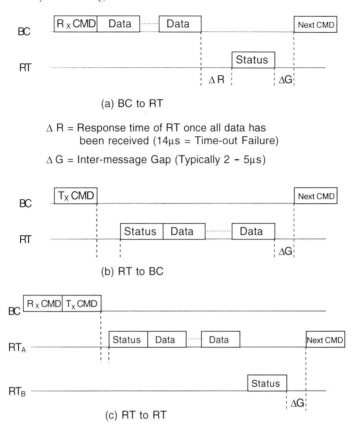

(a) BC to RT

Δ R = Response time of RT once all data has
        been received (14μs = Time-out Failure)

Δ G = Inter-message Gap (Typically 2 - 5μs)

(b) RT to BC

(c) RT to RT

Fig. 9.7 Transfer formats.

- *Word validation* – the terminal checks that each word conforms to the following minimum criteria:
    - Word begins with a valid SYNC field
    - Bits are in valid Manchester II code
    - Information field has 16 bit plus parity
    - Word parity is odd.
  When a word fails to conform to the above criteria the word is considered invalid.
- *Transmission continuity* – the terminal checks the message is contiguous as shown in the formats in Fig. 9.7. Improperly timed data SYNCs are considered a continuity error.
- *Excessive transmission* – the terminal includes a signal time-out which precludes a signal transmission greater than 1 ms plus or minus 0.34 ms.

The data word is deemed valid when the data meet the above criteria and are received in contiguous fashion. The RT responds with a status word when a valid command word and the proper number of contiguous valid data words are detected as specified in the word count field of the command word, or as specified for the mode code contained in the command word.

A fault tolerant configuration of two (or more) identical MIL STD 1553B buses can be connected to the sub-system equipment.

Space constraints limit further discussion of the MIL STD 1553B data bus, in particular the bus operation and organisation to meet the data transfers between the sub-systems connected to the bus in terms of the required words/s. It is hoped, however, that the above coverage will enable the reader to understand and appreciate the basic system and know the questions to ask or look up in the reference literature, if further information is sought.

The '1553' data bus system is a mature system which has now built up a very large operational usage and experience over a period of nearly 20 years since its introduction. It is by far and away the most widely used avionic data bus system and its performance can be summed up in one word – excellent. It is also worth noting that the '1553' bus system and components have also found considerable usage in non-military applications requiring a robust multiplexed data transmission system able to operate in severe environments.

9.2.2 Optical data bus systems

Most readers are probably familiar to some extent with the use of optical fibres to transmit light signals. A brief explanation is set out below for those readers who need to refresh themselves on the subject and also to make clear the difference between multi-mode and single mode optical fibres and their respective applications.

The transmission of light signals along any optical fibre depends on the optical property of total internal reflection. This property is illustrated in Fig. 9.8, which shows four rays of light travelling through a medium of refractive index n_1 to a medium of lower refractive index n_2. Ray 1 is refracted in passing through the second medium, the relationship between the angle the incident ray makes with the normal, i, and the angle the refracted ray makes with the normal, r, being given by Snell's law:

$$\frac{\sin i}{\sin r} = \frac{n_2}{n_1}$$

At the critical incidence angle, i_{crit}, ray 2 is refracted through an angle of 90° and does not pass through the second medium ($i_{crit} = \sin^{-1} n_2/n_1$). All rays with incident angles greater than i_{crit} such as rays 3 and 4 are thus reflected back into the first medium. This condition is known as total internal reflection and is effectively a loss free process.

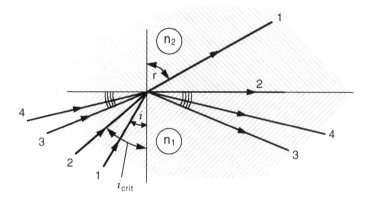

Fig. 9.8 Total internal reflection.

An optical fibre basically comprises a central core of a suitable glass material (for example pure silica) with a very low optical transmission loss and with an outer cladding of a material with a slightly lower refractive index than the core. In multi-mode fibres, as shown in Fig. 9.9, the diameter of the core is large compared with the wavelength of the light being transmitted. For example, a typical core diameter is around 100μm and the operating wavelength around 1μm. A ray entering the fibre at an incident angle θ to the axis of the fibre less than the critical angle, $\theta_{crit} = \cos^{-1} n_2/n_1$, will undergo total internal reflection at the core/cladding

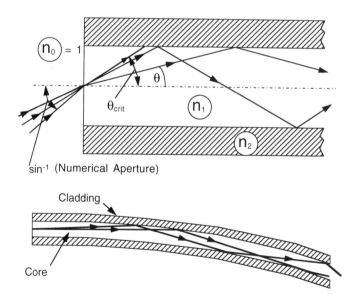

Fig. 9.9 Multi-mode optical fibre.

interface. This ray will then undergo total internal reflection at the lower interface and will thus be guided through the core by repeated internal reflections as shown in Fig. 9.9. There are, however, a large number of different ways or modes by which light can be guided along the fibre depending on the incident angle θ. Hence the term multi-mode fibre. The time taken by a ray to travel along a fibre of length L is thus a function of θ and is equal to $n_1 L/(c \cos \theta)$, where c is the velocity of light. The rays thus travel with different velocities so that a series of light pulses each of width t_1 at the input end of the fibre emerge after transmission through the fibre as a series of pulses of width t_2 (see Fig. 9.10). If the broadening of the pulses due to this time dispersion is large, then ultimately adjacent pulses will overlap at the output and cannot be resolved. This pulse broadening effect is generally tolerable for the current data bus rates of 50 Mbits/s and the relatively short lengths involved in aircraft installations which are generally less than 100 metres.

This pulse dispersion, however, is totally unacceptable for telecommunications applications which require very high data rates and long distances between repeaters to minimise the number of repeaters. This has resulted in the development of highly efficient single mode optical fibres. The major difference between single

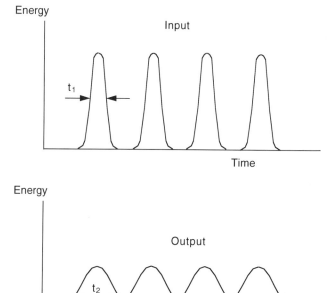

Fig. 9.10 Pulse broadening.

mode fibre and multi-mode fibre is that the core diameter of single mode fibre is of the same order of magnitude as the wavelength of the light source (laser diode)

It can be shown that as the core diameter is decreased and the refractive index difference between the core and the cladding reduced, the number of possible guided modes for transmitting light along the fibre decreases. There is a normalised parameter known as the *waveguide parameter* which is equal to $\dfrac{2\pi \, an_1 (2\Delta)^{1/2}}{\lambda_0}$

where λ_0 = operating wavelength, a = core radius, n_1 = refractive index of core, n_2 = refractive index of cladding and $\Delta = (n_1 - n_2)/n_1$.

When the waveguide parameter is less than a certain critical value (2.4048 for step index fibre) then only one guided mode is possible for transmitting light along the fibre, and the fibre is known as a single mode fibre. Practical single mode fibres have Δ varying from 0.002 to 0.005 and typical core diameters in the range 5 to 10μm. Typical operating wavelength is around 1.5μm.

Single mode means there is only value of θ and hence only one ray path for the light to travel along the fibre by multiple reflections so that there is only one velocity of propagation for a light pulse. The pulse broadening resulting from the differences in velocity for the different ray paths in multi-mode fibre is thus eliminated. It should be noted that there are other smaller sources of dispersion, namely material dispersion and waveguide dispersion which must be minimised in order to achieve very high data rates and very long transmission distances. Material dispersion is the dispersion resulting from the dependence of the refractive index of the fibre material on wavelength. Waveguide dispersion is the dispersion resulting from the spectral width of the source; the different wavelength components experience different refractive indices. These effects can be minimised by techniques such as grading the refractive index profile across the core; for example varying the refractive index linearly from the centre of the core to the cladding. Dispersions of a few pico-seconds per kilometre are attained in modern single mode fibres compared with 50 nano-seconds per kilometre for a typical step index multi-mode fibre. Fourth generation fibre optic communication systems are now coming into service with an information carrying capacity in excess of 10 Gbits/s and repeater spacing of over 100 km; the transmission loss in the fibre is less than 0.3 dB/kilometre.

As stated earlier, multi-mode optical fibre can be used in avionic system applications because of the relatively short lengths involved and the current data rate requirements of 50 Mbits/s. The reason for its use is primarily due to the need for de-mountable connectors in avionic equipment for ease of servicing and replacement of a failed unit. While demountable connectors for single mode fibres are feasible, they present a number of mechanical alignment problems and have not progressed beyond the laboratory stage. (The telecommunications industry uses fusion splicing techniques instead of connectors and allows sufficient extra fibre in a small coil at the end of the fibre to re-splice when necessary.)

The task of designing suitable low loss, robust connectors for multi-mode fibres is eased by the following factors:

(1) Multi-mode fibres have a considerably larger numerical aperture than single mode fibres. The numerical aperture, NA, defines the semi-angle of the cone within which the fibre will accept light and is a measure of the light gathering power of the fibre.

$$NA = n_1 \, (2\Delta)^{1/2}$$

Typical values for a multi-mode fibre are $n_1 = 1.46$ and $\Delta = 0.01$ giving a NA of 0.2. The fibre will thus accept light incident over a cone with a semi-angle of $\sin^{-1} 0.2$, that is 11.5° about the axis. Typical NAs for single mode fibres result in acceptance semi-angles in the region of 4° to 8° as Δ is in the region of 0.002 to 0.005.

The larger NA eases the alignment tolerances of the two halves of the connector.

(2) LEDs can be used for the modulated light source. These approach a Lambertian source with a hemispherical power profile which together with the reasonable NA of multi-mode fibre enables a simple and efficient optical coupling arrangement to couple the light source to the fibre to be implemented.

(3) The larger core diameter eases the mechanical tolerancing problems in aligning the two halves of the connector.

It is in fact possible to adapt existing electrical connectors to incorporate multi-mode optical fibres.

The application of single mode optical fibres in avionic systems is thus currently confined to optical sensors such as the fibre optic gyro. Future requirements for very high data rates, however, could well lead to their adoption. The very high reliability of future avionic systems makes the need for demountable connectors questionable and the use of fusion splicing techniques feasible on the rare occasions when such equipment needs to be removed and replaced.

To summarise, the use of optical fibres to transmit data offers major advantages, such as:

- High data rate capability (>10 Gbit/s using single mode fibre)
- Insensitivity to electro-magnetic interference
- Electrical isolation
- No line capacitance or mutual coupling
- Low crosstalk
- Lower power dissipation
- Reduced weight and volume requirements.

The integration of the various avionic sub-systems to increase mission effectiveness requires an inter-connection network system capable of two way communication of serial digital data at high speed. Present electrical systems, such as the widely used MIL STD 1553B data bus system have limiting restrictions such as:

- relatively slow transmission rate limited by medium (1 Mbit/s)
- restricted number of terminals for communication (max 31)
- restricted number of words transferred per message (max 32)
- central control unit managing all data transfer.

To overcome these problems, a number of higher speed transmission systems are being developed both in Europe and the USA. Table 9.1 summarises their salient features. The systems are described briefly in the following sections.

Table 9.1 Features of transmission systems

Parameter	STANAG 3910 Data Bus	LTPB Linear Token Passing Bus	HSRB High Speed Ring Bus
Data rate	20 Mbits/s	50 Mbits/s	50-100 Mbits/s
Encoding technique	Manchester bi-phase	Manchester bi-phase	4B/5B data encoding
Topology	Bus structure	Bus structure	Point to point linked ring
Max message transfer	4096 data words	4096 data words	4096 data words
Number of stations	31	128	128
Bus control philosophy	Central control	Distributed control	Distributed control
Controlling mechanism	1553 bus control	Token passing control	Token passing control
Bus length	Dependent on 1553 network	1000m	1500m
Interconnect media	Fibre optics	Fibre optics or electrical	Fibre optics or electrical
Standard	STANAG	SAE	SAE
Country of origin	Europe	USA	USA

9.2.2.1 *STANAG 3910 data bus system*

STANAG 3910 is a European data bus with a 20 Mbit/s data rate which has been adopted for the Eurofighter 2000. This advanced fighter aircraft is being jointly developed by the UK, Germany, Italy and Spain for their respective air forces and is planned to enter squadron service around the year 2000.

The bus provides an evolutionary increase in capability by using MIL STD 1553B (STANAG 3838) as the controlling protocol for high speed (20 Mbit/s) message transfer over a fibre optic network as shown in Fig. 9.11.

The optical star coupler is a passive optical coupler which enables light signals from each fibre stub to be coupled into the other fibre stubs and thence to the sub-systems. It is a passive coupler and this inevitably attenuates the signals. For example, a 32 way coupler attenuates the signals by 1/32 and this must be allowed for in the worst case overall signal to noise ratio budget over the temperature range (eg $-55°C$ to $+100°C$).

The basic implementation of a 3910 terminal is illustrated in Fig. 9.12.

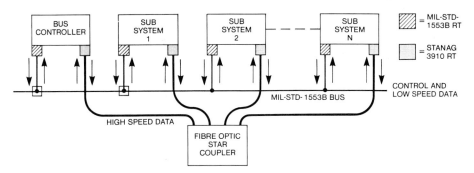

Fig. 9.11 STANAG 3910 data bus system.

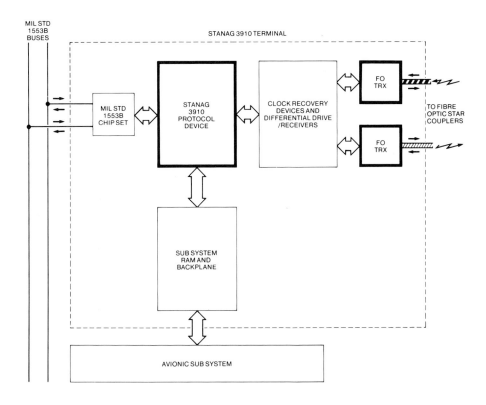

Fig. 9.12 STANAG 3910 terminal (by courtesy of GEC-Marconi Avionics Ltd.).

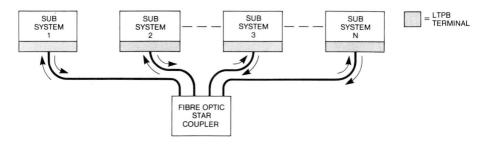

Fig. 9.13 Linear token passing high speed data bus.

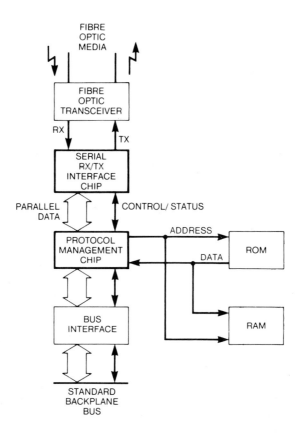

Fig. 9.14 Linear token passing high speed data bus terminal (by courtesy of GEC-Marconi Avionics Ltd.).

9.2.2.2 Linear token passing high speed data bus.

The linear token passing high speed data bus (HSDB) is being developed in the USA for the new generation modular avionic systems as discussed later in Section 9.3. The basic bus configuration is shown in Fig. 9.13. The system uses distributed control by means of a token passing protocol and operates at 50 Mbits/s. The basic implementation of the HSDB terminal is shown in Fig. 9.14. The basic similarities with a 3910 terminal can be seen.

9.3 INTEGRATED MODULAR AVIONICS

The background to IMA has been explained in Section 9.1. The importance of finding a better way of implementing avionic systems can be appreciated when it is realised that avionics currently account for some 30% of the total cost of a new aircraft. Reducing these costs must thus play a major role in containing overall system costs and halting the cost spiral inherent in the current federated architectures as increasing performance and capability are sought. The new IMA architectures offer the best prospects of meeting higher levels of performance and system capability, increased equipment availability and reduced levels of maintenance, so lowering costs right across the system life cycle. Space constraints limit the treatment of this topic, which could occupy several books, to that of an overview explaining the basic philosophy, aims and objectives of the IMA architectures and the implications of their implementation using standardised electronic modules.

The term avionics architecture is a deceptively simple description for a very complex and multi-faceted subject. Essentially, an avionic architecture is the total set of design choices which make up the avionic system and result in it performing as a recognisable whole. In effect, the architecture is the total avionics system design. The complexity of modern avionic systems means that there are very many parts of an avionics architecture and in practice the architecture is best viewed as a hierarchy of levels which comprise:

1. Functional allocation level. The arrangement of the major system components and the allocation of system functions to those components.
2. Communications level. The arrangement of internal and external data pathways and data rates, transmission formats, protocols and latencies.
3. Data processing level. Central or distributed processing, processor types, software languages, documentation and CASE (computer aided software engineering) design tools.
4. Sensor level. Sensor types, location of sensor processing, extent to which combining of sensor outputs is performed.

5.   Physical level. Racking, box or module outline dimensions, cooling provisions, power supplies.

This is not an exhaustive list and there are many other important aspects of the avionics, eg control and displays, maintenance philosophy etc. which are certainly a part of the avionics architecture.

The influence of the architecture also continues down to lower levels of implementation and technology detail. It is the higher levels, however, which are

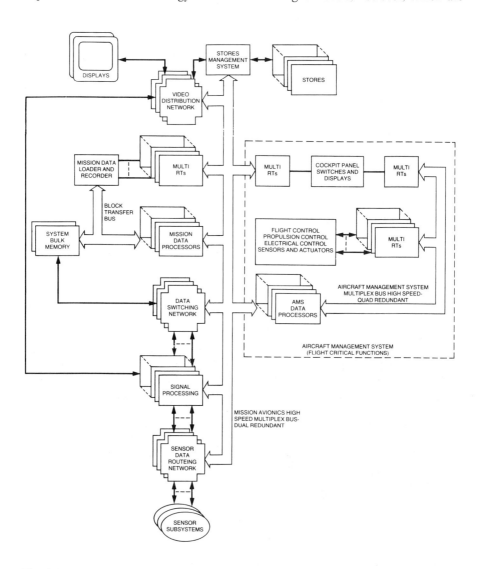

Fig. 9.15 Integrated avionic systems architecture.

most often referred to as the 'architecture' and it is at these levels that integrated modular avionic concepts are most able to influence overall system costs.

An overall avionic systems architecture for a military aircraft and designed for implementation using standard avionic modules is shown in Fig. 9.15. The essential intercommunication system provided by the high speed multiplex data buses can be seen. The grouping together of the systems which carry out flight critical functions, such as flight control, propulsion control, electrical power supply control, sensors and actuators, into the 'aircraft management system' is a noteworthy feature. The diagram is also very broadly representative in concept of future civil IMA architectures with the obvious deletion of the blocks representing 'stores' (ie weapons) and 'stores management system'.

The modular avionics concept relies on the use of a limited range of standard modules which are packaged in a standardised modular format and installed in a small number of common racks. Figure 9.16 illustrates a conceptual rack and modules.

The concept of modular equipment is not new, however, and avionics manufacturers have frequently used modular packaging to seek a competitive cost advantage within their own equipment. What is different with the IMA approach is:

(i)   The proposed use across a range of aircraft platforms (including helicopters) of standard 'F^{3}I' form, that is 'form, fit, function and interface', interchangeable modules procured from an 'avionics supermarket'.

(ii) The proposed integration of data and signal processing across traditionally separate aircraft sub-systems enabling wide scale use of reconfiguration to improve availability.

Fig. 9.16 Integrated modular avionics concept rack and modules (by courtesy of GEC-Marconi Avionics Ltd.).

(iii) The use of sufficient built in test to enable faulty modules to be correctly diagnosed and replaced at first line without additional test equipment.

The use of a small number of standard modules potentially reduces the initial development and procurement cost of equipment through competition and eventually through economics of scale in the production of the modules. It also reduces the maintenance costs by reducing spares holding.

Similarly, the integration of functions across traditional equipment boundaries promises to reduce the proliferation of different module designs across the aircraft and it allows reconfiguration strategies to be adopted economically – it is cheaper to carry a spare set of modules that can be configured to act as different systems than it is to carry complete spare systems. The requirement for a second line avionics intermediate shop is eliminated since the high levels of fault detection and isolation (typically >98%) using built in test circuitry allow maintenance to consist of the replacement of a faulty module on the aircraft with the faulty module being returned to the original manufacturer for repair.

However, with these advantages come many implications for the way equipment is currently contracted for and built. The integrated system design increases enormously both the system complexity and the potential for interactions between sub-systems. At the same time it blurs the traditional lines of responsibility that exist in the industry and it will, therefore, require a very careful and systematic design approach if integrated systems are to be put together successfully. To implement such highly integrated systems, very close collaboration between systems engineers from different avionic, airframe and software suppliers will be required.

Further reading

Ghatak, A.K., and Thyagarajan, *Optical Electronics*: Cambridge University Press

Glossary of terms

Aerodynamic centre The point about which the pitching moment does not change with the angle of incidence (providing the velocity is constant).

Aerodynamic derivative The partial derivative of the aerodynamic force or moment with respect to a particular variable. For small changes in that variable, the resulting incremental force or moment is equal to:
(Derivative) × (Incremental change in the variable).

Adiabatic process A process where no heat enters or leaves the system.

Aliasing The effects from sampling data at a sampling frequency below the frequency of the noise components present in the signal so that spurious low frequency signals are introduced from the sampled noise.

Air density The mass per unit volume of air.

Air density ratio The ratio of the air density to the value of the air density at standard sea level conditions. See 'Standard sea level conditions'.

Altitude The height of the aircraft above the ground.

Attitude The angular orientation of the aircraft with respect to a set of earth referenced axes. This is defined by the three Euler angles – yaw (or heading) angle, pitch angle, bank angle. See 'Euler angles'.

Bank angle The angle through which the aircraft must be rotated about the roll axis to bring it to its present orientation from the wings level position, following the pitch and yaw rotations. See 'Euler Angles'.

Bus controller The unit which controls the transmission of data between the units connected to a MIL STD 1553B multiplexed data bus system.

Calibrated airspeed The speed which under standard sea level conditions would give the same impact pressure as that measured on the aircraft. See 'Standard sea level conditions'.

Category I, II, III landing conditions These three categories define the landing visibility conditions in terms of the vertical visibility ceiling and the runway visual ranges; the visibility decreasing with increasing category number.

Chemosphere The region above the stratopause altitude of 20,000m (65,617 ft) up to 32,004m (105,000 ft) where the temperature is assumed to rise linearly with increasing height at 1.0×10^{-3} °C/m.

Collimation An optically collimated display is one where the rays of light from any particular point on the display are all parallel after exiting the collimating system.

Combiner The optical element of the head up display through which the pilot views the outside world and which combines the collimated display image with the outside world scene.

Complementary filtering The combination of data from different sources through appropriate filters which select the best features of each source so that their dissimilar characteristics can be combined to complement each other.

Coning The motion resulting from a body experiencing two angular vibrations of the same frequency and 90° out of phase about two orthogonal axes of the body.

Consolidation The process of deriving a single value for a quantity from the values obtained for that quantity from several independent sources; for example selecting the median value.

Coriolis acceleration The acceleration introduced when the motion of a vehicle is measured with respect to a rotating frame of reference axes.

Decision height This is the minimum vertical visibility ceiling for a landing to be safely carried out.

Derivative See `Aerodynamic derivative`.

Drag coefficient A non-dimensional coefficient which is a function of the angle of incidence and which is used to express the drag generation characteristics of an aerofoil.

Drift angle The angle between the horizontal projection of the aircraft's forward axis and the horizontal component of the aircraft's velocity vector.

Dynamic pressure The pressure exerted to bring a moving stream of air to rest assuming the air is incompressible.

Ephemeris parameters These comprise 16 parameters which define the GPS satellite orbital position data with respect to earth reference axes.

Euler angles The aircraft attitude is defined by a set of three ordered rotations, known as the Euler angles, from a fixed reference axis frame; the aircraft is assumed to be initially aligned with the reference axes:
1. A clockwise rotation in the horizontal plane through the *yaw angle* about the yaw (or vertical) axis.
2. A clockwise rotation through the *pitch angle* about the pitch (or side-slip) axis.
3. A clockwise rotation through the *bank angle* about the roll (or forward) axis.

Euler symmetrical parameters These four parameters are used to derive the vehicle attitude in a strap-down system and are functions of the three direction cosines of the axis about which a single rotation will bring the vehicle from initial alignment with a reference axis frame to its present orientation, and the single rotation angle.

The four parameters are equal to:
  cosine (half rotation angle)
  (direction cosine 1) × sine (half rotation angle)
  (direction cosine 2) × sine (half rotation angle)
  (direction cosine 3) × sine (half rotation angle)

Exit pupil diameter The diameter of the circle within which the observer's eyes are able to see the whole of the display; the centre of the exit pupil is located at the design eye position.

Fly-by-wire control system A flight control system where all the command and control signals are transmitted electrically and the aerodynamic control surfaces are operated through computers which are supplied with the pilot's command signals and the aircraft state from appropriate motion sensors.

Fluxgate A magnetic field sensor which provides an electrical output signal proportional to the magnetic field.

Free azimuth axes A local level set of axes where the horizontal axes are not rotated in space about the local vertical axis.

Gain margin The amount the loop gain can be increased in a closed loop system before instability results because the open loop gain at the frequency where there is 180° phase lag has reached 0dB (unity)

Gearing A term used in flight control systems to specify a feedback gain in terms of the control surface angular movement per unit angular change, or, unit angular velocity change in the controlled quantity. eg 1° tailplane angle/1° per second pitch rate.

Great circle A circle on the surface of a sphere whose plane goes through the centre of the sphere.

Gyro compassing A method for determining the direction of true north by using inertial quality gyros to measure the components of the earth's angular velocity.

Heading angle The angle between the horizontal projection of the aircraft's forward axis and the direction of true north.

Impact pressure The pressure exerted to bring the moving airstream to rest at that point.

Incidence angle The angle between the direction of the relative wind vector to the aerofoil chord line (a datum line through the aerofoil section). Also known as the 'Angle of attack' in the USA.

Indicated airspeed The speed under standard sea level conditions which would give the same impact pressure as that measured by the air speed indicator (ASI). It is basically the same quantity as the calibrated airspeed but includes instrument errors and static source pressure errors.

Indicated air temperature See 'Measured air temperature'.

Instantaneous field of view The angular coverage of the imagery which can be seen by the observer at any specific instant.

Kalman filter A recursive data processing algorithm which processes sensor measurements to derive an optimal estimate of the quantities of interest (states) of the system using a knowledge of the system and measurement device dynamics, uncertainties, noises, measurement errors and initial condition information.

Knot A measure of vehicle speed; one knot being equal to one nautical mile per hour.

Lapse rate The rate at which the temperature is assumed to decrease with increasing altitude.

Latency The time delay between sampling a signal and processing it so that the processed signal output lags the real signal in time. The resulting phase lag can exert a destabilising effect in a closed loop control system.

Latitude The angle subtended at the earth's centre by the arc along the meridian passing through the point and measured from the equator to the point.

Lift coefficient A non-dimensional coefficient which is a function of the angle of incidence and which is used to express the effectiveness of an aerofoil in generating lift.

Longitude The angle subtended at the earth's centre by the arc along the equator measured east or west of the prime meridian to the meridian passing through the point.

Mach number The ratio of the true airspeed of the aircraft to the local speed of sound.

Magnetic deviation The error introduced by the distortion of the earth's magnetic field in the vicinity of the magnetic sensor by the presence of magnetic materials.

Magnetic dip angle The angle between the earth's magnetic field vector and the horizontal.

Magnetic variation The angular difference between the direction of true north and magnetic north.

Mean aerodynamic chord This is equal to the wing area divided by the wing span.

Measured air temperature The temperature measured by a sensing probe where the air may not be brought wholly to rest.

Meridian A circle round the earth passing through the North and South poles.

Multi-mode fibre An optical fibre whose dimensions are such that there are a large number of ways or modes by which light can be guided along the fibre depending on the incidence angle.

Nautical mile One nautical mile is equal to the length of the arc on the earth's surface subtended by an angle of one minute of arc measured at the earth's centre.

Newton The force required to accelerate a mass of one kilogram at one metre per second per second.

Neutral point The position of the aircraft's CG where the rate of change of pitching moment coefficient with incidence is zero.

Notch filter A filter designed to provide a very high attenuation over a narrow band of frequencies centred at a specific frequency.

Numerical aperture This defines the semi-angle of the cone within which an optical fibre will accept light and is a measure of the light gathering power of the fibre.

Pascal The pressure exerted by a force of one Newton acting on an area of one square metre.

Phase margin The additional phase lag within a closed loop system that will cause instability by producing 180° phase lag at the frequency where the open loop gain is 0dB (unity gain).

Phugoid A very lightly damped long period oscillation in height and airspeed in the longitudinal plane; the angle of incidence remaining virtually unchanged.

Pitch angle The angle between the aircraft's forward axis and the horizontal, being the angle the aircraft must be rotated about the pitch axis following the rotation in yaw to arrive at its present orientation with respect to a fixed reference frame. See 'Euler angles'.

Pitching moment coefficient A non dimensional coefficient which is equal to the pitching moment about the aircraft's CG divided by the product of the dynamic pressure, wing surface area and the mean aerodynamic chord.

Pressure The force per unit area.

Pressure altitude The height above sea level calculated from the measured static pressure assuming a standard atmosphere.

Precession The behaviour of the gimbal suspended spinning rotor of a gyroscope which causes it to turn about an axis which is mutually perpendicular to the axis of the applied torque and the spin axis; the angular rate of precession being proportional to the applied torque.

Prime meridian The meridian passing through Greenwich, England.

Recovery ratio A correction factor to allow for the air not being brought wholly to rest at the temperature probe.

Relative wind Velocity of the airstream relative to the aircraft (equal and opposite to the true airspeed vector).

Remote terminal The interface unit which enables a sub-system to communicate with other systems by means of a time division multiplexed data bus system; in particular the MIL STD 1553B data bus system.

Quaternion A quantity comprising a scalar component and a vector with orthogonal components. The Euler symmetrical parameters are quaternions.

Sagnac effect Two counter propagating coherent light waves experience a relative phase difference on a complete trip around a rotating closed path; this phase difference is proportional to the input rotation rate. The effect is known as the Sagnac effect and the phase difference the Sagnac phase shift.

Schuler period The period of oscillation of a Schuler tuned IN system which is equal to that of a simple pendulum with a length equal to the radius of the earth, that is 84.4 minutes approximately.

Schuler tuning The feed back of the inertially derived vehicle rates of rotation about the local level axes of an IN system so that the system tracks the local vertical as the vehicle moves over the spherical surface of the earth.

Single mode fibre An optical fibre which is specifically designed so that only one guided mode is possible for transmitting light along the fibre. This is achieved by suitable choice of the core radius (of the same order as the transmitted wavelength) and the refractive index difference between the core and cladding.

Spiral divergence A slow build up of yaw and roll motion resulting in a spiral dive which is due to the rolling moment created by the rate of yaw.

Standard sea level conditions These are assumed in the 'standard atmosphere' to be a sea level pressure of 101.325kPa (1013.25mb) and a temperature of 288.15°K (15°C).

Static air temperature The temperature which would be measured moving freely in the air stream.

Static margin The distance of the centre of gravity of the aircraft from the *neutral point* divided by the mean aerodynamic chord.

Static pressure The pressure of the free airstream due to the random motion of the air molecules.

Static source error The error in the measured static pressure due to the effects of Mach number and incidence on the static source.

Stratopause altitude The altitude in the 'standard atmosphere' where the temperature is assumed to start rising linearly with increasing altitude and the region of the chemosphere begins. It is equal to 20,000m (65,617 ft.).

Stratosphere The region between the tropopause altitude of 11,000m (36,089.24 ft) and the stratopause altitude of 20,000m (65,6127 ft) where the temperature is assumed in the 'standard atmosphere' to be constant at 216.65°K (−56.5°C).

Tilt errors The gravitational acceleration component errors introduced by the tilt angles from the horizontal of the input axes of the nominally horizontal accelerometers of an IN system (or derived horizontal acceleration components in the case of a strap-down system).

Total air temperature The temperature that would exist if the moving airstream where brought wholly to rest at that point. It is thus the free airstream temperature plus the temperature rise due to the kinetic heating of the air because of the air being brought to rest.

Total field of view The total angular coverage of the display imagery which can be seen by moving the observer's eye position around.

Total pressure The pressure that would exist if the moving airstream were brought to rest at that point. It is equal to the impact pressure plus the static pressure.

Track angle The direction of the ground speed vector relative to true North.

Tropopause altitude The altitude where the region of constant temperature known as the stratosphere is assumed to start in the 'standard atmosphere'. It is equal to 11,000m (36,089 24 ft).

Troposphere The region from sea level up to the tropopause altitude of 11,000m (36,089.24 ft) where the temperature is assumed to decrease linearly with increasing altitude.

Yaw angle The angle measured in the horizontal plane between a fixed reference axis and the horizontal projection of the aircraft's forward axis. It is the angle through which the aircraft must first be rotated followed by the pitch and roll rotations to bring it to its present orientation. See Euler angles.

List of symbols

Because of the number of different disciplines covered by the blanket name 'Avionics', it is inevitable that there are occasions when the same symbol is used to denote different quantities. Where this is the case, the appropriate chapters are indicated in brackets. In any case, the context should make clear which quantity is being denoted.

A	Local speed of sound (Chapter 2)
A	Area enclosed by closed path of optical gyroscope (Chapter 5)
A	Diameter of HUD crt (Chapter 7)
$\mathbf{A}$	State coefficient (or plant) matrix
A_0	Speed of sound at standard sea level conditions ($A_0 = 340.3$ m/s)
A_1	Constant coefficient in q/η transfer function ($A_1 = M_q / I_y + Z_\alpha / mU$)
A_2	Constant coefficient in q/η transfer function ($A_2 = M_\alpha / I_y + M_q Z_\alpha / I_y mU$)
a	Aircraft acceleration (Chapters 5,6)
a	Core radius of optical fibre (Chapter 9)
a_X	X axis accelerometer output
a_Y	Y axis accelerometer output
a_Z	Z axis accelerometer output
a_0	GPS time correction coefficient representing satellite clock phase offset (Chapter 6)
a_1	GPS time correction coefficient representing satellite clock frequency offset (Chapter 6)
a_1	Magnetic fluxgate error coefficient (Chapter 5)
a_2	GPS time correction coefficient representing satellite clock ageing term (Chapter 6)
a_2	Magnetic fluxgate error coefficient (Chapter 5)
a_3	Magnetic fluxgate error coefficient (Chapter 5)
B	Accelerometer bias error
$\mathbf{B}$	Driving matrix
b_1	Magnetic fluxgate error coefficient
b_2	Magnetic fluxgate error coefficient
b_3	Magnetic fluxgate error coefficient
C_D	Drag coefficient ($C_D = D_W / \tfrac{1}{2} \rho V_T^2 S$)
C_L	Lift coefficient ($C_L = L_W / \tfrac{1}{2} \rho V_T^2 S$)

C_{Lt}	Tailplane lift coefficient
C_{Lmax}	Maximum value of lift coefficient
C_M	Pitching moment coefficient ($C_M = M / \frac{1}{2} \rho V_T^2 Sc$)
C_{M0}	Pitching moment coefficient at zero lift
C_u	Range equivalent of GPS user clock offset
c	Speed of light ($c = 3 \times 10^8$ m/s)
c	Aerodynamic mean chord = wing area/wing span (Chapter 3)
c_p	Specific heat at constant pressure
c_v	Specific heat at constant volume
c_1	Magnetic fluxgate error coefficient
c_2	Magnetic fluxgate error coefficient
c_3	Magnetic fluxgate error coefficient
D	Operator d/dt
D	Diameter of collimating lens (Chapter 7)
D_W	Wing drag force ($D_W = \frac{1}{2} \rho V_T^2 S C_D$)
d_1	Magnetic fluxgate error coefficient
d_2	Magnetic fluxgate error coefficient
d_3	Magnetic fluxgate error coefficient
e_0	Euler parameter ($e_o = \cos \mu/2$)
e_1	Euler parameter ($e_1 = \alpha \sin \mu/2$)
e_2	Euler parameter ($e_2 = \beta \sin \mu/2$)
e_3	Euler parameter ($e_3 = \gamma \sin \mu/2$)
F	Effective focal length of HUD collimating lens
f	Frequency
Δf	Frequency difference
G_q	Pitch rate gearing (or gain)
g	Local value of gravitational acceleration $\left(g = \dfrac{R^2}{(R+H)^2} \cdot g_0 \right)$
g_x	Gravitational acceleration component along aircraft OX (forward) axis
g_y	Gravitational acceleration component along aircraft OY (side slip) axis
g_z	Gravitational acceleration component along aircraft OZ (vertical) axis
g_0	Value of gravitational acceleration at sea level ($g_0 = 9.80665$ m/s^2)
H	Angular momentum of gyro ($H = J\omega_R$) (Chapter 5)
H	True altitude of aircraft above sea level (Chapter 6)
H_G	Geometric altitude - altitude above sea level assuming standard ICAO law and allowing for variation of g with altitude

H_H Horizontal component of earth's magnetic field $\left(H_H = \sqrt{H_1^2 + H_2^2} \right)$

H_I Barometric/inertial height

H_P Pressure altitude - geo-potential altitude above sea level assuming standard ICAO pressure/altitude law and $g = g_0$

H_S Altitude of stratopause (20,000m)

H_T Altitude of tropopause (11,000m)

H_X Earth's magnetic field component along aircraft OX (forward) axis

H_Y Earth's magnetic field component along aircraft OY (side-slip) axis

H_Z Earth's magnetic field component along aircraft OZ (vertical) axis

H_1 Horizontal component of earth's magnetic field along aircraft heading axis

H_2 Horizontal component of earth's magnetic field at right angles to aircraft heading axis

I_G Moment of inertia of DTG gyro gimbal about gimbal axes

I_x Moment of inertia of aircraft about OX (roll) axis

I_y Moment of inertia of aircraft about OY (pitch) axis

I_z Moment of inertia of aircraft about OZ (yaw) axis

J Moment of inertia of gyro rotor about spin axis (Chapter 5)

J Mechanical equivalent of heat (Joule's constant) (Chapter 2)

J_G Moment of inertia of DTG gyro gimbal about polar axis

j $\sqrt{-1}$

K Scalar gain of transfer function KG(D)

K q/η transfer function gain $(K = M_\eta / I_y\, T_2)$

$\mathbf{K}$ Kalman gain matrix

K_G Torsional spring rate of DTG gyro flexural pivots (Chapter 5)

K_p Roll rate error gain

K_ϕ Bank angle error gain

K_ψ Heading error gain

K_0 Gyro output scale factor (Chapter 5)

K_1 Feedback gain

K_2 Feedback gain

K_3 Integral term feedback gain

k_t Tailplane efficiency factor $\left(k_t = \dfrac{\text{Dynamic pressure at tailplane}}{\text{Freestream dynamic pressure}} \right)$

L Lapse rate - rate of change of static air temperature with altitude (Chapter 2)
(Troposphere lapse rate = 6.5×10^{-3} °C/m)
(Chemosphere rise rate = 1.0×10^{-3} °C/m)

L Perimeter of closed light path in an optical gyro (Chapter 5)

L Length of optical fibre (Chapter 9)

L	Distance of observer's eyes from collimating lens of HUD (Chapter 7)
L	Resultant aerodynamic rolling moment (Chapter 3)
L_f	Distance of aerodynamic centre of fin from aircraft CG
L_p	Rolling moment derivative due to rate of roll
L_r	Rolling moment derivative due to rate of yaw
L_v	Rolling moment derivative due to side-slip velocity
L_W	Wing lift force ($L_W = \frac{1}{2}\rho V_T^2 SC_L$)
L_ζ	Rolling moment derivative due to rudder deflection
L_ξ	Rolling moment derivative due to aileron deflection
l_t	Distance of aerodynamic centre of tailplane from aircraft CG
M	Mach number ($M = V_T / A$)
M	Resultant aerodynamic pitching moment
M_q	Pitching moment derivative due to pitch rate
M_u	Pitching moment derivative due to forward velocity increment
M_w	Pitching moment derivative due to vertical velocity increment
$M_{\dot{w}}$	Pitching moment derivative due to rate of change of vertical velocity
M_α	Pitching moment derivative due to incidence change
M_η	Pitching moment derivative due to tailplane/elevator deflection
m	Aircraft mass
m_a	Pendulous mass of accelerometer
N	Resultant aerodynamic yawing moment
N_p	Yawing moment derivative due to rate of roll
N_r	Yawing moment derivative due to rate of yaw
N_v	Yawing moment derivative due to side-slip velocity
N_ζ	Yawing moment derivative due to rudder deflection
N_ξ	Yawing moment derivative due to aileron deflection
n	Phase advance gain (Chapters 4,5,8)
n	Refractive index (Chapters 5,9)
n_1	Refractive index of optical fibre core (Chapter 9)
n_2	Refractive index of optical fibre cladding (Chapter 9)
P	Covariance matrix
P_S	Static pressure
P_{S0}	Static pressure at standard sea level conditions (P_{S0} = 101.325 kN/m^2 = 1013.25 mbar)
P_{SS}	Static pressure at stratopause altitude (20,000m) (P_{SS} = 54.75 mbar)
P_{ST}	Static pressure at tropopause altitude (11,000m) (P_{ST} = 226.32 mbar)
P_T	Total pressure ($P_T = Q_C + P_S$)
ΔP	INS position error (Chapter 6)
ΔP	Angular increment in roll ($\Delta P = p\Delta t$) (Chapter 5)
p	Roll rate – aircraft angular velocity in roll

p	Pressure (Chapter 2)
p_D	Roll rate demand
p_E	Roll rate error ($p_E = p_D - p$)
Q	Dynamic pressure ($Q = \frac{1}{2}\rho V_T^2 S$)
Q_C	Impact pressure
ΔQ	Angular increment in pitch ($\Delta Q = q\Delta t$)
q	Pitch rate – aircraft angular velocity in pitch
R	Radius of Earth
R	Slant range of aircraft from ILS localizer or glide slope transmitters (Chapter 8)
R_a	Gas constant for unit mass of dry air ($R_a = 287.0529$ Joules/°K/kg)
R_1	Range of GPS satellite 1 from user
R_2	Range of GPS satellite 2 from user
R_3	Range of GPS satellite 3 from user
R_4	Range of GPS satellite 4 from user
R_{1p}	Pseudo-range of GPS satellite 1 from user
R_{2p}	Pseudo-range of GPS satellite 2 from user
R_{3p}	Pseudo-range of GPS satellite 3 from user
R_{4p}	Pseudo-range of GPS satellite 4 from user
ΔR	Angular increment in yaw ($\Delta R = r\Delta t$)
r	Yaw rate - aircraft angular velocity in yaw
r	Recovery ratio - constant dependent on the temperature probe installation (Chapter 2)
S	Surface area of wing
S_t	Surface area of tailplane
s	Laplace operator
T	Temperature (Chapter 2)
T	Iteration period (Chapter 6)
T	Time constant (Chapters 3, 4, 5, 6, 8)
T_m	Measured (or indicated) air temperature
T_R	Roll time constant ($T_R = I_x / L_p$)
T_S	Static air temperature - temperature of free air-stream
T_{sp}	Time constant of spiral divergence
T_T^*	Static air temperature at tropopause altitude ($T_T^* = 216.65$°K)
T_T	Total air temperature
T_0	Static air temperature at standard sea-level conditions ($T_0 = 288.15$°K)
T_1	Time constant ($T_1 = I_y/M_q$)
T_2	Time Constant ($T_2 = mU/Z_\alpha$)
t	Time
$t_{s/c}$	Effective GPS satellite time at signal transmission

$t_{o/c}$	GPS epoch time at which polynomial coefficients a_0, a_1, a_2 are referenced
Δt	Time increment
Δt_r	GPS relativistic correction term
$\Delta t_{s/c}$	Time offset between GPS satellite and GPS master time
U	Forward velocity – velocity of aircraft CG along OX (forward) axis in disturbed flight ($U = U_0 + u$)
$\mathbf{U}$	Control input vector
U_A	Forward velocity derived from air data system
U_0	Velocity of aircraft CG along OX (forward) axis in steady flight
u	Forward velocity increment in disturbed flight
V	Air velocity (chapter 2)
V	Side-slip velocity – velocity of aircraft CG along OY (side-slip) axis in disturbed flight
V_A	Side-slip velocity derived from air data system
V_C	Calibrated airspeed
V_E	Velocity along east axis of local north, east, down axis frame
V_F	Horizontal component of aircraft forward velocity
V_G	Ground speed $\left(V_G = \sqrt{V_F^2 + V_S^2}\right)$
V_H	Horizontal component of true air-speed ($V_H = V_T \cos \theta$)
V_N	Velocity along north axis of local north, east, down axis frame
V_S	Horizontal component of aircraft side-slip velocity
V_T	True air-speed
V_W	Wind velocity
ΔV	INS velocity error
v	Side-slip velocity increment in disturbed flight
W	Gyro drift rate (Chapter 6)
W	Vertical velocity – velocity of aircraft CG along OZ (vertical) axis in disturbed flight ($W = W_0 + w$) (Chapters 3,5)
W_A	Vertical velocity derived from air data system (along aircraft OZ axis)
W_0	Velocity of aircraft CG along OZ (vertical) axis in steady flight
w	Vertical velocity increment in disturbed flight
$\mathbf{X}$	System state vector
X_a	Incremental change in aerodynamic force along OX (forward) axis following a disturbance
X_u	Forward force derivative due to forward velocity increment (Chapter 3)
X_u	X axis coordinate of GPS user (Chapter 6)
X_w	Forward force derivative due to vertical velocity increment
X_1	X axis coordinate of GPS satellite 1
X_2	X axis coordinate of GPS satellite 2

X_3 X axis coordinate of GPS satellite 3
X_4 X axis coordinate of GPS satellite 4
x_m Forward axis magnetic fluxgate output

Y_a Incremental change in aerodynamic force along OY (side-slip) axis following a disturbance
Y_u Y axis coordinate of GPS user
Y_v Side-force derivative due to side slip velocity increment
Y_ζ Side-force derivative due to rudder deflection
Y_1 Y axis coordinate of GPS satellite 1
Y_2 Y axis coordinate of GPS satellite 2
Y_3 Y axis coordinate of GPS satellite 3
Y_4 Y axis coordinate of GPS satellite 4
y_m Side-slip axis magnetic fluxgate output

Z_a Incremental change in aerodynamic force along OZ (vertical) axis following a disturbance
Z_m Vertical axis magnetic fluxgate output
Z_q Vertical force derivative due to pitch rate
Z_u Vertical force derivative due to forward velocity increment (Chapter 3)
Z_u Z axis coordinate of GPS user
Z_w Vertical force derivative due to vertical velocity increment
$Z_{\dot{w}}$ Vertical force derivative due to rate of change of vertical velocity
Z_α Vertical force derivative due to incidence change
Z_η Vertical force derivative due to tailplane/elevator deflection
Z_1 Z axis coordinate of GPS satellite 1
Z_2 Z axis coordinate of GPS satellite 2
Z_3 Z axis coordinate of GPS satellite 3
Z_4 Z axis coordinate of GPS satellite 4

α Angle of incidence (angle of attack – USA) (Chapters 2,3,4,8)
α Direction cosine used to specify aircraft attitude and derive Euler parameters e_0, e_1, e_2, e_3 (Chapter 5)
α_{max} Maximum value of angle of incidence
α_T Angle of incidence for trimmed flight

β Side-slip incidence (Chapters 2,3,4,5)
β Direction cosine used to specify aircraft attitude and derive Euler parameters e_0, e_1, e_2, e_3 (Chapter 5)

γ $\dfrac{\text{Specific heat at constant pressure}}{\text{Specific heat at constant volume}}$ $\left(\gamma = \dfrac{c_p}{c_v} \right)$ = 1.4 for air (Chapter 2)

γ	Dihedral angle (Chapter 3)
γ	Direction cosine used to specify aircraft attitude and derive Euler parameters e_0, e_1, e_2, e_3 (Chapter 5)
γ	Doppler radar beam depression angle (Chapter 6)
γ_F	Flight path angle – angle aircraft velocity vector makes with chosen reference axis (usually horizontal axis)
γ_L	Angular displacement of aircraft from ILS localizer beam centre line
γ_V	Angular displacement of aircraft from ILS glide slope beam centre line
Δ	Refractive index difference ratio of optical fibre ($\Delta = (n_1 - n_2) / n_1$)
δ	Drift angle – angle between aircraft track and heading
δ_i	Pilot's stick input
ε	Error signal
ε_X	Vertical gyro X axis output angle
ε_Y	Vertical gyro Y axis output angle
ζ	Angular deflection of rudder from trimmed position (Chapters 3,4)
ζ	Damping ratio of second order system (standard form:- $D^2 + 2\zeta\omega_0 D + \omega_0^2$)
η	Tailplane/elevator angular deflection from trimmed position
η_D	Demanded tailplane/elevator angle
θ	Pitch angle (Chapters 3,4,5,8)
θ	Incidence angle of light ray (Chapters 7,9)
θ_B	Angle between ILS guidance beam centre line and reference axis
θ_{crit}	Critical incidence angle of light ray at which total internal reflection occurs
θ_D	Demanded pitch angle
θ_i	Input quantity
θ_o	Output quantity
θ_p	Angular rotation of stable platform with respect to inertial axis frame
θ_1	Quantity 1
θ_2	Quantity 2
$\Delta\theta$	Tilt angle error
$\Delta\theta_X$	Tilt angle error about X axis
$\Delta\theta_Y$	Tilt angle error about Y axis
λ	Latitude angle (Chapters 5,6)
λ	Wave length (Chapters 5,9)
λ	Root of characteristic equation (Chapter 3)
λ_0	Latitude of initial position

μ	Longitude angle
μ	Single angular rotation used to specify aircraft attitude and Euler parameters e_0, e_1, e_2, e3 (Chapter 5)
μ_0	Longitude of initial position
ξ	Angular deflection of ailerons from trimmed position
ρ	Air density
ρ_0	Air density at standard sea level conditions ($\rho_0 = 1.225$ kg/m^3)
σ	Angle between Doppler radar beams and aircraft centre line (Janus antenna configuration)
τ_1	GPS satellite 1 clock correction
τ_2	GPS satellite 2 clock correction
τ_3	GPS satellite 3 clock correction
τ_4	GPS satellite 4 clock correction
Φ	Bank angle (Chapters 3,4,5,8)
Φ	Phase shift (Chapters 5,8)
Φ_D	Demanded bank angle
Φ_E	Bank angle error ($\Phi_E = (\Phi_D - \Phi)$)
Φ_n	State transition matrix
Φ_s	Sagnac phase shift
ψ	Heading (or yaw) angle
ψ_D	Demanded heading angle
ψ_E	Heading error ($\psi_E = (\psi_D - \psi)$)
$\psi_{G/M}$	Combined gyro/magnetic heading angle
ψ_M	Magnetically derived heading angle
ψ_T	Track angle
ψ_W	Wind direction with respect to true north
$\Delta\psi$	INS heading error
Ω	Earth's angular velocity about polar axis
ω	Angular frequency
ω_R	Angular velocity of gyro rotor about spin axis
ω_0	Undamped natural frequency of second order system (standard form:- $D^2 + 2\zeta\omega_0 D + \omega_0^2$)

List of abbreviations

ADC	Air data computer
ADI	Attitude and direction indicator
AFCS	Automatic flight control system
AHRS	Attitude heading reference system
AMLCD	Active matrix liquid crystal display
ATC	Air traffic control
BHMD	Binocular helmet mounted display
BIT	Built in test
CAA	Civil Aviation Authority
CASE	Computer assisted software engineering
CGH	Computer generated hologram
CRT	Cathode ray tube
DME	Distance measuring equipment
DTG	Dynamically tuned gyro
DR	Dead reckoning
DVI	Direct voice input
EFIS	Electronic flight instrument system
EL	Electro-luminescent
EM	Electro-magnetic
EMC	Electro-magnetic compatibility
EMI	Electro-magnetic interference
EMP	Electro-magnetic pulse
FAA	Federal Aviation Authority
FADEC	Full authority digital engine control
FBW	Fly by wire
FCS	Flight control system
F^3I	Form fit function interface
FLIR	Forward looking infra-red
FMGEC	Flight management guidance and envelope computer
FMS	Flight management system
FOG	Fibre optic gyro
FOV	Field of view
GPS	Global positioning system
GDOP	Geometric dilution of precision
HDD	Head down display
HF	High frequency
HMD	Helmet mounted display
HSI	Horizontal situation indicator
HUD	Head up display
IFOG	Interferometric fibre optic gyro

IFOV	Instantaneous field of view
IKBS	Intelligent knowledge based system
ILS	Instrument landing system
IN	Inertial navigation
INS	Inertial navigation system
IMA	Integrated modular avionics
IR	Infra-red
LCD	Liquid crystal display
LED	Light emitting diode
LLTV	Low light tv
LOS	Line of sight
LRU	Line replaceable unit
LSI	Large scale integration
MCDU	Multi-purpose control display unit
MIMO	Multi-input multi-output
MLS	Microwave landing system
MTBF	Mean time between failures
NA	Numerical aperture
NDB	Non-directional beacon
NVG	Night viewing goggles
OOD	Object oriented design
PFD	Primary flight display
QA	Quality assurance
QFE	Ground pressure
QNH	Mean sea level pressure
RLG	Ring laser gyro
RFOG	Ring resonator fibre optic gyro
SIMO	Single input multi-output
SISO	Single input single output
SMAC	Scene matching area correlation
SSEC	Static source error correction
TACAN	Tactical air navigation
TCM	Terrain characteristic matching
TCN	Terrain contour navigation
TDM	Time division multiplexing
TFOV	Total field of view
TRN	Terrain referenced navigation
UHF	Ultra high frequency
UTC	Universal time co-ordinate
VHF	Very high frequency
VOR	VHF omni directional range
VSL	Vertical speed indication
WDM	Wavelength division multiplexing

Index

Page numbers of figures are shown in **bold**, those for tables in *italic*.